INTRODUCTION TO GEOLOG

Volume 2 Earth History

Part II Later Stages of Earth History

Introduction to Geology

Volume 2 Earth History

Part II Later Stages of Earth History

H. H. READ
F.R.S., F.R.S.E., F.G.S., D.Sc., A.R.C.S.

Sometime Professor Emeritus of Geology,
Imperial College of Science and Technology,
University of London

JANET WATSON
Ph.D., A.R.C.S., D.I.C.

Professor of Geology in the University of London,
Imperial College of Science and Technology,
London

© The Estate of the late H. H. Read
and Janet Watson 1975

First published 1975 by
THE MACMILLAN PRESS LTD
London and Basingstoke
Associated companies in New York Dublin
Melbourne Johannesburg and Madras

SBN 333 17668 5 (hard cover)
SBN 333 17669 3 (paper cover)

Typeset by
PREFACE LIMITED
Salisbury, Wilts
and printed by
COMPTON PRINTING
Aylesbury, Bucks
and bound by
PITMAN PRESS
Bath, Avon

Distributed in the United States by
Halsted Press, a Division of
John Wiley & Sons, Inc., New York

Library of Congress Catalog Card No. 75–501

Contents

List of Figures

List of Plates

Preface

When H. H. Read was asked by Macmillans to write an introductory text-book which would serve as the basis for a first University course in Geology, he originally hoped to cover the whole subject in a single volume. It soon became clear, however, that this could not be done unless much of the factual infomation needed by students was jettisoned. An enlargement of the project became necessary and as I was acting as his Research Assistant he invited me to join him. We decided to think in terms of a two-volume *Introduction to Geology*: Volume 1 (*Principles*) would deal with geological processes and with the rocks and structures produced by them, while Volume 2 (*Earth History*) would illustrate the effects of these processs by reference to the record of geological history. Volume 1 appeared in 1962 and we then began to make plans for Volume 2.

At this time, the scope of historical geology was being dramatically widened by new developments in the earth sciences. It seemed clear to us that if we were to provide the background which would be needed by students in the future we must deal with earth history on a world-wide basis and must cover the full span of geological time. This broad approach made it necessary to give the book a rather unusual balance and to scale down the treatment of some important topics usually favoured by stratigraphers and palaeontologists. The plan that we adopted allowed us to deal with the geological evolution of large crustal units over long time periods. The early stages in the history of these units are covered in Part I; the later stages, from a point in time about a thousand million years ago until the present day, are covered in Part II. Many of the themes introduced in Part I are developed in Part II, and I hope that the two Parts will be regarded for most purposes as an entity. The difficulties of producing a paperback as well as a hardcover version as a single large volume have made it preferable for Parts I and II to be published as separate but integrated books.

Soon after we had begun work on Volume 2, Professor Read suffered a severe illness which restricted his later activities. He pushed on, however, with undiminished enthusiasm and we kept in touch by means of innumerable letters and telephone conversations. About three-quarters of the manuscript had been written when he died in 1970, a few months after his eightieth birthday. Over the next year or so I completed the final chapters, revised the entire text and assembled material for the illustrations and bibliography. It falls to me, too, to thank the many people who have given us advice and information at many stages. In partcular, I must mention my colleagues at Imperial College, many of them former students or colleagues of H. H. Read, who have helped to widen my horizons on many occasions. I also owe a debt of gratitude to Dr P. L. Robinson who has allowed me to use a photograph from her collection, and to many geologists in Europe, Canada, Africa and Australia who have made field excursions both profitable and enjoyable. In writing a book of this scope it is impossible to keep within the limits of one's own experience and although I am, of course, entirely responsible for errors of detail, I know how greatly the book has benefited from the experience and interest of those with whom I have discussed it.

Imperial College of Science   JANET WATSON
and Technology,
London

Acknowledgements

The authors and publishers wish to thank the following who have kindly supplied the originals of the photographs mentioned:

The Director, Grølands Geologiske, Undersøgelse: Plates I and VI.

The Director, Lanmaelingar Islands: Plate V

Aerofilms Ltd: Plates III and IV

Dr P. L. Robinson, University College, London: Plate II

1

New Themes in Earth History

I Introduction

The geological records of the 500–600 m.y. which make up the Phanerozoic eon are not only more detailed than those of earlier periods: they also cover a wider range of geological happenings, notably those connected with the history of the ocean basins, in addition to the continents, and with bodily movements of continental masses as well as relative movements within such masses.

This widening of the scope of historical geology necessitates some changes in approach in the second part of our survey. Still more important is the fact that the periods to be dealt with saw geological developments which were worldwide in their consequences and may even have been unique. During the late Proterozoic and Palaeozoic eras, cycles of mobility were initiated which established the framework of the present structure. By late Palaeozoic times, the supercontinents of Gondwanaland and Laurasia had been assembled and in Mesozoic times they broke up, the continental fragments so formed drifted apart and the widening spaces between them were filled by newly formed crustal material of oceanic type. This latter process of sea-floor spreading was made possible by the activity of a system of mid-oceanic ridges and associated structures. In recounting the history of Phanerozoic times, new geological themes are found to emerge alongside those traced forward from earlier times. Our regional treatment is extended to cover the ocean basins in addition to the continental masses, while some general points not easily dealt with on a regional basis are introduced in this chapter.

II The Hypothesis of Continental Drift

Although several early writers toyed with the possibility that continents now widely separated had once been united, the first serious attempts to examine this possibility were made by Taylor (1910) and by Wegener (1912). Wegener's book, *Die Entstehung der Kontinente*, appeared during the First World War and was not widely discussed until after the publication of an English translation in 1924. With powerful support from du Toit (*Our Wandering Continents*, 1937) and Holmes (*Principles of Physical Geology*, 1944), it began to influence the thinking of many geologists, particularly those familiar with Gondwanaland. Geophysicists, on the other hand, remained generally sceptical until after the Second World War when the tide was turned by new evidence concerning rock magnetism and ocean-floor structure. By the 1960s, the drift hypothesis had become an essential feature of geological thought in most communities of earth-scientists.

Some of the arguments relied on by Wegener and other early proponents of the hypothesis have not stood the test of time and are now only of historical interest. Among the lines of evidence which have proved fruitful, we here summarise the most important, leaving till last the crucial palaeomagnetic studies of recent decades.

1 Fit of the continental masses

Most early suggestions concerning the possibility of continental drift referred to the striking similarities of outline between the west coast of Africa and the east coast of South America (Fig. 1.1). The first comparisons were naturally made on the basis of the coastlines, but when the role of the continental shelf as a submarine extension of the continental mass came to be appreciated, the continental margins rather than the coastlines were matched. Matching of the outlines of the 500- or 1000-fathom contours produces a remarkably good fit which breaks down only where the Niger delta has been built out beyond the original margin of the African continent. The use of arithmetical methods by means of a computer has shown that the 'best fit' has a root-mean-square misfit of only 1½ per cent of the width of the intervening ocean (Bullard *et al.*, 1965). An almost equally close fit between the opposing edges of North America, Greenland and Europe has been demonstrated by similar methods. The only major anomaly here lies in the fact that re-assembly leaves no room for Iceland – an anomaly that is more apparent than real, since Iceland is made almost entirely of igneous material erupted since the date of separation of the continents.

The fitting of continental masses according to a 'jigsaw-puzzle' principle has been tested by investigating the degree of correspondence of the geological structures thus brought together. There is no need to go into this aspect in detail, since it can be assessed from evidence given elsewhere in this book. Remarkably close resemblances exist, for example, between the later Palaeozoic and early Mesozoic successions of southern Africa on the one hand and of South America and the Falkland Islands on the other (du Toit, 1937, Martin, 1961, see also Chapters 6 and 8); between the Palaeozoic fold-belts of north-western Europe and those of eastern North America (see Chapters 2, 3 and 4); and a reasonable fit has been obtained between the Precambrian tectonic provinces of Canada, Greenland and northern Europe (Part I, Chapters 3 and 4).

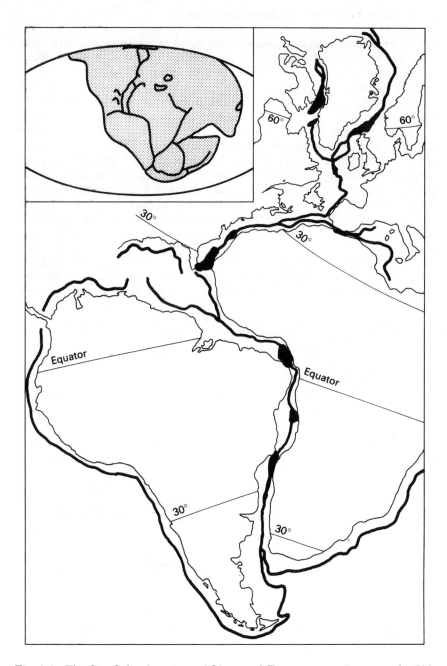

Fig. 1.1. The fit of the American, African and European continents at the 500 fathom line, proposed by Bullard and others (1965); areas of overlap are marked in black. The inset shows Wegener's original reconstruction of Pangea

2 Anomalous distribution of climatic indicators

Indicators of palaeoclimate include certain climate-sensitive sedimentary rocks, most notably glacigene sediments and evaporites: remains of organisms whose climatic preferences can be inferred by analogy with present-day organisms; and remains of organisms showing evidence of seasonal or of uniform growth. Reliable evidence from these indicators is more or less restricted to the Phanerozoic eras for which the stratigraphical or organic records are adequate.

The distribution of Permo-Carboniferous tillites and associated glacigene rocks played a very influential part in early discussions of the drift hypothesis and is still of importance in deciding the most probable relationships of the continents in Gondwanaland. As we see in Chapter 6, the present distribution of glacigene deposits is difficult to reconcile with any likely system of climatic zoning (pp. 175–7). Wegener and du Toit showed that the continents concerned – South America, Africa, India and Australia – could be reassembled in such a way that on the one hand the glaciated area was reduced to a size comparable with that of the Pleistocene ice-sheets while on the other hand the directions of ice-flow indicated by glacial striae fell into a roughly radial pattern. More recently, palaeomagnetic evidence has provided independent indications that the regions subjected to Permo-Carboniferous glaciation lay in high latitudes (Fig. 1.2).

It is instructive also to consider variations in the distribution of climatic environments through successive geological periods. Blackett, for example, points out that the mean (present-day) latitudes at which formations including reef-building corals or evaporites (indicating tropical to sub-tropical and warm–arid conditions respectively) are found, vary with the age of the formations, the older deposits of these types lying to the north of the younger

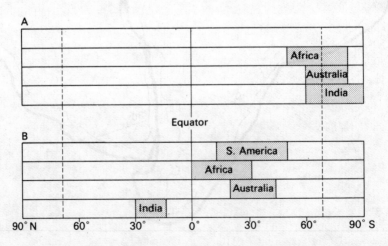

Fig. 1.2. Diagram showing the sites of Permo-Carboniferous glaciation in three continents (A) relative to the palaeopoles determined from magnetic data; and (B) relative to the present pole (based on Blackett, 1961)

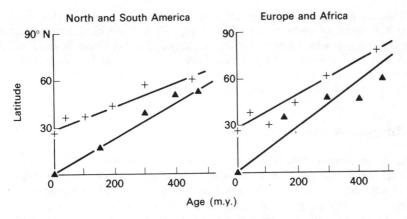

Fig. 1.3. The latitudes of occurrences of corals (triangles) and evaporites (crosses) plotted against age (after Blackett, 1961)

ones (Fig. 1.3). Such a relative shift could be explained by supposing that the climatic zones remained in broadly fixed positions while the continents drifted northward. The changes in palaeolatitude suggested by studies of rock magnetism fit reasonably well with those needed to account for the distribution of the climatic indicators: it is only fair to add, however, that Hill (1957) and Green, to name two among many, have put forward discussions of the distribution of reef-corals and evaporites, respectively, which involve no appeal to drifting of the continents.

3 Other geological anomalies

Under this heading may be recalled some miscellaneous anomalies connected with past distributions of land and sea. There is evidence that land-masses of considerable size have at various times flanked the continents now bordered by the Atlantic Ocean. For example, the late Precambrian Torridonian formation of north-west Britain, close to the western edge of Europe, derived much of its sediment from source-area to the north-west. The Permo-Carboniferous tillites of eastern South America were laid down by ice flowing westward from the region now occupied by the ocean (p. 176). Conversely, marine incursions into the continents from the direction of the present Atlantic and Indian Oceans were lacking over much of the Palaeozoic era, though Jurassic or younger marine deposits are widespread near the present coasts (p. 226).

Anomalies connected with the distribution of organic remains were referred to extensively by Wegener and his assessment of the resemblances between certain faunas and floras of continents now widely separated led him to conclude that the continents of the world remained massed together until at least the late Mesozoic. Interpretation of the evidence involves many biological factors, such as the relationships between the groups concerned, the ecological preferences of both adult and young forms, the possible means of dispersal of adults and young and of spores and seeds. Perhaps because the subject occupies the borderline between geology and biology, it has provoked many disagreements; the best which a recent writer was prepared to say (Westoll, 1965, p. 20) was that the distributions of particular fossils or faunas 'make as much sense and often a great deal more' on continental reconstructions such as those of du Toit, as they do on the continents as at present arranged.

4 Tangential displacements on faults

The demonstration that large rock masses have been moved laterally over the earth's surface for distances measurable in hundreds of kilometres is a comparatively recent achievement. In 1946, Kennedy postulated a displacement of roughly 100 km along the Great Glen Fault of Scotland, basing his conclusions on the matching of the Caledonian orogenic front, of zones of metamorphism and migmatisation and of two granite masses which he regarded as the halves of a single pluton. More recently it has become apparent that the total displacements on such transcurrent faults as the San Andreas Fault of California and the Alpine Fault of New Zealand may reach 500 km. The movements can in some instances be dated accurately enough to allow rough estimates of their average rate to be made. In view of this evidence it seems, as Blackett has remarked (1965, p. ix) that the question is no longer ' "Have or have not the continents drifted?" but the quantitive one "How much have they drifted and when?".'

5 Rock magnetism

Certain rocks common in the geological column possess a remanent magnetization which appears to have remained almost unchanged since an early stage in the formation of the rocks. Rocks possessing such a stable magnetization – notably red sandstones, basic lavas and minor intrusions – provide the data on which the science of palaeomagnetism is founded. When serious studies of rock magnetism began shortly after the Second World War, it was quickly found that although the direction of stable magnetization was more or less constant throughout a single undisturbed rock-unit, it was not as a rule the direction of magnetization appropriate to the position of the unit in the present magnetic field of the earth. If one makes the assumption, for which there is both observational and theoretical support at least as far back as 1000 m.y., that the earth's mean magnetic field has always been that of an axial dipole, this discrepancy is best explained by supposing that the units studied have changed their positions relative to the poles. Systematic measurements of the directions of magnetization of rock-units of different ages situated on several continents have therefore provided new means of investigating both the reality and the effect of continental drift. The results have been accepted by virtually all workers in this field as indicating first, that rafts of continental crust have migrated over distances of thousands of kilometres during at least the past 500 m.y. and, second that the continents have moved relative to each other.

It is important to bear in mind the sort of quantitative information which is supplied by palaeomagnetic studies. The direction of magnetization defines the ancient latitude and the orientation of the rock under investigation, but cannot directly indicate the longitude, 'though Nature has provided the ancient rock with the magnetic equivalent of a Pole Star . . . regrettably it did not also provide it with a chronometer to tell its longitude.' (Blackett, 1965, p. x). Some aberrant directions of magnetization could be accounted for in more than one way – for example, by drift or by rotation. Further difficulties arise from the necessity for dating the samples studied, since the rock-types most often employed – red sediments and basic lavas or minor intrusions – seldom lend themselves to precise stratigraphical dating. Finally, the possibility arises that the magnetization of the sample may not date from the original formation of the rock; it has been suggested, for example, that certain older rocks of Europe and North America were re-magnetized during the Upper Palaeozoic era. When these objections have been admitted, however, the consistency of the results obtained and the extent to which they agree with deductions drawn from other lines of reasoning remain extremely impressive.

The nature of the evidence can be illustrated by reference to the continents of Europe and North America. The directions of magnetization of Pleistocene and Recent rocks from both continents are related to pole positions grouped closely around the present North Pole. Those of Carboniferous and Permian rocks, on the other hand, indicate pole positions far from the present North Pole. The Permian pole positions for European samples fall in the north-west Pacific of the present day, those for American samples fall in eastern Asia, the centres of the two groups being separated by an arc of some 30°. When pole positions for rocks of different ages are plotted, those for each continent are found to form a serial progression with time, defining paths of 'polar wandering' which converge towards the present North Pole (Fig. 1.4). South America, Africa, Australia and India show similar features (Fig. 1.5). These discrepancies suggest that the positions of the continents relative to each other and to the poles have changed drastically since Precambrian times. The evidence suggests that drift relative to the poles has taken place for at least 600–800 m.y. and probably for over 2000 m.y.

The forms in which the results of palaeomagnetic studies are conventionally expressed are illustrated in Fig. 1.4 in which previous pole positions are plotted on a projection of the continents in their present relative positions. Such a presentation involves making few unnecessary assumptions but has disadvantages for geological purposes. An alternative and more evocative form (Fig. 1.6) plots the continents in the successive latitudes and orientations relative to the pole which are indicated by the data for each period. This method involves assigning arbitrary longitudes chosen so as to give a continuous track leading back from the present site of each continent.

An effective northward shift through Phanerozoic time is indicated by data from most continents; this evidence is in reasonable harmony with that supplied by palaeoclimatic indicators already referred to. The apparent northward migration of peninsular India is especially remarkable. Permo-Carboniferous tillites in eastern and northern India indicate a late Palaeozoic position not far from the South Pole. The present mean latitude of the sub-continent referred to

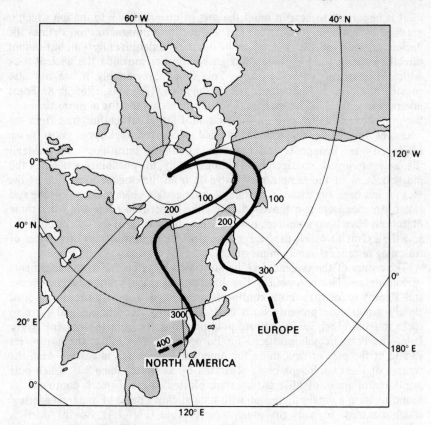

Fig. 1.4. Polar wandering curves for Europe and North America dated in millions of years (after Tarling and Tarling, 1971)

the centrally placed city of Nagpur is 22°N. Palaeomagnetic data suggest that when the Jurassic Rajmahal Traps were erupted, the mean latitude was 49°S and that when the late Cretaceous-Eocene Deccan Traps were erupted it was about 35°S (Radakrishnamurty and Sahasrabudhe, 1965). The average rate of northward drift since Jurassic times suggested by these figures is about 0.5 degrees of latitude per million years.

A second general point concerns the continents bordering the Atlantic. The divergent polar wandering paths for North America and Europe in Mesozoic and Tertiary times have already been mentioned (Fig. 1.4): if the two continents had remained fixed relative to each other, measurements from both should give identical paths, as do measurements from Siberia and Europe, between which there has been little relative movement over the same period. When Africa and South America are compared, the polar wandering paths are similarly divergent, and Creer has shown that when the two continents are fitted together, their polar wandering paths for Palaeozoic times coincide. This coincidence suggests that the continents moved as a single unit during the Palaeozoic, and that the

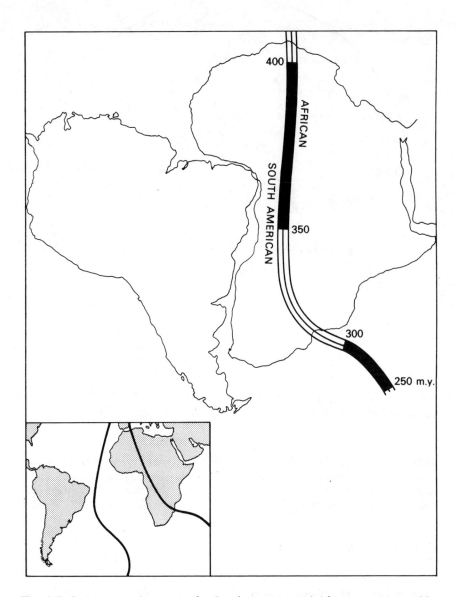

Fig. 1.5. Polar wandering curves for South America and Africa superimposed by restoring the continents to the relative positions occupied before the development of the South Atlantic; in Palaeozoic times both continents migrated together

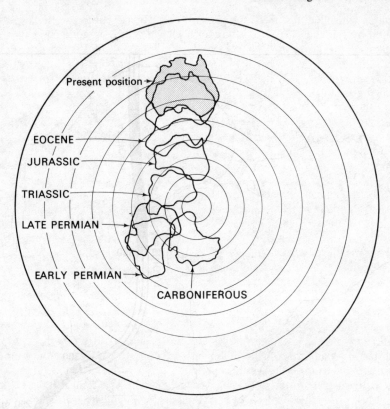

Fig. 1.6. The movement of Australia relative to the South Pole from Carboniferous times onwards, as inferred from palaeomagnetic data

divergence of the paths for Mesozoic and Tertiary times shows the effect of their separation. Palaeomagnetism thus allows one to discriminate between movements *en masse* of the supercontinents and the relative movements of continental fragments to which the term continental drift was originally applied.

III Continental Reconstructions

Several possible reconstructions of the continental masses as they were assembled in late Palaeozoic times have been proposed by geologists and geophysicists on the basis of the evidence discussed in the previous sections. Although these reconstructions differ, certain features common to all seem well-established. All agree in restoring North America and Greenland to positions alongside western Europe, and South America to a position alongside Africa. We have generally adopted for the reconstructions of these continents the computer-controlled fits made by Bullard and others (1965), or the variants favoured by Harland (1965) and Flinn (1971), which provide good bases for the matching of the Precambrian tectonic provinces and of the Palaeozoic mobile belts.

The restoration of North America to Europe has the effect of reconstructing a vast continental mass, *Laurasia,* incorporating North America, Europe and most of Asia. This mass is now bordered on the south by the broad Alpine-Himalayan mobile belt, the main structure of which post-dates the period of continental break-up. To the south of this mobile belt lie, in the west, the continents of South America and Africa and, in the east, the stable continental blocks of Antarctica, Australia and peninsular India. Most authors restore these blocks to positions within a group alongside South America and Africa, thus reconstructing the supercontinent of *Gondwanaland*. There is, however, little agreement as to the relative positions of the components and widely differing proposals have been made by Wegener, du Toit and Tuzo Wilson, to name only three. These divergencies are not surprising, the coastlines which must be matched are short and are often parallel to the structural grain; we follow, in most respects, the grouping adopted by Robinson (Fig. 6.6).

At the present day, fragments of Gondwanaland are in contact with parts of Laurasia in Central America, the Mediterranean region, the Middle East, the Himalayas and, more tenuously, the East Indies. The contact regions are without exception characterised by a complexity of structure which dates from Mesozoic and Tertiary times. The intervention of young mobile belts in which earlier structures are manifestly distorted hinders attempts to reconstruct the former relationships of Laurasia and Gondwanaland. Wegener inferred that in Palaeozoic times the two supercontinents formed a single unit which he termed *Pangea* (Fig. 1.1). Other writers have envisaged arrangements in which they were joined only along a relatively short contact-zone or were altogether separate. This uncertainty may ultimately be resolved by palaeomagnetic evidence; the geological evidence appears to us to favour a close connection between the western parts of Laurasia and Gondwanaland and a wider separation in the east. The great embayment separating the eastern parts of the supercontinents is the former ocean of *Tethys*.

The supercontinents discussed are seen, when examined in more detail, to represent mosaics in which a number of smaller cratonic masses had been welded together by the stabilisation of systems of mobile belts in late Palaeozoic times. Palaeomagnetic studies of the type already dealt with indicate that some at least of the smaller cratons had been widely separated from each other during the early stages of mobility in the enclosing mobile belts — for example, the path of Palaeozoic polar-wandering for the Siberian craton is very different from that for the European craton, suggesting that these units, which are now welded together by the Uralide mobile belt (Fig. 3.1), were moving independantly while the Uralides were developing. Reconstructions of the continental masses for periods earlier than the late Palaeozoic given by Smith, Briden and Drewry (1973) suggest that, whereas the supercontinent of Gondwanaland may already have been in existence at the onset of the Palaeozoic era, the components of Laurasia may have been widely scattered.

IV The Ocean Basins

1 Characters of oceanic crust

Until recently, the greatest unknown factors in earth history have been those concerned with the structure and behaviour of the oceanic crust. The enormous expansion of marine geophysical and geological studies over the past decade has gone some way towards filling this gap and it can now be seen that the history, as well as the structure and constitution, of the oceanic crust has differed fundamentally from that of the continental masses.

The characteristic features of the present ocean basins can be reiterated briefly. They are floored by a crust averaging no more than 5 km in thickness. This crust has geophysical properties consistent with a basic composition and wherever it has been sampled, by dredging, by coring or from oceanic islands, basic igneous material has been found to predominate. The sedimentary cover is usually less than a kilometre in thickness. At the continental margins, the rather abrupt topographical slopes linking the ocean floor with the continental shelf are associated with complex lateral variations in the thickness and constitution of the crust.

Crustal mobility is concentrated along the mid-oceanic ridge systems, which are also zones of high heat flow and of intermittent volcanicity, and in those marginal tracts where the ocean basins are bordered by active orogenic belts. Along these tracts, island arcs and oceanic trenches are developed, along with zones of deep-focus earthquakes (Benioff zones) inclined towards the adjacent continents.

Sampling of material from the true ocean basins has so far failed to reveal the presence of rocks much over 200 m.y. in age. Admittedly, such sampling has been carried out only on a limited scale but, taken in conjunction with other evidence, it suggests that the oceanic crust is geologically young; the most striking confirmation of this proposition has been revealed by magnetic surveys.

Magnetic patterns of the oceanic crust. An early magnetometer survey by Mason and Raff of part of the Pacific Ocean off the North American coast revealed a very distinctive arrangement of magnetic anomalies (1961). The contours of magnetic force define a series of parallel north—south stripes which are offset along a number of transverse lineaments. Similar magnetic patterns have since been found in every ocean basin (pp. 257, 286) and are so extensive that they must be regarded as the expression of a characteristically oceanic structure. The magnetic striping ends at or near the continetal margins and is unlike any magnetic patterns widely developed in continental areas. The magnetic pattern shows a consistent relationship to the oceanic ridges which suggests that it is a growth-structure resulting from the systematic addition of new material to the oceanic crust. In each ocean, the magnetic stripes are parallel to the nearby oceanic ridges and are symmetrically arranged with respect to them. In 1963, Vine and Matthews suggested that the parallel stripes of the magnetic pattern in the North Atlantic marked strips of the ocean floor which were magnetized alternately in normal and reversed directions. The symmetry of the pattern was due to the repeated addition of new crustal material along the mid-oceanic ridge structure. As each increment arrived, it split the previously formed strip into two parts which moved outward on either side. The periodic reversals of the earth's magnetic field have imparted distinctive magnetic signatures to successive strips. The widths of the strips depend on the time-intervals between reversals and on the rate of spreading.

2 Sea-floor spreading

The evidence outlined above suggests that the crust of the ocean basins has been widened by the repeated addition of juvenile igneous material along the active oceanic ridges, while earlier formed parts of the crust were carried outward from the ridges as if by paired conveyor-belts.

The history of the oceanic crust, as seen at the present day, is dominated by this phenomenon of *sea-floor spreading* initially proposed by Holmes in 1929 and first stated in the current form by Dietz in 1961. As a result of the localisation of crust-forming processes along the ridge structures, the youngest strips lie nearest to the ridge and progressively older rocks and structures are encountered as one goes outward from it — geological history is therefore to be read horizontally rather than vertically. The periodic reversals of the earth's magnetic field provide time-markers which allow the correlation and dating of growth-stages in widely separated areas.

3 Sea-floor spreading and continental drift

The oceans which have been opened by dispersal of the fragments of Laurasia and Gondwanaland — the Atlantic, Indian and Arctic Oceans and the Red Sea — record phases of sea-floor spreading which are closely related to the process of continental drift; the generation of new crustal material in the ocean basins took place at approximately the rate required to fill the gaps opened up by separation of the continents (Fig. 1.7).

The Pacific and Tethys, on the other hand, are old oceans. The Tethys has been almost eliminated by collision of the continents on its northern and southern sides. The Pacific appears to have been narrowed by the westward advance of the Americas. Yet the Pacific floor exhibits a magnetic pattern which, from the width of the stripes, suggests sea-floor spreading at rates at least equal to those of the Atlantic. The addition of new crustal material along the ridge-systems of the Pacific has not been compensated for by retreat of the adjacent continents, but, instead, the spreading edges of the oceanic crust appear to have been over-ridden by the advancing continents. The zones of deep-focus earthquakes extending down to depths of several hundred kilometres beneath the continents appear to represent fundamental dislocations (Benioff zones) along which surplus oceanic crust is returned to the mantle. The mobility of the circum-Pacific belt, and of the Alpine-Himalayan belt bordering the former Tethys may be seen as a response to the collision of converging crustal plates. The rates of convergence are controlled by the growth-rate of the new ocean

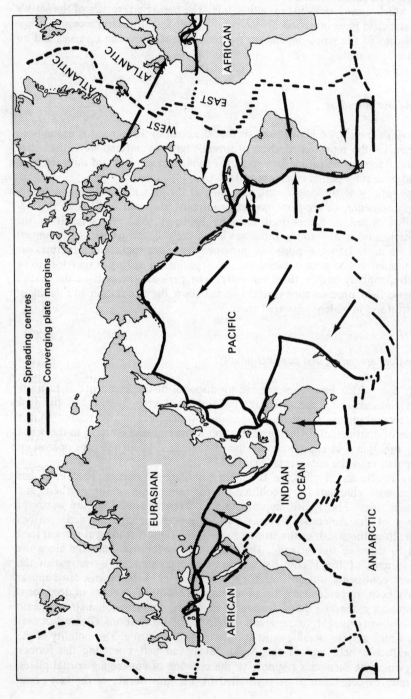

Fig. 1.7. The principal crustal plates of the earth today. The arrows (based on Le Pichon) indicate motions resulting from sea-floor spreading

Table 1.1. DIFFERENTIAL MOVEMENTS BETWEEN CONVERGING CRUSTAL BLOCKS
(according to calculations of Le Pichon, 1968)

Plates	Location	Rate (cm/yr)
E. Asian: Pacific	Kuriles trench	7.9
	Kuriles trench	8.5
	Japan trench	8.8
	Japan trench	9.0
	Mariana trench	9.0
	Mariana trench	8.9
Indian: Pacific	N. Tonga trench	9.1
	S. Kermadec trench	4.7
	S. New Zealand trench	1.7
	New Guinea	11.0
American: Pacific	E. Aleutian trench	5.3
	W. Aleutian trench	6.2
	W. Aleutian trench	6.3
Indian: Eurasian	Turkey	4.3
	Iran	4.8
	Tibet	5.6
	W. Java trench	6.0
	E. Java trench	4.9

basins from which the advancing plates move and have been calculated on this basis by Le Pichon at values of 5–10 cm per year (Table 1.1).

The ocean basins with their relatively simple structure are by this means adjusted with remarkable precision to complex and changing patterns of the continental masses. The integrated movements whereby relatively rigid lithospheric plates made up of continental and oceanic crust, together with portions of the underlying mantle, move relative to one another in response to the processes of sea-floor spreading, and 'consumption' of crust at destructive plate margins, defines a style of crustal activity known as plate tectonics. The characteristic arrangement of continental and oceanic units seen today has been produced by plate movements whose effects can be traced back for almost 1000 m.y. In the following chapters we shall deal first with the early stages, which were dominated by the evolution of a network of mobile belts and then with the later stages involving the break-up of the supercontinents and the dispersal of their fragments.

2

The Caledonides and their Forelands

I Preliminaries

The last few hundred million years of Precambrian time saw the initiation of a world-wide network of mobile belts most of which were to remain active until well into the Phanerozoic eon. In Laurasia (Fig. 2.1) this system was built around a fairly small number of cratons; most parts of the system remained mobile at least until the end of Lower Palaeozoic times and some are still in being at the present day. In Gondwanaland, the mobile belts enclosed a rather larger number of smallish cratons; much of this southern network was effectively stabilised well before the end of the Lower Palaeozoic and only a few branches remained active in later Phanerozoic times.

The late Precambrian and early Palaeozoic mobile belts of Laurasia will be considered in two groups: first, those which were stabilised by the end of the Palaeozoic era (the Caledonides of Europe and Greenland (this chapter), the Hercynides of Europe and Asia (Chapter 3) and the Appalachians of North America (Chapter 4): and, second, those in which activity continued through Mesozoic and Tertiary times (Chapters 7 and 11).

The adjective *Caledonian*, derived from the Roman name for a part of northern Britain, was originally applied by Suess to the north-east–south-west fold-belt which was developed in Scotland before the deposition of the Old Red Sandstone. This orogenic belt, the *Caledonides*, extends into England, Wales and Ireland on the one hand and Norway and Sweden on the other (Fig. 2.1). The Caledonian orogenic cycle is therefore the geological cycle which led to the formation of the Caledonides of these regions and, by extension, of the contemporaneous orogenic belts of other areas.

The beginning of the Caledonian cycle has not been precisely defined. Tectonic, metamorphic and depositional events can be recognised at least as far back as 800 m.y. and we regard these early events as phases of a long and

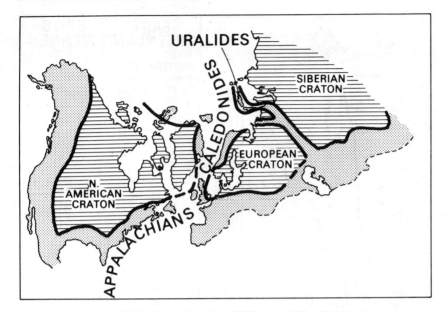

Fig. 2.1. The mobile belts and cratons of Laurasia in mid-Palaeozoic times

complex Caledonian cycle. Some authors, however, recognise two cycles in the evolution of certain Caledonian belts, the later being taken as the Caledonian cycle. Russian geologists, for example, distinguish a *Baikalian* cycle preceding a Caledonian cycle, and in Brittany and adjacent regions a *Cadomian* orogeny terminating at about 500 m.y. has been distinguished. These restrictions seem to us to obscure the essential unity of the long cycle responsible for building the Caledonides.

II The Main Geological Units

Orogenic activity in the Caledonian belts was preceded by the development of basins of deposition in which thick supracrustal successions accumulated. The centres of deposition shifted from time to time, with the result that it is often possible to distinguish a number of zones derived from the infilling of different basins which may show characteristic styles of tectonics and metamorphism.

In western Europe (Fig. 2.1), the principal craton was that made by the *Baltic shield and Russian platform*. This craton was flanked on the east by the *Ural Mountains*, a belt which remained mobile until the end of the Upper Palaeozoic and is most appropriately dealt with in a later chapter. The *Caledonides of Britain and Scandinavia* which flank the Baltic shield to the north-west have remained substantially unmodified since the end of the Lower Palaeozoic as have the *Caledonides of East Greenland and of Spitsbergen* which are represented in Fig. 2.1 as portions of the same belt. The craton which formed the *western*

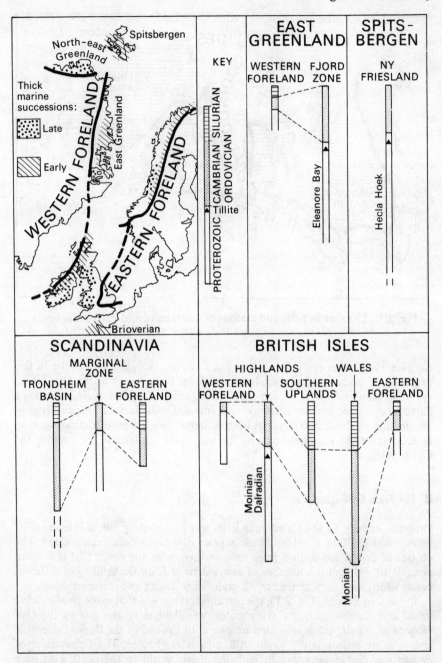

Fig. 2.2. The major components of some basin-successions deposited in the Caledonian mobile belt of north-west Europe, East Greenland and Spitsbergen; the columns are only roughly to scale

foreland in north-west Europe and Greenland has been fragmented by continental drift, a small part remaining along the western seaboard of *north-west Scotland*, and a larger part forming *central and western Greenland*.

III The Caledonian Cycle in Outline

The supracrustal cover-successions in the mobile belts range in age from late Precambrian to Silurian, collectively spanning a period of more than 400 m.y. and reaching thicknesses of 15 km or more (Fig. 2.2). Some units of the belt, which we may call the *early basins*, received the bulk of their fill before the end of the Cambrian period. The *late basins* received deposits through most of Lower Palaeozoic times. *Platform successions* on the cratons to east and west consist mainly of Lower Palaeozoic rocks; a conspicuous unconformity at the base of these successions records a widespread marine transgression in Cambrian times. The platform successions contain little or no volcanic material.

Figure 2.2 shows that deposition related to the cycle did not begin everywhere at the same time. Although dating and correlation of the Precambrian groups are difficult, broad variations in timing are demonstrated by the relationships of time-markers provided by late Precambrian glacigene deposits. Assemblages of turbidites and volcanics – the characteristic assemblages of unstable environments – are concentrated in the late basins and towards the top of the sequences in the early basins. These deposits are, in a broad sense, syn-orogenic. Post-orogenic deposits of non-marine facies are represented by the Old Red Sandstone, of Devonian age. The change of environment which led to the onset of molasse-type deposition took place at much the same time – around 400 m.y. – in every section of the belt, evidently reflecting a fundamental change in the crustal regime.

The tectonic history of the mobile belt began towards the end of Precambrian times with the definition of the early basins of deposition (Fig. 2.3). The boundaries of these early basins against the adjacent forelands may be taken to mark the original limits of the mobile belt. These limits appear to have shifted only slightly during the later stages of the cycle and coincide approximately with orogenic thrust-fronts developed towards the end of the Lower Palaeozoic.

Episodes of orogenic compression and metamorphism began in the early basins of deposition well before the end of the Precambrian and continued over a long time-span. In most of the late basins, metamorphism reached only low grades, whereas in some of the early basins high-grade metamorphism and migmatisation took place on a regional scale.

Towards the end of the Silurian period, the cycle approached its terminal stages; regional uplift was accompanied by deep erosion and the complementary accumulation of the Old Red Sandstone. The final stages of compression led to rupture and the development of thrust-zones at the orogenic fronts. The emplacement of late-orogenic granites overlapped with the period of molasse-formation and was completed in most regions by mid-Devonian times. These terminal events heralded the stabilisation of the Caledonian mobile belts throughout northern Europe, Greenland and Spitsbergen. In the more southerly portions of the European network and in America, stabilisation was delayed

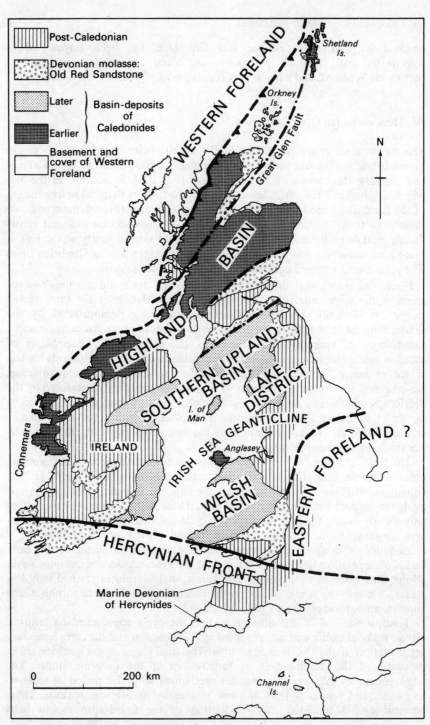

Fig. 2.3. The Caledonian mobile belt in Britain

until late Palaeozoic, Mesozoic or Tertiary times; in these regions, there was no molasse-formation and only limited granite-emplacement at the end of the Lower Palaeozoic.

IV The Early Basins of Deposition

The early basins of deposition in the Caledonides will be dealt with in the following order:

(1) Deposits of the Highland basin and north-west foreland in Britain
(2) Deposits of the southern basin in Britain and Armorica
(3) Deposits of the Caledonian basin and western foreland in East Greenland
(4) Deposits of Spitsbergen
(5) Deposits of the Sparagmite basins of Scandinavia
(6) The Infracambrian Glacial Deposits

1 The Highland basin and north-west foreland in Britain

In the Highland area of Scotland and its continuation in Ireland (Fig. 2.4), the cover-succession rests unconformably on a Lewisian basement. From whole-rock Rb–Sr dating of metasediments, it has been inferred that deposition of the *Moine Series*, the lower division, took place within the period 1000–800 m.y. The occurrence of tillites at a low level in the *Dalradian Series* and of Cambrian fossils in a limestone intercalated in greywackes at a high level suggests that deposition of this upper division spanned the Infracambrian – Cambrian boundary. The Moinian reaches a thickness of at least 7 km near the western border of the basin. The main tract of Dalradian rocks, in which their thickness probably exceeds 8 km, lies to the south-east of this portion of the Moinian outcrop and it is by no means certain that a Dalradian sequence was ever laid down over the thickest portion of the Moinian.

The equivalent *foreland succession* which fringes the western margin of the Highland basin consists of three divisions separated by angular unconformities of about 20°. The two lower divisions, grouped together as the Torridonian Series, reach a total thickness of up to 4 km. Whole-rock dating of shales from these divisions has given apparent ages of 1000–800 m.y. The uppermost unit, less than a kilometre in thickness, is of Cambrian and early Ordovician age.

Taken as a whole there are conspicuous differences between the basin and foreland successions. The foreland succession is entirely of piedmont, fluviatile and transgressive marine orthoquartzite facies. The thicker basin-succession begins with deposits of similar facies but passes up into groups characterised by turbidites and basic volcanics, some of which appear to be the time-equivalent of the shallow-water Cambro-Ordovician of the foreland. The coincidence in space between the basin-margin defined in the period 800–900 m.y. and the orogenic thrust-front which was not established in its present form until about 400 m.y. is worth noting.

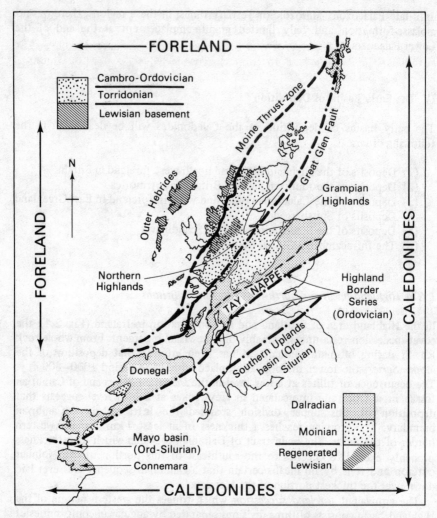

Fig. 2.4. The Highland area of the British Caledonides

The Torridonian sediments of the foreland succession rest on an irregular surface with a relief of at least 600 m. They are predominantly feldspathic sandstones, containing abundant debris derived from the gneissic Lewisian basement, together with pebbles of acid volcanics probably derived from late Laxfordian volcanics no longer preserved in situ. Current bedding in the sandstone divisions which overlie the lowest groups indicate a system of palaeocurrents radiating eastward from a number of foci, and Williams (1969) has suggested that these divisions represent alluvial fans laid down by rivers issuing from a mountain-tract in the foreland.

The Infracambrian glacigene horizons are absent in the foreland, and it appears that the episode of glaciation fell in the time-gap represented by the Torridonian–Cambrian unconformity. The Cambro-Ordovician *Durness succes-*

sion forms a transgressive succession which recurs in an almost identical form on, or within, the western border of the Caledonides in East Greenland, Spitsbergen and north-eastern America. A thin quartz-pebble conglomerate is followed by orthoquartzites which pass up from current-bedded rocks into the intertidal 'pipe-rock' penetrated by vertical worm-burrows. Thin shales are followed by the main division of the sequence, the dolomitic Durness limestone. Lower Cambrian faunas (including *Olenellus* which is scarcely represented elsewhere in Britain) extend up to the lower parts of the limestone, Middle and Upper Cambrian faunas are absent and the upper parts of the limestone carry Lower Ordovician (Canadian) fossils.

In the *Highland basin*, the Moine Series, which forms the lower part of the succession, resembles the Torridonian in that it is dominated by badly sorted psammites but carries a higher proportion of pelitic material and appears to rest on a basement surface of low relief. The direction of transport indicated by current bedding appears to be from south to north, roughly parallel to the margin of the basin. The monotonous lithology, the abundance of shallow-water depositional features and scarcity of carbonate rocks would be consistent with a deltaic environment of deposition.

Deposition of the Moine Series was followed, before the end of the Precambrian, by folding and metamorphism affecting at least the western part of the basin-fill (p. 45). It might be expected that the deposition of the overlying Dalradian Series would have been preceded by an important break, but no structural or metamorphic unconformity is known. Such an unconformity may have been obliterated by later folding, but it seems more probable to us that deposition was essentially continuous and that the early deformation was confined to the deeper parts of the succession. The lower divisions of the Dalradian (Table 2.1) form orthoquartzitic assemblages of quartzites, limestones (some oolitic or stromatolitic), pelites and black pelites which are remarkably consistent along the length of the mobile belt. A change of facies is heralded by the incoming of turbidites in the Middle Dalradian and the bulk of the Upper Dalradian, several kilometres in thickness, consists of graded greywackes and greywacke-pelites with pillow-lavas and basic ashes.

Table 2.1. THE DALRADIAN SUPERGROUP OF THE HIGHLAND BASIN IN SCOTLAND

UPPER DALRADIAN	Greywackes and pelites of turbidite facies with arenaceous limestones (Loch Tay Limestone etc.) near base. Spilitic pillow-lavas (Tayvallich lavas) occur in the south-west Highlands and a horizon of ashes (Green Beds) extends through most of the outcrop. In the centre of the belt the Leny Limestone, in the highest sub-division, has yielded a Lower Middle Cambrian fauna.
MIDDLE DALRADIAN	The division begins with a tillite horizon (Schichallion or Portaskaig conglomerate) which is followed by current-bedded quartzites (Islay Quartzite etc.) and by pelites, black pelites, calcareous pelites, thin limestones, quartzites and slide-conglomerates.
LOWER DALRADIAN	Impure current-bedded psammites and pelites pass up to a black pelite-limestone group often ending with a conspicuous limestone (Islay Limestone etc.)

2 The southern basins in Britain and Armorica

Fragmentary remnants of thick supracrustal groups including clastic sediments and acid volcanics with high-level acid intrusions underlie the Lower Palaeozoic basin-fill in Wales, south-eastern Ireland and the Welsh Border. They appear to date from late Precambrian times and may be tentatively regarded as deposits of one or more early Caledonian basins. The majority of these remnants – for instance, the sequence of acid volcanics (Uriconian) followed by grey and red clastic sediments (Longmyndian) at the eastern margin of the Welsh basin – show evidence of deformation prior to the deposition of the Cambrian. The *Monian* which forms the basement of the Irish Sea land-mass (p. 40) is exceptional in showing the effects of late Precambrian metamorphism and migmatisation, and has much in common with the middle and upper parts of the Dalradian Series. In the Armorican massif of Brittany a late Precambrian succession of eugeosynclinal aspect, the *Brioverian*, overlies a crystalline basement. Some details of this region are given in Chapter 3 (p. 63).

3 East Greenland

The cover-succession in the Caledonides of East Greenland consists of a very thick (up to 16 km) pile of Precambrian, Infracambrian and Cambro-Ordovician strata corresponding roughly in time-span with that of the Highland basin (Fig. 2.2). The deposits of the mobile belt crop out at intervals across a tract well over 200 km in breadth (Fig. 2.5). Within a central tract (the 'central metamorphic complex') the grade of metamorphism is high and details of the stratigraphy are obscure. The *fjord zone* which flanks this tract on the east reveals a well-preserved supracrustal assemblage which provides the standard succession given in Table 2.2. The less accessible *nunatak zone* on the western side of the central metamorphic complex displays a number of isolated groups (the Petermann and Gregory Series near Kejser Franz Josephs Fjord, the Krummedal and Charcot Land sequences further south) whose relationships are obscure. The *western foreland* is largely hidden by the inland ice, but portions of a thin cover-succession, ranging from Cambrian to Silurian in age, are exposed locally near the thrust-front and more widely in northern Greenland where a salient of the craton projects into the angle between the Caledonides of eastern and northern Greenland.

The basin-sequence of the fjord region includes three principal divisions which are itemised in Table 2.2. The Eleanore Bay Group which forms the main part of the succession is wholly Precambrian and consists mainly of psammites, pelites and dolomites (Plate I). The Morkebjerg Formation (roughly equivalent to the 'Tillite Group' of Haller) is a glacigene group assigned to the Infracambrian. It is separated by minor unconformities or disconformities both from the Eleanore Bay Group and from the overlying Cambro-Ordovician sequence of quartzites and dolomites. Neither turbidites nor volcanic rocks are present in significant amounts. The Cambro-Ordovician division in the fjord region bears a remarkable resemblance to the Durness succession. Apart from the fact that the Greenland sequence amounts to some 3 km instead of less than a

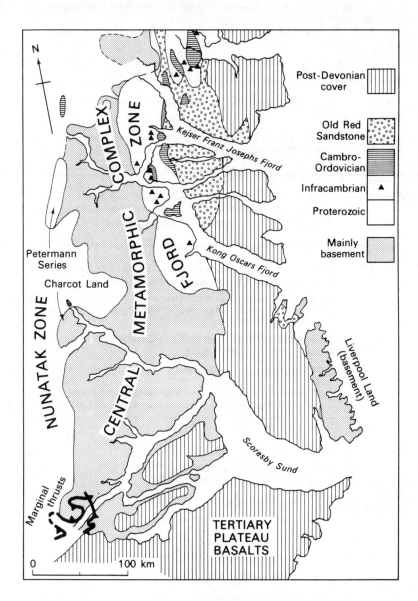

Fig. 2.5. The Caledonides of East Greenland (mainly after Haller, 1971)

Table 2.2. THE CALEDONIDES OF EAST GREENLAND

The succession in the fjord zone (based on Haller, Katz and others)

LOWER AND MIDDLE ORDOVICIAN	Cass Fjord (oldest), Cape Weber, Narhvalsund and Heimbjerge formations: dolomitic limestones. *c*. 2100 m

←——————————————— *non-sequence* ———————————————→

MIDDLE CAMBRIAN	Hyolithus Creek and Dolomite Point formations: dolomites. *c*. 600 m
LOWER CAMBRIAN	Kløftelv (oldest) Bastion and Ella Ø. formations: sandstones passing up to interbedded limestones and shales. *c*. 300 m

←—————————— *often a non-sequence or minor unconformity* ——————————→

EOCAMBRIAN	Morkebjerg formation: tillites interbedded with arkose, sandstones, conglomerates, shales, often followed by dolomite, black shale, limestone, sandstone and mudstone. maximum *c*. 500 m

←—————————— *local non-sequence or minor unconformity* ——————————→

PRECAMBRIAN	Nökkefossen formation: a carbonate-formation of limestones and dolomites, with stromatolites. maximum *c*. 1000 m
	Brogetdal formation: multicoloured series consisting mainly of red and green pelites and marls interbedded with dolomitic limestones. maximum *c*. 1000 m
ELEANORE BAY GROUP	Agardhsberg formation: quartzites and feldspathic psammites with minor pelites, maximum *c*. 2000 m
	Alpefjord formation: alternating quartzites and pelites with a central sub-division consisting of dolomitic limestones, pelites and quartzites. maximum 6500 m, no base in fjord zone, but in Gaaseland, apparently equivalent strata overlie a tillite horizon resting on crystalline basement.

kilometre, even minor lithological sub-divisions of the two sequences can be matched. Trilobites, gastropods, cephalopods, stromatolites and worm-traces found in both regions reveal affinities with a 'Pacific' faunal province extending to Spitsbergen, Arctic Canada, western North America and the western foreland of the Appalachians.

4 Spitsbergen

The original position of the Spitsbergen group of islands prior to the onset of continental drift was probably rather nearer to the north-east of Greenland than the present position of the group (Fig. 2.1) and it is therefore no surprise to find that the history of deposition in the Caledonides followed a course not unlike that discussed above. The *Hecla Hoek succession* ranges from late Precambrian through Infracambrian to Cambro-Ordovician and in some localities appears to provide a record of almost continuous deposition over this time-range. In spite

Plate I. The Upper Eleanore Bay Group in Berzelius Bjerg, Southern Lyellsland, East Greenland. The section (highest point about 1800m) shows gently dipping Quartzite Series, Multicoloured Series and Limestone-dolomite Series. (Reproduced by permission of the Director, Greenland Geological Survey.)

of this apparent continuity, there is radiometric evidence of an early period of metamorphism taking place before the final end of deposition. The maximum thickness of the Hecla Hoek has been estimated at 20 km, the bulk of this being made by the lower and middle divisions (Table 2.3). The succession is exposed at intervals over a tract some 200 km in width, measured across the tectonic trend. Considerable lateral variation has led some authors to think in terms of deposition in two basins intermittently separated by a north–south ridge.

5 Scandinavia

The distribution of stratigraphical groups and tectonic elements in the Caledonides of Scandinavia presents, in many ways, a mirror-image of that of comparable units in the British Caledonides. In Scandinavia, the Baltic shield provides an eastern foreland separated from the mobile belt by westward-dipping thrusts. The oldest deposits of the mobile belt, of Precambrian to early Ordovician age, occupy basins immediately alongside the orogenic front, while thick younger accumulations of Cambrian to Silurian age lie further from the front in one or more internal basins. The succession of the foreland ranges from Cambrian to Silurian. The outer boundaries of the geosynclinal basins against the eastern foreland which were established in the late Precambrian appear to have coincided rather closely with the position of the thrust-front formed in the late

Table 2.3. THE CALEDONIDES OF SPITSBERGEN

The Hecla Hoek Group Succession in Ny Friesland (western basin)
(based on Harland and others)

Age	Division	Description
LOWER ORDOVICIAN	UPPER HECLA HOEK	Oslobreen Series: thin sandstones passing up into dolomites and limestones, dolomites carry L. Cambrian–L. Ordovician faunas. c. 1000 m
LOWER CAMBRIAN		
INFRACAMBRIAN		Polarisbreen Series: dark shales with one or more tillites. c. 1000 m
LATE PRECAMBRIAN	MIDDLE HECLA HOEK	Upper division, predominantly dolomites and limestones, c. 2300 m
		Lower division, predominantly quartzites, impure psammites and pelites, c. 3800 m
	LOWER HECLA HOEK	Predominantly detrital, psammites, semi-pelites and pelites, with many basic layers, possibly representing lavas and pyroclastics, possible acid pyroclastics. 7–8000 m

stages of orogeny, and the marginal deposits of the basin have therefore been disrupted by thrusting.

The deposits of the cycle fall into two principal divisions (Table 2.4). The underlying *Sparagmite Group* is Precambrian and Infracambrian and includes a tillite formation comparable with those already mentioned. It is widely developed in the western part of the mobile belt where it reaches thicknesses usually considerably less than 6 km; but does not extend out onto the foreland (Fig. 2.6). The Lower Palaeozoic sequence forms a relatively thin cover on the foreland, where it rests directly on the basement-rocks of the Baltic shield. In the marginal parts of the mobile belt, Lower Palaeozoic of foreland facies rests, with only a minor discordance or non-sequence, on the Sparagmite Group. In the central parts of the belt the Lower Palaeozoic succession is of considerable thickness and is of a eugeosynclinal facies.

The name of the Sparagmite Group (Greek, *sparagma* = fragment) gives an indication of the dominant lithology. The bulk of the group consists of impure feldspathic sandstones with which are associated impure pelites, a few

Table 2.4. SUCCESSIONS OF THE SCANDINAVIAN CALEDONIDES

		Bergen arcs (Kolderup)	Trondheim region	Eastern marginal region, Lake Mjøsa	Oslo Region
SILLURIAN	U				sandstones and shales, shelly limestone
	M	Phyllites, sandstones, conglomerates	*Horg Group*: black pelites, psammites quartzite-conglomerates	shales, sandstones, limestones	
	L				shales, sandstones conglomerates
ORDOVICIAN	U	Phyllite, limestone	*Upper Hovin (and Røros) Group:* schists, psammites, tuffs, polymict conglomerate.	(no deposits)	shales and limestones with shelly faunas
	M	greywacke, greenschist, limestone	*Lower Hovin Group* schists, psammites, limestones, green-stones, conglomerates.	shales and limestones	
	L	schist and phyllite	*Støren Group:* pillow-lavas, greenstones, keratophyres, pyroclastics		black shales
CAMBRIAN	U	schist and phyllite	*Gula Schist Group:* micaschists, lime-stone, quartzite-conglomerate	Alum Shales	*Alum Shales:* black shales M–U Cambrian
	M				
	L			sandstones and shales	Basal conglomerate
INFRACAMBRIAN			(Infracambrian and Precambrian probably represented in psammitic pelitic meta-sediments)	Vemdel formation (quartz-sandstone) Ekra shale Moelv tillite Moelv sparagmite ⎫⎬⎭ Upper Sparagmite	
PRECAMBRIAN				Biri shales and limestone Biri conglomerate Brøttum shale and limestone Brøttum sparagmite ⎫⎬⎭ Lower Sparagmite	

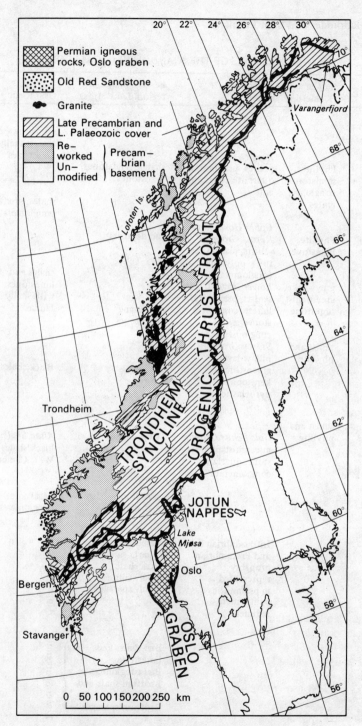

Fig. 2.6. The Caledonides of Scandinavia

conglomerates and occasional dolomitic limestones. Many of the psammites are current-bedded, the limestones include stromatolite-bearing and oolitic types and the assemblage generally appears to be of shallow-water facies.

The Lower Palaeozoic, which follows the Sparagmite in the marginal parts of the mobile belt, extends upward to the Lower Ordovician and consists mainly of shales with thin sandstones, limestones and dolomites. There is no evidence to show whether later Ordovician or Silurian were deposited in this zone. On the foreland, on the other hand, a full, though thin, Lower Palaeozoic sequence is represented (p. 42).

6 The Infracambrian glacial deposits

Repeated reference has been made in the last few pages to rocks interpreted as tillites occurring in and on the margins of the Caledonides. Most of these rocks appear to lie at roughly the same stratigraphical position, that is, below the base of the fossiliferous Cambrian but separated from this neither by a major unconformity nor by thicknesses of sediments exceeding about 1000 metres. They are therefore assigned to a period immediately preceding the Cambrian for which we adopt the name Infracambrian (many Scandinavian geologists use *Eocambrian* in the same sense). Glacigene rocks of similar age have been recorded from numerous localities in Asia, Australia, Africa and western North America. The general question of a *world-wide Infracambrian glaciation* is therefore more appropriately considered after these occurrences have been mentioned (pp. 156–8). Meanwhile we can adopt Kulling's term *Varanger Ice-age* for this episode as recorded in northern Laurasia and use the adjective Varangian (Varegian of some authors) for the corresponding deposits; both names are derived from Varangerfjord in northern Norway.

The Varangian tillites of northern Laurasia usually overlie dolomitic limestones (often oolitic or stromatolitic), and are followed by shallow-water psammites. Within the mobile belt, they typically form intercalations in shallow-water marine successions and for this reason have been interpreted as marine tillites, perhaps laid down by shelf-ice. Along the eastern margin of the Scandinavian Caledonides, on the other hand, the tillite group is often unconformable on the Lower Sparagmite, or the basement. Here it sometimes rests on a surface of irregular relief, perhaps terrestrial, and in the type locality of Varangerfjord a planed-off and striated glacial pavement of Lower Sparagmitian underlies it.

In structure and lithology many of the Varangian rocks are typical tillites displaying an assortment of irregularly distributed and poorly rounded rock fragments in an unbedded matrix. Boulders of dolomites and other types derived from underlying members of the succession are common in the lower tillites whilst the upper tillites contain exotic crystalline rocks, including pink alkaline granites, granophyres, porphyries, gneisses and metasediments which remain remarkably consistent from Spitsbergen to Ireland, a distance of some 3000 km. A source for this suite in the Baltic shield has been proposed by Scandinavian geologists. We may perhaps envisage shelf-ice and icebergs riding out into the Caledonian basins from a major ice-sheet over the eastern foreland.

V The Base of the Cambrian System

In some early basins in the Caledonian mobile belts, as we have seen, deposition went on without significant interruption from the latest Precambrian into the Cambrian. The minor unconformities or non-sequences between the fossiliferous Lower Cambrian and the underlying strata are of no greater magnitude than breaks at other levels in the succession; and in many localities the succession is essentially continuous. On the other hand, where a Lower Palaeozoic cover extends onto the stable cratons its basal members, almost without exception, rest with spectacular unconformity on the underlying basement.

Since the stratigraphy of the older cover-formations of the Caledonides remained for many decades almost unknown, it is hardly surprising that successive generations of European geologists have attached great significance to the unconformity which separates the Lower Palaeozoic strata from the underlying Precambrian on the cratons. This unconformity does indeed record a marine transgression of tremendous extent which led to the flooding, among other regions, of considerable parts of the Baltic shield and Canadian shield. A transgression of such dimensions certainly provides a geologically important datum, but the Cambrian transgression was not unique in this respect and, moreover, was not recorded in Gondwanaland where the Cambrian was generally a time of regression. If we accept the long-term orogenic rhythm as a basis for the division of geological time, we must admit that the Cryptozoic–Phanerozoic boundary is of comparatively minor significance. This view is reflected in the common use of such terms as 'Eocambrian' or 'Infracambrian' which imply a close connection between the Cambrian and the underlying rocks. The real importance of the Infracambrian–Cambrian boundary is biological.

VI Early Orogenic Activity

Evidence showing that orogenic activity began even before the complete infilling of the early basins has already been mentioned. More widespread episodes of deformation, accompanied in some areas by metamorphism and migmatisation, followed the ending of deposition in the early basins (see p. 48). In northern and western Britain, these important episodes appear to have taken place in late Cambrian to early Ordovician times. Detritus derived from metamorphic rocks is abundant in Ordovician and Silurian clastic sediments of the Southern Uplands and Mayo basins, which were flanked by crystalline massifs made of folded Dalradian rocks. In a broad sense, therefore, the Lower Palaeozoic sequences of the later basins in the Caledonides may be regarded as syn-orogenic deposits laid down after the onset of orogenic deformation. A somewhat different history was followed by the southern branch of the mobile belt extending into the Channel Islands and Brittany. In this branch, the orogenic phase which ended the accumulation of the Brioverian – the *Cadomian orogeny* dated at 600– 550 m.y. – was followed by granite-emplacement and stabilisation (see later, p. 64).

VII The Later Basins of Deposition

1 Distribution

The changes outlined in the last section, overlapping to some extent with the final stages in the history of the early Caledonian basins, led to marked changes in the siting of zones of subsidence. A number of relatively narrow basins were defined within or alongside the early basins, separated from each other in some places by positive areas subject to erosion or receiving only thin successions (Fig. 2.7). The forelands remained, for the most part, as low-lying areas subsiding relatively slowly: but more vigorous subsidence defined one or two cratonic basins such as the Ludlovian trough which extends southward from southern Sweden.

Upper Ordovician and Silurian strata are, with trifling exceptions, absent from the orogenic regions of East Greenland and Spitsbergen and from the Highland zone of Britain. They are not represented among the pebbles of the molasse-conglomerates of these regions and we conclude that thick successions of later Lower Palaeozoic rocks did not accumulate in many of the northern parts of the Caledonian system. Indeed, the occurrence of detritus derived from the Highland zone in Ordovician and Silurian sediments of adjacent younger basins in Britain indicates that the zone stood at least partly above sea-level and was subject to denudation.

Much of the south-eastern part of the Caledonian belt in Britain was occupied by subsiding Lower Palaeozoic basins elongated parallel to the length of the belt. The *Southern Uplands Basin* of Scotland and north-east Ireland may have been developed on a basement of folded Dalradian. The *Mayo basin* of western Ireland evidently developed as a downwarp within the early fold-zone. In both these basins the oldest strata of the cover-sequence are Arenigian (Fig. 2.8). To the south, a thick sequence possibly including Cambrian exposed in the *Lake District*, the Isle of Man and south-east Ireland, is separated by a NE–SW zone of restricted deposition (the 'Irish Sea land-mass'), from the Welsh basin. Deposition in the *Welsh basin* appears to have begun in Infracambrian times, after folding and metamorphism had already taken place in the Monian (p. 40).

In Scandinavia there is no obvious metamorphic unconformity between the Sparagmitian and Cambro–Ordovician on the one hand and the later Lower Palaeozoic sequences on the other. Two facies zones are distinguished in the Ordovician strata, a more easterly zone passing through Trondheim characterised by detrital sediments with volcanics and a more westerly coastal zone by limestones. Beyond the orogenic front, cratonic cover-successions of Lower Palaeozoic age spread widely along the southern border of the Baltic shield and over the stable block of central England.

Taken as a whole, the sequences which accumulated in the later basins of deposition in the Caledonides differ from those of the early basins in several respects. They include bulky and varied volcanic contributions and contain thick turbidite-groups indicating repeated instability. Rapid variations of thickness and facies express the increasing complexity of the mobile belts both in terms of topography and of structure.

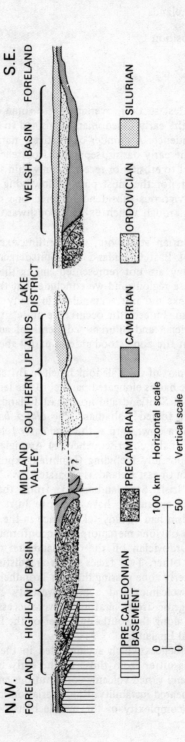

Fig. 2.7. Diagrammatic cross-section illustrating the inferred structure of the British Caledonides after the infilling of the later basins of deposition but before the terminal stage of end-Silurian folding and uplift

2 Stratigraphical nomenclature

The regions dealt with in this chapter include those in which the three Lower Palaeozoic systems were first distinguished and in which the classic studies of Lower Palaeozoic stratigraphy were made. All three systems were defined by reference to the succession in and on the borders of the Welsh basin, after a long tussle in which Sedgwick, Murchison and Lapworth were the main protagonists. Initially, only Cambrian and Silurian were recognised. The separation of the Ordovician system was proposed by Lapworth largely as a means of resolving a dispute between Sedgwick and Murchison as to the position of the boundary between Cambrian and Silurian. The independent status of the Ordovician was not immediately accepted and many early publications classify as Lower Silurian strata now referred to the Ordovician.

There are still discrepancies between the conmmon systems of sub-division adopted in different countries (Table 2.5). British geologists have been almost alone in classifying the Tremadocian as Cambrian rather than Ordovician: a

Table 2.5. STRATIGRAPHICAL DIVISIONS OF THE LOWER PALAEOZOIC

	EUROPE		NORTH AMERICA
SILURIAN	Downtonian (usually classified with Devonian)		
	Ludlovian		Cayugan
	Wenlockian		Niagaran
	Llandoverian		Alexandrian Medinan
ORDOVICIAN	Ashgillian	} Bala	Cincinnatian
	Caradocian		
	Llandeilian		Champlainian
	Llanvirnian		
	Arenigian		Canadian
	Tremadocian (British usage places this division in Cambrian)		
CAMBRIAN	Upper		Croixan
	Middle		Albertan
	Lower		Waucoban

difference in usage which can confuse discussions of the early Caledonian basins of Europe and North America. The position of the Downtonian is also in doubt; originally assigned to the Silurian, with which it is conformable in the type area, it is now more often classified as partly or wholly Devonian.

The principal zone-fossils employed in sub-division of the Lower Palaeozoic systems are graptolites, trilobites and brachiopods. The graptolites are the principal fossils of off-shore deposits in the geosynclinal basins, whereas trilobites and brachiopods occur mainly in the shelf-sea deposits of the foreland and the inshore deposits of the basins. The successions of these contrasted environments are therefore often zoned on different lines and, in Britain especially, distinctions have been drawn between a *graptolitic facies*, broadly geosynclinal, and a *shelly facies* of marginal or shelf-sea environments.

3 The western foreland

Along the western margins of the Caledonides in Britain and Greenland, no Lower Palaeozoic strata younger than mid-Ordovician are preserved; but in north-east and north-west Greenland, the cratonic cover-succession includes strata ranging from early Cambrian to late Silurian. This succession reaches total thicknesses of not more than about 3000 m and is interrupted by non-sequences and minor discordances. The bulk of it consists of limestones and shales carrying shelly faunas, as well as sporadic graptolites and cephalopods. Similar successions extend into Arctic Canada, suggesting that shelf-seas may have spread widely over the craton.

4 The basins of Mayo and the Southern Uplands

The termination of deposition in the Highland basin was rapidly followed by folding and metamorphism of the contents of the early basin. The later basins, to be considered first lie within, or on the flanks of, the area occupied by the Highland basin; subsidence appears to have begun in early Ordovician times and a large volume of erosion-debris from the Highland zone was trapped in the new basins. The Mayo basin of western Ireland is a narrow east—west structure flanked on the north by the main mass of Dalradian rocks and on the south by the Connemara massif of Dalradian. The Southern Uplands basin, with its continuation in north-east Ireland, is separated from the Highland zone by a tract some 70 km broad which carries only a thin platform succession of Silurian: this tract is the *Midland Valley of Scotland*, a graben of middle Palaeozoic age. There are indications that a massif standing above sea level, or subsiding relatively slowly, at times separated the Southern Upland basin from those of the Lake District and Wales.

Several major faults which traverse the region under consideration should be mentioned at this point. The *Highland boundary fault* forms the northern limit of the Mayo basin and continues into Scotland as a lineament separating the uplifted massif of the Highland zone from the Midland Valley graben. The *Southern Uplands fault* defines the southern side of the Midland Valley graben and the north-western margin of the Southern Uplands basin. Both faults were probably active from early Ordovician times and are characterised by the occurrence of serpentine pods. They are evidently deep fractures and the Southern Uplands fault is regarded by some authorities as a suture of major importance (p. 50).

The filling of the Mayo trough (Fig. 2.3) began in the Arenigian, was interrupted by an episode of late Ordovician folding, and thereafter continued up to Wenlockian times. The first episode was volcanic, leading to the eruption of basic pillow-lavas. A similar Arenigian volcanic episode of ophiolitic type is widely recorded in Britain, not only in the Mayo, Southern Uplands and Welsh basins but also at a number of localities beyond the basin-limits in north-western Ireland and along the Highland boundary region (Table 2.6). It has been identified as a critical stage in the evolution of the mobile belt, some authors envisaging a tapping of mantle material as a result of orogenic deformation and others (Dewey, 1969) assigning the volcanism to an oceanic trench environment. This activity coincided roughly with the great tectonic changes involved in the elimination of the Highland basin and initiation of subsidence in new troughs. Overlying the pillow-lavas of the Mayo basin are marine and coastal-plain clastic sediments representing detritus poured into the basin from the adjacent massifs. The Ordovician is said to reach maximum thicknesses of some 12 km, the Silurian rather less than 2 km. The Lower Ordovician sediments are mainly turbidites, the later Ordovician mainly badly sorted deltaic sandstones. The Silurian, which follows unconformably, begins with thin non-marine and littoral conglomerates and sandstones and continues with better sorted neritic sediments.

The distribution of bouldery slide-rocks in the turbidite groups suggests that material was contributed from both flanks, though axial flow of turbidity currents is indicated by sole-markings. The later Ordovician delta-flat sediments wedge out northward and appear to have been derived largely from the south. Volcanic activity in the basin, after the initial basic episode, is recorded mainly by numerous beds of ash and felsitic welded tuff emitted by explosive activity from acid volcanoes near the margins of the basin (Table 2.6).

In the Southern Uplands basin (Fig. 2.8, Table 2.6) Arenigian pillow-lavas marking the basic igneous episode already mentioned form the lowest units of the succession. The *Ballantrae complex* associated with these lavas on the west coast of Scotland has been interpreted as an upthrust wedge of oceanic crust (see p. 50). During the period of some 25–35 m.y. from Llandeilo or Caradoc to late Wenlock times, successive wedges of turbidites were laid down in the basin, thinning abruptly towards the south-east and sometimes passing laterally in this direction into condensed successions of graptolitic black shales. Although the orientation of sole-structures shows that turbidity currents frequently moved axially along the basin, the bulk of the detritus appears to have been derived from sourcelands to the northwest, that is, from the rocks of the Highland zone. Volcanic activity was of little importance after the initial Arenigian episode, but andesitic–rhyolitic lavas and pyroclastics of Llanvirn or Llandeilo age reach thicknesses of about 3 km (the Borrowdale volcanics) in the Lake District, further south.

At the north-western side of the Southern Uplands basin a platform

Table 2.6. VOLCANIC ROCKS IN BASIN-FILLINGS OF THE BRITISH CALEDONIDES

		Highland zone	Southern Uplands and Girvan	Mayo basin	Lake District	Wales
SILURIAN	U			? Sp (AR)		A (eastern border)
	M					
	L					
ORDOVICIAN	U					S R
	M		S A T	R A	(R)	S R
	L			R A / S Sp	A B	S A R
CAMBRIAN	U					
	M					
	L	Highland border (S Sp) / S B ? Sp	S Sp			
PRECAMBRIAN						R

A andesites   B basalts
R acid volcanics   T trachytes
S spilites
Sp serpentine

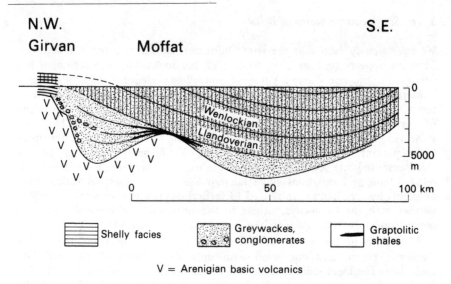

Fig. 2.8. Schematic section illustrating variations of thickness and facies in the Lower Palaeozoic basin of the Southern Uplands (based on Williams, George and others)

succession equivalent to that of the basin is exposed near the west coast. This succession of the *Girvan area* is only some 1000–2500 m in thickness and consists of conglomerates, sandstones, mudstones and thin shelly limestones. It thickens abruptly south-eastward at the margin of the basin where fault-movements appear to have controlled subsidence and passes into the monotonous greywackes of the clastic wedges. The lateral variations from the margin to the basin, and within the basin, were demonstrated as long ago as 1878 by Charles Lapworth who was for some years the master of a village school near Moffat. His zoning of the succession by means of graptolites provided the key to both the succession and the structure.

The tract of Upper Ordovician black shales in the Moffat region was for long considered to mark the deepest part of the basin. The identification of the greywackes in the clastic wedge to the north-west of this tract as turbidites deposited by axial currents had modified this concept, since axial turbidity currents must be assumed to follow the lowest part of the sea-floor. The black shales are now regarded as starved sediments accumulating on a rise to the south-east of the axis of the trough.

Variations in the Silurian successions follow a similar pattern. The greywacke turbidites of Middle Silurian age, however, extend south-eastward over the site of the Ordovician black shales and appear to be derived in part from the south-east. These Silurian greywackes reach thicknesses of up to 8 km. The equivalent deposits of the platform area in the Midland Valley are only a few kilometres in thickness. They pass up without a break into red sediments with fish and eurypterid remains, heralding the onset of terrestrial deposition.

5 The South-eastern basins of Britain

We have already seen that the later Ordovician and Silurian basin in southern Scotland appears to have been flanked on the south-east by a region of less profound subsidence. There is yet more compelling evidence that a similar region bordered the Welsh basin on its north-west side. Portions of this resistant region – the *Irish Sea Land-mass* or Irish Sea geanticline – appear in the island of Anglesey and in inliers to the south-east of the Lake District where the Lower Palaeozoic successions are thin, usually of shallow-water facies and interrupted by important unconformities. Within the *Welsh basin*, variations in the thicknesses indicate repeated shifting of the zones of maximum deposition while minor lithological variations and stratigraphical breaks suggest tectonic unrest. Some of these variations are related to vertical displacements on faults roughly parallel with the basin-edge, others to the occurrence of lenticular volcanic groups some of which appear to have been built up into islands.

Deposition in the Welsh basin began at or before the beginning of the Cambrian period and continued with only short interruptions until late Ludlovian. The basin-sediments are almost entirely detrital and are characterised by the recurrence at many levels of groups of turbidites. Interbedded with the turbidites are siltstones, flags, shales and black pelites, as well as shallow-water sandstones near the margins of the basin and around local rises within it. The scanty faunas of graptolites, trilobites and brachiopods contrast with the much more varied and abundant faunas of the south-east foreland (pp. 41–2). The maximum thickness of Lower Palaeozoic sediments in the basin is estimated by Jones (1938) as well over 12 km. At the end of the period of deposition, the surface of the basement appears to have sloped down from the margin towards the depths of the basin at a maximum angle of about 15°. The maximum rates of accumulation over the three Lower Palaeozoic periods as estimated by Jones, are between 30 cm in 5000 years and 30 cm in 3700 years.

Volcanic rocks reach an aggregate thickness about 2 km in some parts of the basin. They are mostly of Ordovician age, although there are some minor Silurian groups and, in North Wales, a considerable thickness of Infracambrian volcanics (Table 2.6). A remarkable variety of rock-types is represented, including spilitic pillow-lavas, basalts, andesites and many kinds of acid rocks. These acid types predominate in Cader Idris (Arenigian) and the Snowdon ranges (Caradocian). Pyroclastic rocks including welded tuffs are often as abundant as lavas and bodies of granophyre, felsite or rhyolite interpreted as high-level intrusions are not uncommon. The extrusion of volcanics and updoming by the associated intrusions built up several centres into volcanic islands which contributed erosion-debris to the surrounding sediments.

6 The Scandinavian basins

The later members of the Lower Palaeozoic sequence of the Caledonides in Norway are interfolded with and metamorphosed along with the underlying rocks. Fossils are scarce and details of the stratigraphy are not well known. Table 2.4 summarises the succession and indicates some useful markers. In the *coastal belt*, volcanics low in the succession are followed by pelites associated with carbonate groups in which Lower Ordovician faunas comparable with those of the western foreland (p. 36) have been identified. The youngest fossils found in this zone are of mid-Silurian age. In the Bergen area, two arcuate belts of strongly metamorphosed supracrustals – volcanics, psammites, polymict conglomerates and marbles – contain Upper Ordovician and Lower Silurian corals, crinoids, brachiopods and graptolites.

In the *Trondheim region*, the bulk of the later basin-fill is made up of detrital sediments and basic volcanics. An important group of Lower Ordovician greenstones carries many small copper-bearing sulphide ore-deposits. It is followed by varied detrital metasediments, including tuffaceous material, which are tentatively assigned to the Middle and Upper Ordovician. Rapid variations of lithology, with the occurrence of local unconformities and of polymict conglomerates, point to instability in the basin and there are indications of a fold-phase (the *Trondheim disturbance*) in early Ordovician times.

The instability of the mobile belt is still more clearly illustrated by the rocks of the *Jotunheim* to the south-east of the Trondheim zone (Fig. 2.6) where the great Jotun nappes of basement igneous rocks rest with thrust contacts on Cambro-Ordovician of foreland facies. The basement rocks of the nappes carry a cover of Cambro-Ordovician pelites, conglomerates and greenstones laid down before the period of thrusting. Sandwiched between lower and upper Jotun nappes is a group of coarse feldspathic sandstones, pelites and conglomerates containing erosion-debris from the igneous rocks of the nappes. This formation, the *Valdres Sparagmite*, reaches about a kilometre in thickness and has been tentatively assigned to the late Ordovician; it is regarded as a syn-orogenic deposit of flysch type laid down during the emplacement of the nappes and over-ridden by the upper nappe. Very similar pelites and polymict conglomerates are sandwiched between nappes of basic and ultrabasic igneous rocks in the Shetland Islands.

7 The south-eastern foreland

The cover-sequences laid down on the south-eastern foreland are exemplified by two well-known successions, those of Shropshire in Britain and of the Oslo region in Norway. In both, a transgressive Cambrian sequence rests with profound unconformity on older rocks. Infracambrian strata such as those represented in the adjacent Caledonian basins are missing. The bulk of the sediments deposited are of neritic facies, turbidites are almost absent and volcanics very restricted. The total thickness of Lower Palaeozoic is generally less than 3 km.

The foreland succession of England is incomplete in many places, the Ordovician being absent over wide areas. Well sorted basal sandstones are followed by shales alternating with thin sandstones and limestones. Shoreline facies are seen at several levels near Precambrian inliers such as those of the Longmynd and the Wrekin which formed islands in the shelf-seas. The thin limestones of the foreland carry richer and more varied faunas than the rocks of

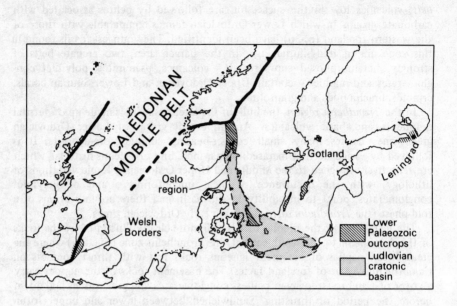

Fig. 2.9. The principal outcrops of Lower Palaeozoic on the eastern foreland of the Caledonides. The accompanying diagram (after Stormer, 1967) shows lithological variations in the Lower Palaeozoic of the Oslo region in relation to the supply of detritus from the rising mobile belt

the basins; some are packed with brachiopods, others contain abundant corals, trilobites and echinoderms.

Towards the top of the Ludlovian shales, marine faunas become increasingly restricted until finally *Lingula* remains almost alone. *The Ludlow Bone Bed*, a thin but persistent winnowed bed made largely of phosphatised fish teeth and scales and eurypterid remains, marks the upper limit of indications of marine deposition. It is followed conformably by red or brown shales and cornstones which represent the first deposits of Old Red Sandstone facies (p. 55).

The foreland successions of Scandinavia are best seen in the Oslo region (Fig. 2.9, Table 2.4) and in Jemtland in Sweden. Shallow-water sandstones and limestones alternate with shales, many of which are carbonaceous. These black shales, exemplified by the Alum Shales of the Oslo region, were remarkably extensive in southern Norway and Sweden during the Middle and Upper Cambrian and Lower Ordovician, suggesting the persistence of anaerobic bottom-conditions. The later Ordovician and Silurian deposits of the Scandinavian foreland are characterised by a number of incursions of coarser psammitic sediments, derived mainly from the west, which appear to represent material eroded during periods of uplift in the mobile belt. Such psammites are prominent in the Llandeilian, Ashgillian, Lower Llandovery, Lower Wenlock and late Ludlovian. A late Silurian cratonic basin extending southward from the Oslo region through southern Sweden and across the Baltic received almost 2000 m of shales and mudstones mainly of Ludlovian age. The uppermost Silurian deposits are often non-marine, indicating a gradual emergence of the foreland in the final period before the phase of orogenic uplift.

VIII Structure and Metamorphism

1 The British Caledonides

In Britain the tectonic and metamorphic styles developed in the early and later basins are very different and it is inferred that the contents of each basin behaved more or less independently. In the Highland basin, deformation and metamorphism began in Precambrian times and the principal structural and metamorphic patterns were established by early Ordovician times, though stabilisation was delayed until the end of the Silurian. The structure is characterised by the occurrence of large recumbent folds (involving regenerated basement at deep structural levels) which are distorted by sets of younger folds. The metamorphic grade ranges up to amphibolite facies and is associated with regional migmatisation. A well-defined marginal thrust-zone (the *Moine thrust-zone*) follows the junction of the Highland zone with its western foreland.

In the later basins of deposition, episodes of disturbance recurring throughout the Lower Palaeozoic are indicated by angular unconformities and by the occurrence of polymict conglomerates and wedges of coarse sediments. The main structural patterns in these basins date from the end of the Lower Palaeozoic. The dominant fold-structures were developed on steeply inclined axial planes. The metamorphic grade was low, no regional migmatites were developed and the basement was not extensively regenerated.

2 The Highland basin: structure

In spite of the undoubted evidence of polyphase deformation and metamorphism spread over a span of several hundred million years, the rocks of
the Highland basin exhibit a fundamentally simple arrangement of tectonic and
metamorphic structures. The characteristic outcrop-pattern of lithological units,
and of the principal axial planes and dislocations, is arcuate, curving from
north—south to east—west as one goes south-westward. Near the north-western
border of the belt the earliest folds, which are almost recumbent, face westward

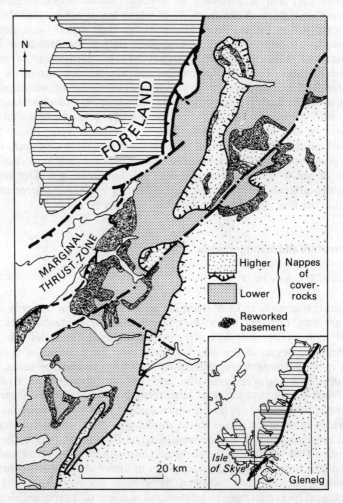

Fig. 2.10. Basement-cover relationships in the Highland zone of the British
Caledonides: simplified map of part of north-west Scotland showing outcrops of
regenerated basement interleaved with the Moinian cover as fold-cores and
thrust-slices

toward the foreland (Fig. 2.10). The structures which refold these nappes are developed on steeper axial planes which generally dip eastward and the dislocations of the Moine thrust-zone, the youngest structures of the western region, also dip eastward. In the central parts of the Highland belt, the character of the earliest folds is difficult to establish owing to a lack of marker horizons and to the complexities resulting from later refolding: some authors doubt the existence of major recumbent structures in this central region. Towards the south-eastern border of the Highland zone, very large recumbent folds which face south-eastward are developed in the Dalradian Series, in association with slides which tend to replace the normal, rather than the reversed, fold limbs (these dislocations are the *lags* of Bailey). The largest recumbent structure in the Dalradian is the anticlinal *Tay nappe* whose inverted limb extends for 400 km along the axial direction. Refolding on steep axial planes, generally with north-westerly dips, turns the anticlinal nappe-hinge downward along the Highland Border as a downward-facing synform.

The crudely symmetrical arrangement of early folds facing outward from a central tract and of later folds on axial planes fanning towards the margins of the Highland zone is found, when analysed in more detail, to incorporate structural elements of different styles and ages. There is considerable disharmony between the folds involving the basement and the lower parts of the Moinian and those involving the Dalradian. The flat-lying early structures of the Glenelg and Morar regions near the western margin of the zone (Fig. 2.10) incorporate anticlinal cores and thrust-nappes of regenerated basement gneisses. Kennedy and Ramsay have shown that some of these folds decrease in amplitude at higher structural levels. On the other hand, the great nappes of Dalradian near the south-eastern margin of the zone appear to be rootless structures with amplitude decreasing downward and may result from gravity-sliding.

In the north-west the flat folds and thrust-slices involving basement rocks, and the earliest structures which refold them, appear to have been associated with metamorphic episodes dated by Rb–Sr methods at 800–750 m.y. If the dating is reliable, these structures must have been formed before the bulk of the Dalradian Series had been deposited and therefore long before the formation of the symmetrically disposed Tay nappe and associated folds affecting Dalradian rocks. The remarkable consistency in the alignment of structures and in the location of certain isograds from the earliest episodes until the terminal stages of Caledonian orogeny, lead us to conclude that successive episodes were controlled by essentially similar tectonic and thermal regimes and are best assigned to a single long and complex geological cycle.

3 The Highland basin: metamorphism

Metamorphism within the Highland zone shows regional variations in grade and type which plot out to give a rather simple thermal structure (Fig. 2.11). A thermal gradient defined by a westward decrease of metamorphic grade marks the junction with the foreland. A similar gradient defined by a south-eastward decrease of grade marks the present south-eastern boundary of the Highland zone in Scotland but is not preserved in Ireland. In the south-west Highlands and

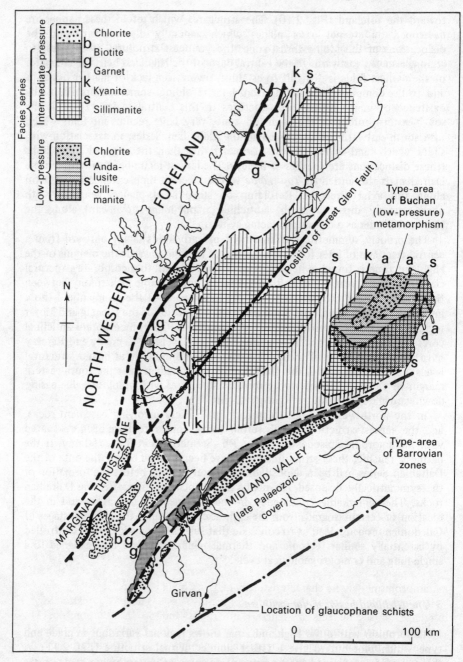

Fig. 2.11. Metamorphic map of the Highland zone of the Caledonides in Scotland; allowance is made for post-metamorphic sinistral displacement on the Great Glen Fault (based on compilations by Kennedy, 1948, Dewey and Pankhurst, 1970)

in northern parts of Ireland, low grades of metamorphism characterise the entire width of the zone, but in western Ireland and north-east Scotland, the central part of the zone is occupied by a core of highly metamorphosed rocks ranging up to sillimanite grade and associated with regional migmatites. The high-grade core of north-eastern Scotland was interpreted by Kennedy (1948) as the centre of an *antiform of thermal surfaces* aligned in the tectonic grain and plunging south-westward. The scale of this thermal structure is of a different order from that of the tectonic units mentioned above and Kennedy related the antiform to a 'tectogene' underlying the entire Highland zone.

The characteristic index minerals of successive metamorphic zones over the greater part of the Highland zone are those whose development can be related, according to modern nomenclature, to an intermediate facies series. Metamorphic zoning of this type was first described by George Barrow (1893) from part of the Dalradian outcrop and is sometimes referred to as the *Barrovian type*. By a remarkable coincidence, zoning of a different type characterises the easternmost part of the Highland zone in Scotland where Horne at a very early date described regional development of andalusite and cordierite. This *Buchan type* of metamorphism (Read, 1952) was for long regarded as aberrant. It can now be assigned to a low-pressure facies series developed in response to a steep geothermal gradient at relatively shallow depths — the rocks characterised by Buchan type metamorphism lie mainly in the upper limb of the Tay nappe (the highest structural unit of the belt) close to the culmination of the great thermal antiform (Fig. 2.11). Sodic migmatites are developed within the areas of high-grade metamorphism in several broad belts and sheets roughly concordant with the lithological boundaries.

The full history of metamorphism in the Highland belt still remains to be established, but certain episodes can be identified with some confidence (Table 2.7). Precambrian metamorphism, with pegmatite intrusion, affected the basement and the lower members of the cover in the western part of the belt (see p. 45). The western limit of metamorphism lay within the tract later disrupted by development of the marginal thrust-zone. The eastern limit is unknown. There is no indication that a 'metamorphic unconformity' separates the upper portions of the Moine Series or the Dalradian Series from the rocks of the west in which the Precambrian metamorphic episodes are recorded, although the earliest metamorphic events affecting the Upper Dalradian cannot be earlier than late Cambrian. These anomalies suggest that metamorphism began in the lower part of the cover before deposition had ceased, and intermittently extended its domain into the younger cover-rocks. As Shackleton has put it 'Such apparently continuous deposition and continuously waxing and waning metamorphism' may be characteristically geosynclinal (1969, p. 21).

The episodic nature of the metamorphic processes has been demonstrated by reference to several sets of igneous intrusions and by the relationships of successive mineral-assemblages to successive fold-systems and tectonite fabrics. Evidence of this kind suggests a remarkable stability in the thermal structure in that the distribution of relatively higher- and lower-grade zones conformed to similar patterns during successive episodes. The overall simplicity of the metamorphic zoning reflects a thermal structure which remained broadly consistent throughout the long period of metamorphism.

The date at which the final metamorphic pattern revealed in Fig. 2.11 was

Table 2.7. POLYPHASE CALEDONIAN METAMORPHISM IN THE HIGHLAND ZONE
OF SCOTLAND

Low structural levels (Moinian, basement)	High structural levels (Dalradian)	
Late-orogenic granites (410–390 m.y.) emplaced before complete cooling: local met^m along Moine thrust-zone. Mineral ages, 440–410 m.y., define period of cooling?	Late-orogenic granites	
	Mineral ages 470–430 m.y., define period of cooling?	
Lamprophyre dykes recrystallised during or after phase of emplacement		
	Retrogression associated with major folding	
Polyphase met^m and migmatisation associated with repeated folding – no precise correlation with episodes recognised in Dalradian rocks: Cambro-Ordovician rocks involved in folding	Gabbros of NE- Scotland, 500 m.y.	
	Post-kinematic met^m } climax of met^m with	
	Syn-kinematic met^m } migmatisation	
	Folding f$_2$	
Early granites c. 550 m.y.	Early granites?	
	Early met^m associated with recumbent folding, f$_1$: Cambrian (possibly Arenigian) rocks involved	
Pegmatites 740 m.y.		
Polyphase met^m and migmatisation associated with repeated folding involving regenerated basement		

established is controversial. In western Ireland, the Connemara massif of Dalradian rocks is flanked by Lower Palaeozoics of the Mayo basin which do not share its high-grade metamorphism and migmatisation. Clastic sediments of mid-Ordovician and younger ages contain metamorphic debris believed to come from Dalradian sources. In Scotland, Arenig supracrustal rocks along the south-east border of the Highland zone appear to be of much the same metamorphic grade as the adjacent Dalradian and may even form part of the Tay nappe, but late Ordovician clastics of the Southern Upland basin contain metamorphic detritus. These relationships suggest that the folded rock-mass of Moinian and Dalradian began to rise and to cool at about the end of the Cambrian period, but that tectonic and metamorphic activity were not halted everywhere at the same time. K–Ar ages obtained from Dalradian metamorphic minerals and rocks, especially those of low metamorphic grades, fall mainly in the range 470–430 m.y., those for Moinian rocks in the range 440–410 m. This contrast has been interpreted by several writers as being related to the effects of uplift and cooling subsequent to the cessation of active metamorphism in early Ordovician times, that is, at about 480 m.y. The survival of residual thermal gradients till late Silurian times is indicated by variations in the contact-relations of late-orogenic granites dated at about 400 m.y., and in the extent of recrystallisation of minor intrusions associated with these granites. Taking these terminal effects into account, we may say that the history of metamorphism in the Highland zone extended from about 750 m.y. to 400 m.y.

4 The younger basins in Britain

Although earth-movements powerful enough to produce angular unconformities took place several times during the infilling of the younger basins, the principal fold-structures date from the closing orogenic episodes which took place towards the end of the Silurian period. Since Wenlockian (and in some instances Ludlovian) strata are represented, and since folding was completed before the intrusion of late-orogenic granites dated at about 400 m.y., the period of orogenic deformation must have been little more than 20 m.y. in duration. Despite this short time-span, polyphase deformation involving the superposition of two or three fold-sets and associated minor structures is indicated by evidence from many localities, notably in the Southern Uplands, the Isle of Man and Mayo. In the regions flanking the later basins, the relatively thin Ordovician and Silurian cover-rocks tend to be less strongly folded than those of the basins themselves, a contrast which suggests that each basin was deformed as an independent unit.

The tectonic patterns defined by the axial planes of major folds and by associated cleavages and dislocations tend to parallel the margins of the basins. Complications are introduced by refolding and, especially in Wales, by the effects of superposition of folds on pre-existing volcanic domes. The tectonic style is dominated by the effects of folding on axial planes with moderate or steep dips and by the abundance of associated dislocations. In the Southern Uplands, the principal folds are monoclinal structures with steep limbs facing north-westward, which may be associated with steep thrusts dipping to the north-west. In the Mayo basin the axial planes are almost vertical. In Wales, the axial planes are steep and usually dip westwards. The principal folds thus face the south-eastern foreland: folding dies out towards the foreland and there is no clearly defined thrust-front comparable with that at the western border of the Caledonides in Scotland.

Metamorphism in the younger basins is generally confined to the deeper parts of the cover and is usually only of greenschist facies; the basement rocks, where exposed, show evidence of retrogression and dislocation and competent horizons within the cover are often extensively fractured. A feature which has attracted interest in recent years is the local occurrence of glaucophane, with some other minerals suggesting metamorphism of blueschist facies. Glaucophane is recorded only at two localities. At the northern border of the Southern Uplands basin, it is developed in the ophiolitic assemblage of serpentines, eclogites and pillow-lavas which forms the Ballantrae complex; the metamorphism here is probably early Ordovician, though some of the rocks involved may be Precambrian. In Anglesey, glaucophane occurs in a narrow zone within Monian rocks whose metamorphism is generally of Barrovian type and probably of latest Precambrian age. The juxtaposition of the Highland zone with its extensive high-temperature metamorphism and the more southerly basins with their low-temperature metamorphism has been compared with the development of the paired metamorphic belts characteristic of the Pacific margin (see p. 295). Dewey for example (1969) suggests that the Ballantrae complex represents a wedge of oceanic crust emplaced on the site of a former oceanic trench and that the blueschist metamorphism resulted from the consumption of oceanic crust at the margin of a continental plate advancing south-eastward. The Southern Uplands fault (Fig. 2.3) may on this hypothesis be identified as a suture near the line of union of two separate plates.

5 The Caledonides in Greenland and Spitsbergen

Although the infilling of the basins of deposition appears to have been completed by early Ordovician times and although orogenic activity began even before this date (p. 27), Caledonian folding and metamorphism in Greenland and Spitsbergen appear to have continued until the late Silurian. Isotopic dates of 420–400 m.y. have been yielded by metamorphic minerals and may, like those of the British Caledonides, record the final stages of cooling.

On a broad scale, the Caledonides of East Greenland exhibit a core of migmatitic and metamorphic rocks (the *central metamorphic complex* p. 24), overlain by a non-metamorphic zone. The front of metamorphism transgresses the stratigraphical divisions to produce a dome-like thermal culmination: at its highest points, it reaches the Infracambrian Morkebjerg tillite-group (p. 26). Above the metamorphic front, folding on north–south axial planes is broad and simple, and effects only a moderate amount of east–west shortening. In the metamorphic complex, the structures become tighter and more complicated, involving both N–S and E–W fold trends. Metamorphism follows the Barrovian trend and migmatites form tongues, domes and sheets whose migrations are believed to be responsible for the structural complexity. The crystalline basement is reconstituted with a Caledonian fabric and rises into the cover-rocks in regenerated migmatite-bodies. The contrasting styles of the non-metamorphic *Oberbau* and the migmatitic *Unterbau* produce a conspicuous disharmony between the structures above and below the transitional zone of regional metamorphism. The resulting patterns define an upper and lower 'storey' in the tectonic architecture, the phenomenon which Wegmann termed *Stockwerk-tectonics*.

6 The north-western orogenic front

Both in Greenland and in Britain, a zone of eastward-dipping thrusts separates the western foreland from the Caledonian orogenic belt. Many thrusts in this zone carry wedges of basement rock and it seems clear that thrusting originated at depth and did not involve simply a stripping-off of the cover. The minimum displacements proved on several individual thrusts are of the order of 15–30 km, and the real total of displacement across the orogenic front may well be much greater. The spectactular Moine thrust-zone of north-west Britain provided the material for the first detailed study of an orogenic thrust-zone towards the end of the nineteenth century by Lapworth and by Peach, Horne and others.

The major thrust-nappes of the orogenic front were formed towards the end of the Caledonian cycle. In Britain they disrupt the principal fold-systems and displace the isograds of metamorphism but are cut by granites emplaced at about 400 m.y. (Table 2.7). In Greenland they rest on a foreland cover-series ranging up to the Wenlockian. The relatively late thrust-front, however, coincides over long distances with marginal features of very much earlier date and appears to follow a crustal feature established early in the Caledonian cycle. In Britain, as we have seen, the initial border of the Highland basin and the western limit of Precambrian folding and metamorphism in that basin fall in or close to the Moine thrust-zone. Some discordance between the late orogenic front and the earlier marginal structures is suggested by the fact that the transgressive Cambro-Ordovician succession, which is confined to the foreland and thrust -zone in Britain, extends eastward for about 200 km across the Caledonian belt in East Greenland and Spitsbergen.

7 The Scandinavian Caledonides and the eastern orogenic front

In Scandinavia, the eastern orogenic front is marked by a zone of large low-angle thrusts which carry basement rocks and deposits of the marginal basins, Sparagmites with overlying Cambro-Ordovician, over the almost undisturbed Lower Palaeozoic of the foreland. This thrust-front lies close to the south-eastern margin of the area in which Infracambrian and Precambrian sediments are preserved. Over most of its length, it lies some distance east of the front of metamorphism and hence the rocks of the lower and more easterly nappes show little or no Caledonian recrystallisation.

The scale of the nappes of the marginal zones is remarkable. In the region north-west of Lake Mjøsa, erosion has dissected the warped thrust-mass resting on the foreland for many tens of kilometres in a direction perpendicular to the front. In northern Scandinavia the great *Seve nappe* which overlaps one or more lower nappes to rest almost directly on the foreland can be traced parallel to the front for over 1000 km. Oftedahl, following Kautsky, estimates that a displacement of 160 km on the basal thrust of this nappe is likely and one of 240 km possible, though these estimates are not accepted by all geologists familiar with the area.

In the southern part of the Norwegian Caledonides, the lowest thrust-sheets are overlain by two great crystalline nappes, the *Lower and Upper Jotun nappes.* The lower parts of these nappes are made of basement rocks largely of granulite facies, including basic, anorthositic, mangeritic, syenitic and granitic varieties. The lines of thrust are marked by mylonites but the bulk of the nappes retain their pre-Caledonian characters and resemble material of the Proterozoic belt of southern Norway. The Jotun nappes are associated with the flysch-like Valdres Sparagmite (p. 41) which is thought to be of late Ordovician age.

In the interior of the orogenic belt, the grade of metamorphism rises and the complexity of folding increases. The effects of recrystallisation and deformation make it difficult to separate basement and cover or to recognise major dislocations. It has been suggested that great thrusts such as that at the base of the Seve nappe extended into the interior and were later annealed and repeatedly folded. The structure is complicated by interference patterns which give steep-sided domes and basins, especially in the north, or arcuate structures such as the 'Bergen arcs' in which two strongly curved tracts of Lower Palaeozoic metasediments are sandwiched between tracts of reworked gneisses including anorthosites. In the Trondheim region, recumbent folds in which thrusts partially replace certain limbs are refolded to form the broad 'Trondheim syncline'.

Gneissose or granitic rocks tentatively assigned to the basement play a larger part in the Scandinavian Caledonides than do the corresponding rocks in the British Caledonides. Near the eastern border of the orogenic belt, the basement-gneisses are readily recognisable and show little other than cataclastic modifications. In the interior, on the other hand, the basement rocks are extensively reworked, their unconformable relationship with the cover is largely destroyed and, in some instances, they appear to pass into Caledonian intrusive granites. Ramberg has regarded the mobilisation of buoyant granitic material from the basement as an important factor in the development of the characteristic dome-structures of the belt.

The dating of orogenic episodes in the Scandinavian Caledonides is still somewhat uncertain. The cover-succession involved in folding ranges at least up to Lower Silurian and the principal recumbent folds and slides, as well as the great marginal thrusts, appear to be younger than the whole of this succession. The occurrence of several polymict conglomerates and other coarse clastics in the the Trondheim region has long been regarded as evidence that earlier orogenic episodes interrupted deposition. A few localities in the west reveal indications that folding and local low-grade metamorphism took place in early Ordovician times. A few radiometric dates of about 490 m.y. appear to be related to this *Trondheim* or *Trysil* disturbance.

IX Terminal Stages of the Caledonian Cycle

1 Preliminary

Over much of the Caledonian system in Europe and the north Atlantic, the end of the Silurian period was marked by conspicuous changes in orogenic activity which may be summarised as follows:

(a) Within the zones which had been subjected to intermittent metamorphism, crustal temperatures fell to levels below those at which minerals could behave as open systems with respect to radiogenic elements. The setting of the radiometric 'clock' is indicated by the strong peak of K–Ar age determinations at about 450–410 m.y. in Britain, Scandinavia, East Greenland and Spitsbergen.

(b) Folding associated with the development of tectonite fabrics ceased in most areas, though certain younger formations suffered extensive warping and more localised folding and fracturing.

(c) Changes in the pattern of sedimentation led to the accumulation in a number of newly developed basins of thick sequences of Old Red Sandstone of continental facies. These late-orogenic sequences of molasse type were connected with regional uplift and erosion.

(d) Intrusive granites were emplaced in great numbers in the folded rocks of the orogenic belts, recording a substantial upward migration of mobile granitic material. The majority of these granites post-date the regional patterns of folds and metamorphic zones; some cross the basal unconformity of the Old Red Sandstone and most fall within the range 410—380 m.y. These terminal changes — cooling, uplift and erosion, beginning of molasse-deposition and granite-intrusion, took place within a span of 30—40 m.y. which represents no more than 10 per cent of the entire Caledonian cycle. Moreover, they were broadly contemporaneous over the whole system of mobile belts discussed in this chapter although, as has been seen, there had been conspicuous variations in timing during earlier stages of the cycle. Both phenomena suggest a fundamental change in the controlling mechanism at depth.

2 Late-orogenic granites and associated rocks

The majority of intrusive granites in the Caledonides appear from their geological relationships to date from the terminal stages of the orogenic cycle. In Britain, they form more than fifty separate bodies, while intrusives known to have been emplaced before about 450 m.y. amount to only about a tenth of this number. In Scandinavia, Greenland and Spitsbergen, there is a similar preponderance of late-orogenic granite intrusions. Some members of the late-orogenic suite are older than the earliest molasse-deposits, others penetrate the Old Red Sandstone, or are associated with andesitic or rhyolitic volcanics in this group.

The late-tectonic granites of Britain include both forcefully emplaced plutons and ring-complexes emplaced by cauldron-subsidence (among the latter being the ring-complex of *Glencoe*, the first such structure to be recognised). They spread across almost the whole width of the Caledonides, occurring in the Highland zone and in all the later basins except that of Wales. Many show well developed contact-aureoles and were evidently emplaced in relatively cool rocks. Some of those in the central tract of the Highland zone, which represented the last part of the structure to cool (p. 48), are surrounded by contact-migmatites and appear to have been emplaced in relatively hot country-rocks.

The majority of late-orogenic intrusions of Britain have a sodic cast and are granodioritic in character. Hundreds of small pipe-like appinites and allied rocks are clustered around some plutons; some are igneous breccias which appear to have been emplaced by mechanisms involving volatile fluxes. Felsite, porphyry and lamprophyre dyke-suites are widespread and form swarms centred on some ring-complexes. A few alkaline complexes of syenites and basic syenites are sited in or near the western marginal zone in Scotland.

3 Molasse-formation: the Old Red Sandstone

The terminal orogenic changes in the Caledonides were associated with regional uplift which led to the emergence of a land-area incorporating much of northern

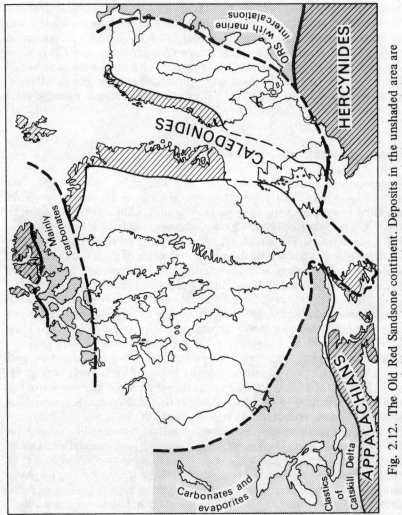

Fig. 2.12. The Old Red Sandsone continent. Deposits in the unshaded area are non-marine and of Old Red Sandstone facies; around the borders of this area deposits of similar facies interfinger with marine sediments

Europe, Spitsbergen and Greenland. This *Old Red Sandstone continent* (Fig. 2.12) has remained for the most part above or just below sea level to the present day.

The formations of continental facies which make up the Old Red Sandstone are thickest in and at the margins of the Caledonian belt, where they accumulated in intermontane basins and piedmont tracts. Evidence that similar formations were originally more widespread is provided by the occurrence of a few outliers on the crystalline rocks of the adjacent shields. At the southern margin of the Baltic shield, Devonian sediments of Old Red Sandstone facies interdigitate with marine sediments laid down in basins within the Hercynian mobile belt (see Chapter 3). Within the Caledonian orogenic belt, the formation rests with profound unconformity on rocks folded and often metamorphosed during the Caledonian cycle. At the margins of the belt it commonly follows conformably on the Lower Palaeozoics of the foreland cover-succession.

The sediments of the main intermontane and piedmont basins consist largely of red sandstones, conglomerates and shales, together with dark-coloured lacustrine sandstones, shales and bituminous limestones. Some of the earliest known terrestial floras (including the well-known flora of the Rhynie chert in north-eastern Scotland) and a great range of aquatic vertebrate remains are contained in these formation. The bulk of the detrital sediments represent fluviatile or lacustrine deposits: neither eolian sandstones nor evaporites are present and there are no indications of extreme aridity. Volcanic groups in the succession include basaltic, andesitic and rhyolitic layers and pyroclastics.

In *Spitsbergen*, the principal outcrop of Old Red Sandstone is contained in a NNW tract less than 100 km in breadth. This basin is bounded on the west by a normal fault and on the east by a high-angle thrust. The Old Red Sandstone, ranging up to the top of the middle division, is strongly warped and violently folded along the line of the thrust, indicating late-Devonian (*Svarlbardian*) disturbances. The basin-fill is almost entirely sedimentary, including red and grey sandstones, conglomerates and siltstones with maximum thicknesses of 4 km. The lowest deposits are Downtonian and, from the dates assigned to rocks of the Hecla Hoek Group, Gayer *et al*. have concluded that molasse deposition began within 5 m.y. of the termination of metamorphic activity.

In *East Greenland* a north–south tract of Old Red Sandstone crossing Keyser Frans Josephs Fjord includes red sandstones, conglomerates, breccias and lacustrine marls and shales, together with acid lavas and pyroclastics and other volcanic materials. This assemblage, reaching maximum thicknesses of about 8 km, is almost entirely of Middle and Upper Devonian age. It shows considerable evidence of late-orogenic disturbance. Angular unconformities separate successive sub-divisions and, especially towards the north, thrusts and folds of some magnitude are seen; syn-kinematic granites form sheets and irregular intrusions associated with some of these structures.

In *Scandinavia*, scattered outliers of Old Red Sandstone along the coastal zone between Bergen and Trondheim mark the remains of intermontane basins. These outliers consist almost entirely of breccias, conglomerates and sandstones belonging to the Middle Old Red Sandstone and, like those of Greenland, are extensively disturbed and traversed by several thrusts.

In *Britain* large outcrops of Old Red Sandstone represent the contents of at least one marginal and two intermontane basins. The marginal basin – the

Table 2.8. THE OLD RED SANDSTONE OF THE INTERMONTANE BASINS IN BRITAIN

OLD RED SANDSTONE	THE MIDLAND VALLEY BASIN	THE ORCADIAN BASIN (Caithness succession)	Stages of Marine Devonian
UPPER	Dura Den Beds etc., mainly sandstones and shales	Sandstones of Dunnet Head etc., local basalts 1000 m	Famennian
	(unconformity)		Frasnian
		John o'Groats Sandstone Group c. 600 m	Givetian
MIDDLE	Strathmore Group; 500 m: sandstones, marls, shales	Thurso Flagstone Group / Achanarras Beds (impure limestone, with fish) / Passage Beds Group / Wick Flagstone Group — cyclothems of sandstone, flagstone, shale, limestone, c. 5500 m	Eifelian
	Garvock Group, 1500 m; andesites, basalts, conglomerates (2500–3000 m)	(unconformity)	Emsian
LOWER	Crawton Group; andesites, basalts (2500–3000 m)	'Barren Group' 500 m; mainly sandstones, conglomerates, breccias	Siegenian
	Dunottar Group; conglomerates, volcanics (2500–3000 m)	(unconformity)	
	Stonehaven Beds (Downtonian), 900 m sandstones, shales, volcanic conglomerates	Metamorphic rocks of Moine Series with intrusive granites	Gedinnian

Anglo-Welsh basin — appears to mark the site of a coastal plain sloping southward from the newly elevated Caledonian mountains of Wales to the marine troughs of the Hercynian mobile belt. The succession in this tract follows conformably on the platform succession of Lower Palaeozoics, the change from marine to terrestrial conditions being a gradual one (p. 43). The sequence is almost entirely sedimentary and consists predominantly of fine red sandstones, siltstones and shales with calcareous layers.

The principal intermontane basins — the *Orcadian basin* extending from north-east Scotland to Orkney and Shetland and the basin of the *Midland Valley of Scotland* — are fault-bounded structures each containing up to about 6 km of Old Red Sandstone (Table 2.8). Considerable thicknesses of volcanic material, mainly basaltic or andesitic in character, are incorporated in the succession in some regions and similar volcanics extend beyond the limits of the basins over the eroded Dalradian rocks of the Highland zone. The basal divisions are strongly warped, especially in the vicinity of marginal faults such as the Highland Boundary Fault, and are separated by angular unconformities from the less disturbed middle and upper divisions. Conglomerates and breccias representing screes and piedmont deposits form thick wedges near the margins of the basins, especially in the lower parts of the successions. They pass laterally and upwards into finer-grained detrital sediments. The Middle Old Red Sandstone of the Orcadian basin includes a distinctive lacustrine assemblage of pale green or grey flagstones, mudstones and bituminous limestones in which fish remains are locally abundant.

3

The Hercynides and Uralides with their Forelands

1 Introductory

Over much of the network of mobile belts established in Laurasia at about 800 m.y. (Fig. 2.1), orogenic activity came to an end in the later part of the Lower Palaeozoic era. Those parts of the network which remained active through Upper Palaeozoic times display a complex structure resulting from the superposition of later patterns on Caledonian or older foundations. These long-lived belts include the Hercynides of western and central Europe, the Uralides, the Altaids of northern and central Asia and the Appalachians of North America.

Our present concern is with the Hercynides and Uralides, together with the cratonic area which flanks them to the north and west. In western Europe, the Hercynides diverge markedly from the Caledonian branch of the orogenic system which was stabilised at the end of the Lower Palaeozoic and it is therefore not surprising that the Hercynian cycle has traditionally been regarded as an independant affair by geologists in central Europe. We shall see, however, that the origins of the Hercynides and Uralides go back to much the same late Precambrian time as those of the Caledonides.

There has been a conflict of usage among geologists of two terms, *Hercynian* and *Variscan* with respect to the European fold-belts under discussion. In the *Explanation of the Tectonic Map of Europe* (1964), these terms seem almost interchangeable, though earlier authorities have favoured restrictions of various kinds. Out preference is for Hercynian and we shall not make use of the term Variscan in this chapter.

The Hercynides of western and central Europe were developed as a broad tract running east and west at the southern boundary of a cratonic region including the Baltic shield, the Russian platform and, from Devonian times onwards, the stabilised Caledonides of Britain (Fig. 3.1). The remnants of this belt are exposed in two situations. In the southern and Mediterranean areas, they form anticlinal wedges or fold-cores or larger massifs within the younger Alpine mobile belts and have been extensively modified by Alpine orogenic activity. To the north of the Alpine front, the Hercynian rocks have not suffered later orogenic disturbance, but their continuity has been broken by Mesozoic and Tertiary faulting and basin-formation on the craton. Hercynian complexes are exposed here in uplifted massifs isolated between broad stretches of Mesozoic and Tertiary cover-successions. This isolation of the complexes, together with the large number of national boundaries traversed by the Hercynian belt as a whole, makes it exceptionally difficult to gain a synoptic view of the belt. The key massifs which together provide a cross-section of the structure are the classic areas of the *Ardennes–Eifel-Rhenish Schiefergebirge* astride the middle Rhine and the *Bohemian massif* and its satellites centred on Prague. In Western Europe, the massifs of south-west England, Armorica, central France and the Iberian peninsula provide a second traverse.

In very broad terms we may distinguish at the outset a central *Moldanubian* tract including the Bohemian massif (Fig. 3.2) which is dominated by pre-Hercynian crystalline rocks. To the south of this tract, the Palaeozoic cover is well represented in a Mediterranean facies. To the north lies the broad geosynclinal tract containing the classic Devonian and Carboniferous sequences

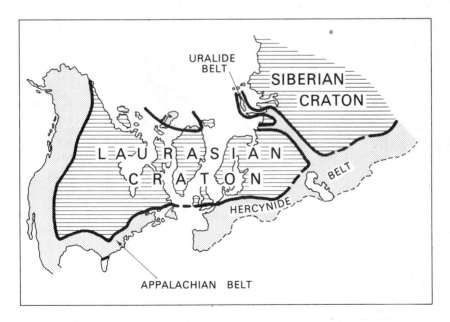

Fig. 3.1. The Hercynian mobile belts in Europe and western Asia

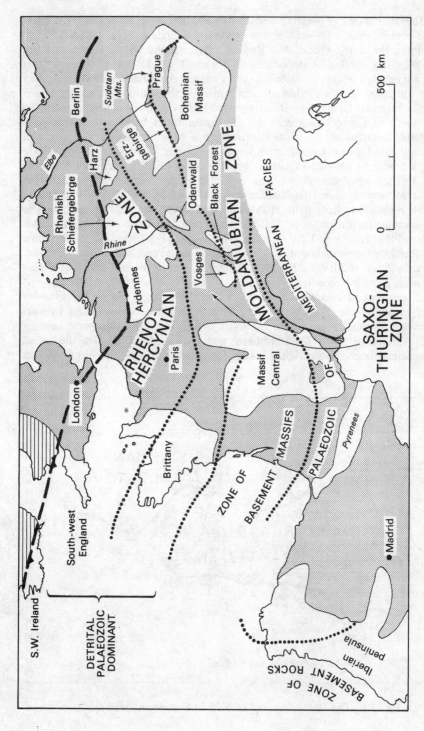

Fig. 3.2. Structural zones of the Hercynides. Post-Palaeozoic cover shaded

of the Rhine and Ardennes. An unstable *foredeep* bordering the northern craton persisted until late Carboniferous times.

The geological evolution of the Hercynian tract in Europe lasted some 500 m.y. from the late Precambrian to the end of the Palaeozoic. The Hercynian cycle proper occupied the later part of this life-span, from Devonian to Permian times.

The *Uralides of western Asia* form a conspicuous north—south lineament separating the Russian platform from the Siberian platform and extending northward to the Arctic Ocean via the peninsula of Novaya Zemlya. As in the Hercynides, the evolution of the Urals mobile zone can be traced from late Precambrian times forward to orogeny and stabilisation soon after the end of the Upper Palaeozoic era.

II The Foundations of the Hercynides

1 Precambrian basement and cover

The making of the Hercynides began with the development of subsiding basins on a crystalline basement in late Precambrian times. Although the relationships of these lower units in the Hercynian structure are frequently obscured by later orogenic or plutonic transformations, some details of their original characters can still be assembled. Precambrian rocks, as already noted, are predominant in the massifs which fall within the Moldanubian zone and it is largely with these massifs that we are concerned at present.

The Bohemian core. The central quadrilateral of the Bohemian massif (Fig. 3.3) consists of a basement block of high-grade metamorphic rocks within and around which basins of subsidence were formed at intervals from late Proterozoic to Tertiary. The oldest of these basins, that south-west of Prague, contains a succession of late Proterozoic to Devonian rocks collectively known as the *Barrandian.*

The rocks of the crystalline basement are generally grouped together as the *Moldanubian.* They include pelitic, psammitic, graphitic and calcareous meta-sediments, together with basic and ultrabasic volcanics and intrusives, all deformed, migmatised and metamorphosed in amphibolite or granulite facies; kyanite-gneisses and eclogites are recorded among these assemblages.

The Barrandian cover itself falls into two portions — a Proterozoic sequence known by the somewhat unfortunate term *Algonkian* (derived from an early American terminology) and an uncomfortable Palaeozoic sequence. The Algonkian division is of eugeosynclinal type, consisting of some 10 km slates, greywackes and conglomerates followed by the 3 km Spilite Horizon of lavas, tuffs, cherts and black shales and then by further greywackes and slates. Tillites have been doubtfully identified among these rocks. The entire succession shows intense folding related to the '*Assyntian orogeny*' which brought the Proterozoic era to a close. The first sediments which overlie them are unfossiliferous molasse-like conglomerates which are regarded for convenience as Lower Cambrian and are succeeded by transgressive Middle Cambrian shales.

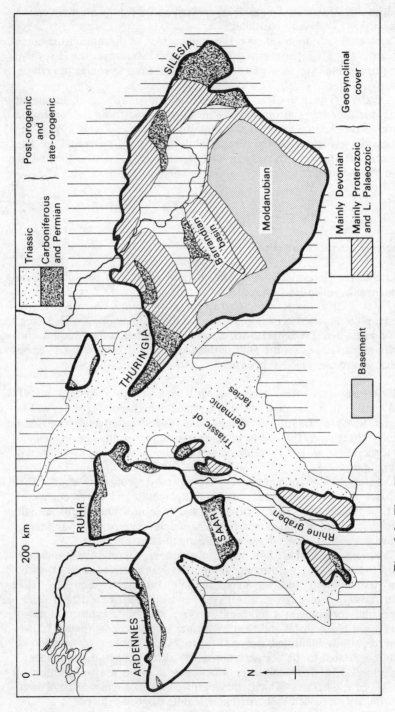

Fig. 3.3. The Hercynian massifs of central Europe showing units of the Palaeozoic succession and important coalfields (capital lettering); granites are omitted

Towards the west, the Moldanubian zone is continued in the massifs of the *Schwarzwald* and *Vosges* which stand on either side of the Rhine graben, and in the small *Odenwald* and *Spessart* massifs. Crystalline basements of Moldanubian type occur in these massifs, but are almost swamped by Hercynian migmatites and granites. In the *Massif Central* of France a core of crystalline Precambrian rocks in the north-east is bordered by wide complexes of schists whose metamorphism is assigned by some French geologists to a *Cadomian* episode of very late Precambrian age (compare with the Assyntian of central Europe).

Armorica. The Precambrian components of the Armorican massif (Fig. 3.4) include an ancient crystalline basement and a supracrustal cover. A complex of

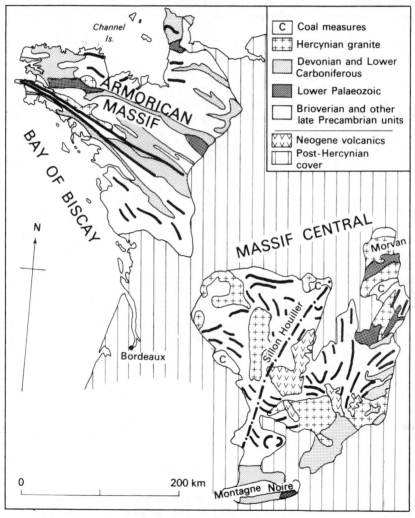

Fig. 3.4. The Hercynian massifs of western Europe

gneisses, amphibolites and granitoid rocks has been recognised in Brittany and the Channel Islands. This *Pentevrian* basement has yielded a few isotopic ages up to 2600 m.y. and probably has a polycylic Precambrian history. It is covered unconformably by the *Brioverian* succession referred to in Chapter 2.

The Brioverian supracrustals are seen to have been folded and eroded prior to the deposition of conglomerates marking the local base of the Cambrian. This deformation was related to the *Cadomian orogeny* named from Caen in Normandy. It was accompanied by metamorphism ranging up from greenschist to almandine—amphibolite facies, by restricted migmatisation and by the emplacement of a suite of granites. Radiometric ages ranging from 690 m.y. (Gneiss de Brest) to about 500 m.y. suggest that granite-emplacement spanned the Precambrian—Cambrian boundary; we have already noted that the terminal Cadomian granites are considerably older than those of the British Caledonides.

The Iberian peninsula. The Iberian peninsula (Fig. 3.8) is a large block consolidated during the Hercynian orogeny, onto which the Alpine fold-zones of the Pyrenees and the Betic cordillera were moulded. This great massif is isolated from the remainder of the Hercynides and its place in the regional structure is doubtful. Many geophysicists believe that it originally stood in closer proximity to the south of the Armorican massif and moved to its present position during the opening of the Bay of Biscay in Mesozoic or Tertiary times (p. 274). Such a hypothesis might bring the north-western corner of the peninsula into line with the north-westward projection of the Moldanubian zone. A Precambrian basement emerges from beneath the Palaeozoic cover in this north-western area, comprising parts of Galicia and of northern Portugal. Here, a highly metamorphosed supracrustal series of greywackes, arkoses, pelites and calcareous rocks is associated with garnet-amphibolites, peridotites, pyroxenites and eclogites. Migmatitic gneisses, kyanite-gneisses and charnockites are derived from these parent-rocks while granitic gneisses and riebeckite-gneisses represent deformed intrusive bodies. The youngest of the basement intrusives have been dated radiometrically at about 500 m.y. East of Cape Finisterre, the basement is traversed by a remarkable north—south lineament extending for more than 150 km — the *fosse blastomylonitique* — along which repeated movements appear to have taken place in Hercynian times. Alkaline pre-Hercynian intrusions are emplaced in this zone which den Tex believes to have originated as a tensional feature in earliest Palaeozoic times.

2 The Lower Palaeozoic story

At the close of the Proterozoic, a positive belt of folded late Proterozoic rocks with their accompanying plutonics formed as a result of Cadomian or Assyntian orogeny appears to have traversed the Hercynian domain. This belt contributed sediment to the first Palaeozoic basins and had a more lasting effect in separating differing lithological and faunal provinces. The Palaeozoic successions to the north exhibit some common features from the Rhenish Schiefergebirge to Armorica and south-west England. Those to the south represent a Mediterranean facies, which accumulated largely in shelf-seas receiving only small amounts of

Table 3.1. THE LOWER PALAEOZOIC OF GORY SWIETOKRZYSKIE
(Holy Cross Mountains)

SILURIAN (max. about 3000 m)	Ludlovian (2–2600 m)	shales and limestones greywackes and shales shales	
	Wenlockian	shales	
	Llandoverian	shales	
ORDOVICIAN (max. about 400 m)	Ashgillian	siltstones and marls	northern area; often missing elsewhere
	Caradocian	graptolitic shales	
	Llandeilian	limestones, ironstones	
	Llanvirnian	limestones, ironstones	central areas; often missing elsewhere
	Arenigian	sandstones, conglomerates	
	Tremadocian	(missing)	
CAMBRIAN	Upper	shales, sandstones	
	Middle	sandstones	
	Lower	shales, siltstones, sandstones	

detritus. These seas, for much of the time, were open to the true ocean of *Tethys* which appears to have separated Europe from Africa (p. 11); their deposits present a somewhat disjointed story, since many are entangled in the Alpine fold-belts and others – for example, those of Spain and Corsica – are situated in blocks which have suffered rotation or bodily displacement in post-Palaeozoic times.

The first Lower Palaeozoic sediments of central and western Europe were deposited in transgressive seas. In broad terms, sedimentation began with clastic deposits which gave way to pelitic and carbonate types. Regions of subsidence were thus initiated in Cambrian times, but their subsequent evolution was by no means uniform. Lateral variations were related especially to the availability of detritus and the incidence of volcanic activity; vertical variations to episodes of epeirogenic or orogenic movments. These variations are illustrated in Tables 3.1 to 3.3 and are amplified in the next few paragraphs.

With the transgression of early *Ordovician* times, the coherent history of the Hercynian domain really started. The principal geosynclinal basin defined at this time occupied the northern part of the domain in the Ardennes and Rhenish Schiefergebirge. The sedimentary facies of the Ordovician here is detrital; limestones are scarce, though ironstones are conspicuous in Bohemia, Thuringia and Armorica. Thick basic volcanic groups occur low in the sequence in Bohemia and both basic and acid volcanics in the Upper Ordovician of Armorica. It is difficult to make out widely developed movement-phases: the phase styled

Table 3.2. THE LOWER PALAEOZOIC OF THE BARRANDIAN BASIN
(*Regional Geology of Czechoslovakia*, Vol. 1, 1966)

SILURIAN (*c.* 600 m)	Upper	graptolitic shales red sandstones with eurypterids Lochkov limestones: flaggy limestones often with cherts Budnanian: basic volcanics and shelly limestones, followed by bituminous limestones
	Lower	graptolitic shales followed by limestones with basic lavas and tuffs
ORDOVICIAN (maximum 2000 m)	Ashgillian	shales and sandstones
	Caradocian	ironstones, shales, sandstones, basic tuffs
	Llandeilian	shales, quartzites, ironstones, tuffs
	Llanvirnian	shales, ironstones, basic volcanics
	Arenigian	shales, ironstones, basic volcanics
	Tremadocian	sandstones, acid tuffs, followed by shales and cherts
CAMBRIAN (maximum 6000 m)	Upper	conglomerates, acid volcanics
	Middle	conglomerates, sandstones, arkoses, porphyries and tuffs
	Lower	molasse-like marine and continental sandstones and conglomerates

'Bohemian' is pre-Tremadoc in Bohemia, post-Tremadoc elsewhere; that styled 'Taconic' is pre-Caradocian in the Ardennes and Montagne Noire, post-Ashgillian in central Poland. Such appellations imply a unity where, in our opinion, none is adequately demonstrated.

The *Silurian* period began with a widespread transgression and ended in many places with shallowing and uplift which was locally accompanied by folding assigned to a Caledonian phase (see below). Typically, the early Silurian transgressive sandstones gave place to dark graptolitic shales and these in turn to shallow-water limestones, often of reef-form. Exceptions to this general pattern are found in the Barrandian basin and the north-eastern part of Armorica where no Caledonian disturbance or uplift is recorded.

The *Mediterranean facies* of the Lower Palaeozoic is developed in the successions of the Montagne Noire (Table 3.3) at the southern tip of the Massif

Table 3.3. THE LOWER PALAEOZOIC OF MEDITERRANEAN TYPE:
THE MONTAGNE NOIRE

Ludlovian Wenlockian Llandoverian	black shales followed by shelly limestones
Caradocian	sandstones followed by limestones
Llandeilo, Llanvirn	(not represented)
Arenigian	shales and sandstones
Tremadocian	sandstones, shales, greywackes
Upper	shaly sandstones
Middle	calcareous shales
Lower	sandstones followed by Archaeocyathid limestones

Central, in the Pyrenees and in western Spain. A similar facies is especially well displayed in the Moroccan Meseta, a plateau region floored by Hercynian complexes to the south-east of Rabat and Casablanca. The successions of these regions are relatively thin and are usually lacking in volcanic material; they lack also the rapid variations and numerous stratigraphical breaks that characterise the northern domains. The predominant rocks are rather fine detrital sediments, shales interbedded with fine sandstones. Limestones occur repeatedly as beds and lenticles, often richly fossiliferous and occasionally of reef facies.

Archaeocyathus is a characteristic fossil of Lower Cambrian limestones in the Montagne Noire and elsewhere. The Silurian of the Mediterranean zone is frequently dominated by black shales which provide a valuable marker-horizon in the strongly deformed complexes of the Pyrenees.

3 Caledonian orogeny in central Europe

In a zone embracing the Ardennes and the Rhenish Schiefergebirge, and in Thuringia, Saxony and parts of Poland, there are indications of late Silurian to early Devonian disturbances which have been referred to the Caledonian orogeny. These disturbances seem best regarded as simply episodes in the long Palaeozoic cycle. In the Sudetes and adjacent regions of south-western Poland, Lower Palaeozoic sequences were folded, metamorphosed to low or medium grades, locally granitised and invaded by bodies interpreted as syntectonic and post-tectonic granites, before deposition was resumed in Lower Devonian or later times. In the Ardennes and the Brabant massif, several somewhat local phases of movement are recorded; the main phase of the Ardennes (the *Ardennic phase* of Stille), fell at the Devonian—Silurian boundary, and was marked by folding on steep axial planes. In other parts of the Hercynian belt, there are no strong indications of crustal disturbance at the end of the Silurian. In some instances, elevation took place, with consequent erosion of Silurian sediments and non-deposition of early Devonian, while in others, deposition continued without interruption.

The connection between the Caledonides of Britain and the longer-lived mobile zone of central Europe in which 'Caledonian' disturbances are sometimes recorded has been much discussed. The rocks of the Brabant massif disappear northward beneath a younger cover and their original connections are obscure. To us, it seems most probable that a link with the British Caledonides may be found via Armorica in the west of the system and that by Lower Palaeozoic times the main part of eastern and central England was occupied by a stable salient of the Baltic shield.

III The Hercynian Cycle Proper

1 The Hercynian geosyncline

The stabilisation of Caledonian belts in north-west Europe created a new situation in which the developing Hercynian zone was flanked by a craton forming the entire northern part of the continent. Much of this northern foreland was elevated in late Silurian or early Devonian times to form the Old Red Sandstone continent from which detritus was intermittently supplied to the Hercynian troughs. The fluctuating shore-line ran roughly east and west from northern Poland to the Low Countries and on into southern Britain and Ireland, defining the northern boundary: the *Hercynian geosyncline*. A positive area, the Franco—Allemanian—Bohemian island roughly on the site of the Moldanubian zone, separated it from the Tethyan ocean and its fringing seas in southern Europe. Two contrasting facies groups were thus re-established: a northern detrital facies frequently associated with volcanics, and a deeper-water Mediterranean facies.

The main events in the Hercynian cycle are here summarised and the major stratigraphic and orogenic terms employed in the descriptive sections that follow are summarised in Tables 3.4 and 3.5.

Table 3.4. UPPER PALAEOZOIC STRATIGRAPHICAL DIVISIONS

Stages and Series in Europe		System		Stages and series in North America
Tartarian Kazanian }	Upper		Upper	Ochoan
Kungurian Artinskian Sakmarian Asselian }	Lower	PERMIAN	Lower	{ Guadalupian Leonardian Wolfcampian
Stephanian Westphalian }	Upper (Silesian)	PENNSYLVA-NIAN		{ Virgilian Missourian Desmoinesian
Namurian Visean Tournaisian }	Lower (Dinantian)	CARBON-IFEROUS		{ Atokan Morrowan
		MISSISSI-PPIAN		{ Chesterian Meramecian Osagean Kinderhookian
Famennian Frasnian }	Upper		Upper	{ Bradfordian Cassadagan Cohoctonian Fingerlakesian
Givetian Eifelian }	Middle		Middle	{ Taghanican Troughoriogan Cazenovian Onesquethawan
Emsian Siegenian } Coblencian Gedinnian }	Lower	DEVONIAN	Lower	{ Espusian Derparkian Helderbergian

Table 3.5. STILLE'S TECTONIC PHASES OF THE HERCYNIAN CYCLE

Phase	Date
Pfalzic	between Permian and Triassic
Saalic	Lower Permian
Asturic	uppermost Carboniferous (pre-Stephanian): the main phase in the foredeep
Erzgebirgic	Upper Carboniferous (pre-Westphalian)
Sudetic	between Lower and Upper Carboniferous: the main phase in the interior of the belt
Bretonic	between Devonian and Carboniferous

I The Devonian transgression that opened the geosynclinal phase was followed by regression towards the close of the period. Unrest during regression culminated in a number of movement-phases affecting especially the southern parts of the geosyncline and grouped as the *Bretonic disturbances.*

II The Lower Carboniferous or *Dinantian* transgression advanced onto the Old Red Sandstone continent. In the geosyncline, sandy–shaly sediments of Culm facies were deposited on the foreland limestones and deltaic sediments of shelf facies. About the end of the Dinantian, most of the northern geosyncline was obliterated during the powerful *Sudetic phase* of deformation and metamorphism.

III During the *Namurian* and *Westphalian stages*, paralic sediments accumulated on the foreland and in a narrow belt – the foredeep – situated between the foreland and the newly risen Hercynides to the south. Towards the close of the Westphalian stage, they were deformed during the *Asturic phase* and the contents of the foredeep were added to the Hercynides.

IV The last stage of the Carboniferous period, the *Stephanian*, was marked by deposition of coal-measure type in fault-controlled troughs, such as the Saar, let into the Hercynian massifs.

V Sedimentation in the Hercynian cycle was concluded by the deposition of *Permian and Triassic formations* of molasse and continental types, collectively making the *New Red Sandstone.* The dying phases of the Hercynian orogeny are recorded by minor disturbances and by varied volcanicity.

2 Devonian sedimentation and vulcanicity

The Northern Geosyncline. The full width of the geosyncline is displayed in the Rhenish Schiefergebirge (Fig. 3.5) where the pattern of sedimentation indicates irregular subsidence. Four tectonically defined units reflect the sedimentation-history. These units are:

(a) The Sauerland Synclinorium in the north, immediately south of the external coal-basin of the Ruhr.

(b) The Siegerland Block, the main positive unit.

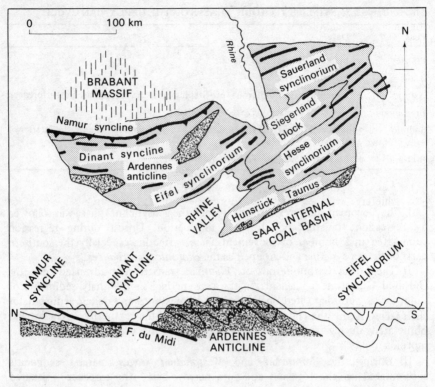

Fig. 3.5. Outline map of the Ardennes and Rhenish Schiefergebirge with diagrammatic section across the eastern Ardennes. Pre-Devonian rocks are shown by horizontal lines; broken lines indicate pre-Devonian rocks obscured by a post-Hercynian cover

(c) The Hesse Synclinorium, showing deep-water sediments in the lower Moselle and Lahn valleys.

(d) The Hunsruck-Taunus quartzite ridges that dominate the internal coal-basin of the Saar to the south.

At the beginning of the Devonian, before the Siegerland block began to rise, sedimentation was controlled by the proximity of land on both geosynclinal margins, littoral and continental Old Red Sandstone (see p. 54) accumulating in Sauerland and shallow-water types in the Hunsruck-Taunus unit to the south. Thick greywackes and shales accumulated in a central subsidence on the site of the future Siegerland block uplift. This block began to rise during the final (Emsian), stage of the Lower Devonian. In the Hesse trough the Emsian is complete, and consists of semipelites and pelites with a few soda-rich lavas; north of the uplifted block, in Sauerland, the lower Emsian is missing and the stage begins with a basal conglomerate followed by shallow-water shales and coral-limestones that wedge out southwards against the Siegerland block.

The same palaeogeographic controls continued during the Middle and Upper Devonian. Littoral deposits were formed on the emergent Siegerland block and on its borders to north and south. Deep-water shales and limestones in the centre of the Sauerland trough gave place northwards to reef-limestones and sandstones and, at the continental margin, to coarse clastics including some of continental type. South of the Siegerland block, geosynclinal conditions developed in the Hesse trough where great thicknesses of deep-water shales accumulated, with shallow-water clastics at the margins. Extensive vulcanicity, begun in the Emsian, continued throughout the Middle Devonian and lasted until the early Fammenian stage, giving rise to keratophyric and spilitic lavas and tuffs (Schalsteine), together with hematitic ironstones formed by emanations from the volcanics. During the Middle and Upper Devonian deformations of differing magnitudes were recorded, the main disturbances coming at the close of the period.

The Schiefergebirge are continued westwards as the *Ardennes* where only the northern part of the geosyncline is exposed. In this area successive beds of the Devonian overstep northwards onto the Caledonian *Brabant massif*. The line of successions given in Fig. 3.6 brings out the main features, notably the spanning of the whole Devonian by diachronous basal conglomerates and the passage of the Lower Devonian from dominantly deep-water shales in the centre of the geosyncline to a littoral facies of conglomerates and red shales and sandstones at the northern edge. Similar lateral changes are seen in the higher Devonian stages. In the *Ferques inlier* of the Boulonnais, Givetian rests unconformably on folded Silurian and the Devonian rocks have a mixed marine, littoral and continental lithology recalling similar horizons in Belgium.

In South-west England, the marine Devonian of the type region has two contrasted developments, one to the south and the other to the north of a wide synclinorium of Carboniferous rocks. The southern assemblage is almost entirely marine and includes, towards the top, fossils from an open-sea environment; pelitic and calcareous rocks predominate and basic volcanic material is abundant. In the northern belt marine and continental facies alternate, the fossils are mostly of shallow-water organisms and volcanic rocks are scanty — the rocks of this northern belt were laid down along the margin of the Old Red Sandstone continent and pass northwards into the Old Red Sandstone of the Anglo-Welsh basin (pp. 56–7).

Massifs south of the Northern Geosyncline. The relative uniformity of Devonian sedimentation in the northern geosyncline is lost in the isolated Hercynian massifs to the south. These massifs can, however, be arranged in two groups reflecting the distribution of Caledonian orogenic disturbances (p. 67).

The contrast is well exemplified in the Bohemian massif. In the Barrandian basin, the Lower Devonian is fully represented as a calcareous facies of Mediterranean affinities, but sedimentation ended with a regression at the close of Middle Devonian. On the north-western borders of the massif, in Thuringia and Saxony, a late Middle Devonian transgression from the north onto an uplifted landmass of the Munchberg and Erzgebirge was recorded and basic vulcanism was widespread. Similar contrasts are revealed in the Massif Central, between the southern (Montagne Noire) and Morvan areas, and in Armorica, between the northern and southern regions thereof.

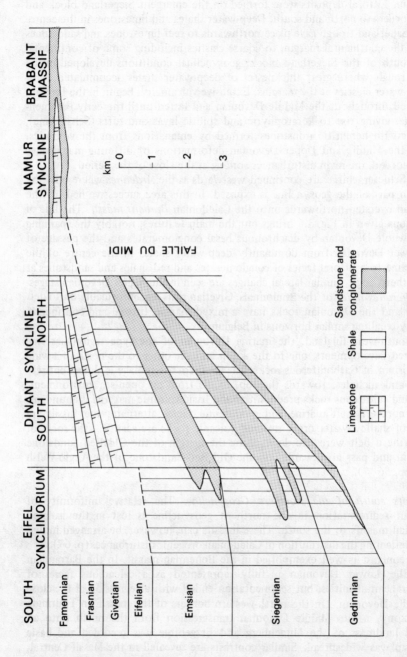

Fig. 3.6. Variations of thickness and facies in the Devonian of the Ardennes (based on Fourmarier)

3 The Bretonic disturbances

Several of the crustal disturbances which took place during Devonian and early Carboniferous times have been awarded special names with implied stratigraphical significance. The extent and nature of these disturbances are not always apparent and their dating is often difficult. Accordingly, we employ only one term, the *Bretonic disturbances*, to cover the dozen or so pulses of movement proposed for late Devonian and early Carboniferous times. In much of the Hercynian belt, structures produced during the Bretonic movements were almost obliterated by those resulting from more powerful later movements.

In the southern portion of the Rhenish Schiefergebirge, well documented Bretonic structures increase in intensity towards the south, attaining an Alpine style in the Hesse basin and in the Taunus where the earliest Carboniferous is transgressive on deformed Devonian. Rather similar relationships are seen west of the Rhine in the southern Eifel. No plutonic activity of Bretonic age is recorded here or in the Ardennes. In south-west England, late Devonian movements were perhaps responsible for the displacement of the *Lizard complex* – an assemblage of metamorphic rocks associated with a large serpentine-gabbro body – which rests on a tectonic mélange (the Meneage crush-zone) separating it from phyllitic or slaty Devonian rocks of the southern facies belt (p. 71). Isotopic dating suggests that some phases in the sequence of igneous intrusion and low-grade metamorphism in the area were late Devonian.

Bretonic disturbances are recorded on a grander scale in the *Bohemian massif* where the Moldanubian core as a whole was overthrust onto its bordering complexes. The eastern margin of the Moldanubian block is a well defined thrust-zone running roughly north and south for 200 km. The Moldanubian is displaced eastwards on this zone and its cover-succesion shows overfolds and thrusts of Alpine style. Similar thrust structures, and a number of granites probably of Bretonic age, are seen in the Vosges and the Schwarzwald.

4 Dinantian sedimentation and vulcanicity

In the Hercynian mobile belt, two important orogenic events, the *Sudetic* and the *Asturian* (=Asturic of Stille) punctuated the accumulation of Carboniferous sediments (Table 3.5). The succession therefore falls into three divisions: the Dinantian, the Namurian–Westphalian and the Stephanian.

The transgressive sea, which advanced at the opening of the Carboniferous period, traversed two regions which were to be characterised by different facies. In the rapidly subsiding troughs above the old Devonian geosyncline a thick series of dominantly detrital sediments akin to flysch – lumped together as the *Culm facies* – was laid down, usually resting discordantly on disturbed Devonian. In the shallow transgressive seas on or near the foreland, limestones accumulated, passing landward into deltaic and lagoonal sediments. It is convenient to describe this assemblage as the *shelf-sea facies*.

Continued crustal instability in the mobile belt is reflected by the occurrence of beds of coarse detritus, by gaps in the sequence, and by volcanic intercalations. In the northern and eastern border-regions of the mobile belt, profound subsidence took place in Dinantian times, leading to accumulation of pelagic sediments without accompanying volcanics. In the Ruhr and in Sauerland, the deep-water Culm facies is characterised by cherts, siliceous and flaggy limestones and black shales; these pelagic sediments are interleaved at many levels with sandstones and other clastic deposits. Somewhat similar associations occur in the East Sudetes and Moravia.

In the western part of the border-zone, the Culm of south-west England and south-west Ireland includes Dinantian, Namurian and lower Westphalian components. The 'Lower Culm', equated with the Dinantian and following without a break on the Devonian, is made up of some thousand metres of black shales, cherts, black limestones and sandstones. Pillow-lavas and tuffs are abundantly developed in the southern part of the outcrop.

Classic developments of the Dinantian shelf-sea facies appear in northern France and adjacent parts of Belgium, where the type localities of Dinant, Tournai and Visé are all situated, as well as in central and northern England. All these regions are characterised by limestones of varied types, some rich in shells, some characterised by stromatolites and some by corals.

In the Franco-Belgian area, the Dinantian follows on the underlying marine Devonian with a transitional junction. It reaches a maximum of about 700 m in the Dinant basin, thinning northward against the Brabant massif and eastward towards Aachen and the northern Ruhr. Among the dominant limestones, a reef facies, the Waulsortian facies, with abundant bryozoa and a distinctive molluscan fauna is well developed in a belt south of Dinant.

In Britain, the shelf-sea facies begins immediately north of the Culm geosyncline of south-west England and extends irregularly northward to Scotland (Fig. 3.7). A low-lying island extending westward from the Midlands to eastern Ireland (*St George's Land*) stood out of the shallow seas and clean neritic limestones were deposited around it. These constitute the *Carboniferous Limestone* of British geologists. Near the northern shoreline in the Midland Valley of Scotland and north-east England, the equivalent deposits are detrital sediments of deltaic, lagoonal and swampy facies associated with alkaline olivine-basalts, trachybasalts, trachyandesites, trachytes and rhyolites reaching thicknesses of up to 1000 metres.

The foreland Dinantian successions in Britain show conspicuous variations in thickness and lithology which illustrate especially clearly the effects both of topographical irregularities and of differential movements in the basement. (Fig. 3.7). In particular, a number of blocks and uplifts were defined, towards which the succession is interrupted or thinned. The east—west island of St George's Land, which has already been mentioned, stood consistently above sea level. The north—south *Usk anticline* separating the coalfields of South Wales and the Forest of Dean is marked by thin sandy deposits which were folded towards the end of Dinantian times. In the *Pennine region* of northern England more stable blocks, on which accumulated conglomerates followed by massive limestones, were flanked by basins which received thicker sequences of thin-bedded limestones and shales; reef-knolls are situated close to the 'hinges' linking block and basin. Still further north, limestones interfinger with detrital

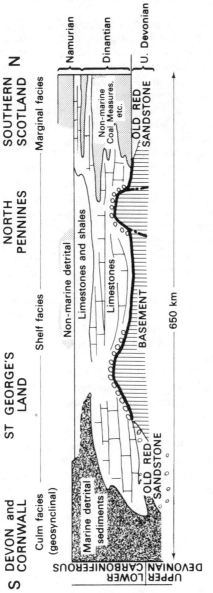

Fig.3.7. Facies variations in the Lower Carboniferous of the Hercynian foreland in Britain (based on Gignoux, and others)

sediments, locally containing evaporites, which mark the complex of lagoons and coastal flats fringing the northern landmass.

5 The Sudetic orogenic phase

In and on the borders of the principal regions of deposition, the Dinantian disturbances recorded by breaks in deposition or by local unconformities culminated towards the close of Viséan times in the principal orogenic episode of the Hercynian cycle. By these movements, grouped as the Sudetic phase, the axial parts of the Hercynian belt were raised into mountain-tracts and the geosynclinal stage was virtually terminated.

The fragments of the Hercynian belt which remain in central Europe fall into a number of tectonic zones which can be somewhat doubtfully extended westward to the Atlantic margin. These zones, originally defined by Kossmat in 1927 but modified by later workers, reflect the older zonation illustrated in Fig. 3.3; from south to north, they are the *Moldanubian zone,* the *Saxo–Thuringian zone* and the *Rheno–Hercynian zone.* The first is derived largely from the basement and was the site of extensive Hercynian metamorphism and granite-emplacement. The second reveals Palaeozoic successions folded over domes and cores of basement and invaded by abundant granites. The third and most northerly is occupied mainly by folded Palaeozoic supracrustals of low metamorphic grades. We shall deal with the zones in this order and then turn to the outlying areas of the west and south.

The Moldanubian zone. The Bohemian massif, the Schwarzwald and the Vosges form the principal units of the Moldanubian zone proper. In these massifs and in the north-east of the Massif Central, the basement rocks were partially regenerated and the remnants of cover-rocks were folded between or overridden by blocks of basement.

A conspicuous feature of the basement structure is the occurrence of great zones of movement, often transverse to the trend of pre-Hercynian folds, which appear to define a system oriented roughly north-west. The history of these fracture-belts is long. Some were pre-Hercynian fault-zones which provided channels for the ascent of Hercynian granites and hydrothermal solutions and continued moving sporadically up to Tertiary times. The most remarkable in the Bohemian massif is the *Bavarian Pfahl*: a straight west-north-west lineament which is traceable for 140 km. The Moldanubian gneiss is mylonitised for some hundreds of metres on either side of a central fracture-zone marked by quartz-breccias.

The main *Hercynian metamorphism* of the mobile belt may be mentioned here although it spreads beyond the Moldanubian zone and cannot everywhere be assigned to the Sudetic phase. Schists and migmatites carrying andalusite and cordierite, or sillimanite and cordierite, are widely developed in Bohemia, the Schwarzwald and the Vosges. These rocks are generally regarded as products of regional Hercynian metamorphism and a few isotopic dates of about 340 m.y. have been obtained from localities in the Bohemian massif. Palaeozoic rocks up to Devonian in age are locally recognisable in the metamorphic terrains and in basement rocks, andalusite or cordierite-bearing assemblages are seen to be overprinted on earlier high-grade assemblages some of which contain kyanite.

The massifs of the Moldanubian zone exemplify the regional Hercynian metamorphism which is developed throughout the central parts of the mobile belt. Voluminous migmatites and granites are associated with a low-pressure metamorphic facies series distinguished by the occurrence of andalusite and cordierite, not only in central Europe but also in the Massif Central, the Pyrenees and the Iberian peninsula (Fig. 3.8). Many parts of the basement retain remnants of earlier assemblages in which the presence of kyanite and of eclogites suggests a Precambrian high-pressure facies series. In a study of the entire belt, Zwart (1967) points out that the Hercynian metamorphic zones are thin and that where stratigraphical control is adequate the evidence suggests that the overburden at the time of metamorphism was small. In the Pyrenees, Zwart estimates that the top of the andalusite zone rose to within 4000 m of the earth's surface, suggesting a geothermal gradient of about 150°C/km. Gradients of up to 200°C/km have been inferred by other authors from evidence taken from the Massif Central and from Silesia.

The abundance of Hercynian granites not only in the Moldanubian zone but in the Iberian peninsula, the Pyrenees and some more northerly regions (Fig. 3.8) is in accord with this picture of the development of high temperatures in the crust.

The late Precambrian and early Palaeozoic cover-rocks which occupy the Barrandian basin south-west of Prague and other areas on the margins of the basement massifs are not only folded but also share in the disruption produced by transverse fractures of the Pfahl system. Overthrusts marked by mylonites are developed along the junctions of different units in the structure, some of which include Hercynian granites.

The rocks of the Vosges strike south-westward towards the Auvergne section of the Massif Central and from the western side of this sector the strikes turn sharply north-westward towards the southern part of Armorica. Precambrian rocks folded, metamorphosed and invaded by granites in pre-Hercynian episodes form much of the Massif Central. Cover-successions ranging up to Devonian or early Carboniferous are represented mainly in the east of the massif and are commonly separated from the older rocks by intrusive Hercynian granites.

The time-relations of orogenic events in the Massif Central as a whole are extremely controversial. Early tectonic episodes are assigned by French geologists to Cadomian or Caledonian cycles. In the Montagne Noire, far to the south, cover-rocks from Cambrian upwards are unmetamorphosed, and granites dated at 560 m.y. invade folded late Precambrian supracrustals. Certain meta-sediments incorporated in the massif have yielded isotopic ages of over 400 m.y. On the other hand, Devonian and possibly Dinantian rocks are seen in a metamorphic condition in the Série de la Brevenne west of Lyons, and the youngest migmatites of the massif have given ages in the range 300–325 m.y. Hercynian granites fall mainly in the Carboniferous age-range and the oldest post-orogenic sediments are generally Stephanian. The distribution of metamorphic zones in the Massif Central has been expressed by Roques and

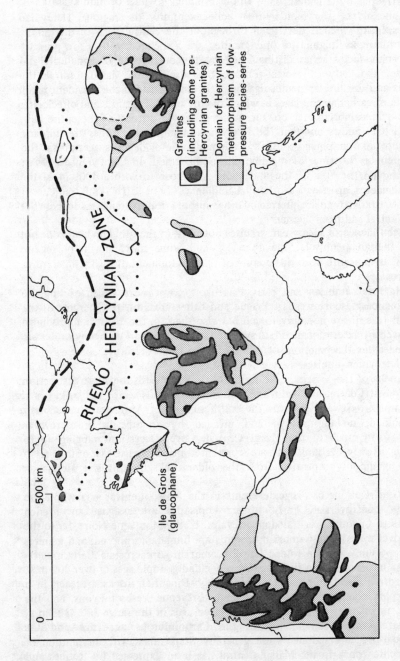

Fig. 3.8. Hercynian plutonism; the map shows the distribution of Hercynian granites and of metamorphic assemblages of the low-pressure facies series (modified after Zwart, 1967)

RHENO - HERCYNIAN ZONE

Ile de Grois (glaucophane)

Granites (including some pre-Hercynian granites)

Domain of Hercynian metamorphism of low-pressure facies-series

500 km

0

Jung, among others, in remarkably simple terms. A zonal sequence of *ectinites* not affected by migmatisation was considered to pass upward from gneisses through schists to phyllites, the zone boundaries being more or less flat. Transgressing this sequence, a migmatite front, beneath which migmatitic gneisses of various types are developed, rises from north to south. This picture was related by Jung and Roques essentially to metamorphism proceeding *pari passu* with sedimentation and burial. The complexity of the time-relationships and the evidence of repeated deformation and metamorphism outlined above seem to us to rule out such a simple concept.

The Saxo-Thuringian zone. The type areas of the zone lie to the north-west of the Bohemian Massif and are characterised by a SW—NE tectonic trend. Towards the north-east, the zone is terminated by the old-established Elbe lineament coursing north-westward across the Hercynian belt. In the large block of the West Sudetes on the eastern side of the Elbe line, Hercynian structures of north-westerly trend are superimposed on Caledonian structures and are traversed by conspicuous north-westerly movement-zones, many of which remained active in post-Hercynian times. Some small high-level granites are emplaced in the Palaeozoic cover.

In Thuringia and Saxony, the geosynclinal filling is folded into a series of wide troughs and saddles of SW—NE trend. The great central basin of Thuringia, occupied by Devonian and Culm, is flanked by anticlinal massifs revealing Lower Palaeozoic and Algonkian rocks, associated with high-grade gneisses. These massifs are in some places bordered by thrusts and the entire structure is traversed by belts of faulting and disruption parallel to the Elbe lineament; like the latter, these belts moved repeatedly up till Tertiary times.

Among the anticlinal massifs of the zone, the celebrated *Granulitgebirge* of Saxony reveal a core of granulites derived from sedimentary and volcanic parents surrounded by gneisses and by an envelope of mica-schists and phyllites which pass out, it is thought, into early Palaeozoic rocks. The Saxon granulites are characterised by the wide distribution of hypersthene, the occurrence of cordierite, and by remarkable fabrics in which quartz forms leaf-like plates parallel to the foliation. Hercynian retrogression has lowered the metamorphic grade in places, the envelope-rocks are disrupted by thrusts, and post-tectonic granites and pegmatites invade both core and envelope. Somewhat similar assemblages are seen in other dome-like massifs of the zone. Dome structures in the gneisses are attributed largely to magmatic granite-emplacement; and high-level granites associated with tin and lead rise into the envelope in some areas.

The Armorican massif far to the west lies in a position very broadly equivalent to that of the Saxo—Thuringian zone. Much of the massif is occupied by Brioverian rocks folded and metamorphosed during the Cadomian cycle, but Palaeozoic strata up to Lower Carboniferous are preserved in bundles of synclinoria coverging north-westward. In northern Brittany and the Channel Islands, the Brioverian and their Pentevrian basement suffered little rejuvenation, but towards the south, Hercynian metamorphism and migmatisation are recorded. Although andalusite and cordierite are developed on a regional scale, glaucophane-schists, suggesting a high-pressure type of metamorphism, occur at one locality on the Isle de Grois and are regarded as Hercynian. Granites yielding ages of about 340 m.y. and about 300 m.y. invade the structure. Those in the tightly folded southern area have a linear north-westerly outcrop, those further north are usually elliptical. Those of south-west England, which give dates of about 290 m.y., appear to rise as a line of bosses from a buried batholith of ENE trend.

The Rheno-Hercynian zone. The Rheno-Hercynian zone, sited on the Upper Palaeozoic geosyncline (pp. 68–71), includes the Ardennes-Schiefergebirge and Harz massifs. Bretonic disturbances affected the southern parts of the Schiefergebirge but died out northward, so that on the border of the geosyncline the Dinantian passed up into the Namurian without break. The main movements in all Hercynian phases in the key-region were northerly, the tectonic style being expressed by steep isoclinal folding, complicated by imbrication and thrusting. The metamorphic grade is low, except in the Harz where plutonic activity is recorded and ottrelite-schists are developed from Devonian parent-rocks.

The Pyrenees, the Montagne Noire and the Iberian peninsula. At the southern border of the Massif Central, the Hercynian structure is dominated by the development of gravity-nappes made up of cover-rocks ranging from Cambrian to Lower Carboniferous, which appear to have slid southward from a rising dome of crystalline rocks in the Montagne Noire. Almost on strike from this dome rises the northern portion of the *axial zone of the Pyrenees,* which constitutes an east–west Hercynian massif elevated along powerful border faults in Tertiary times. Although this massif is flanked both to north and to south by zones in which Mesozoic and early Tertiary sediments are quite strongly folded, it retains its Hercynian structure and metamorphic features almost unmodified. Stephanian sediments resting unconformably on folded sequences ranging up to early Carboniferous indicate that the main orogenic disturbances were Sudetic. The principal features of the Pyrenees are summarised in Table 3.6.

The structure of the axial zone in the Pyrenees is dominated by the occurrence of elongated domes within which are exposed metamorphic rocks representing the lower portions of the cover, with early granitic intrusions and, locally at least, a Precambrian basement. De Sitter, Zwart and their colleagues from Leiden have drawn a distinction between the metamorphic *infrastructure* revealed in the domes and the low-grade, or non-metamorphic, *suprastructure* made largely of Devonian and early Carboniferous strata, which occupies the intervening tracts. In the infrastructure, the schistosity and axial planes of small folds are flattish and two or more tectonite fabrics are superimposed on each other. In the suprastructure, the axial planes of large-scale folds are steep and are associated with a steep east–west axial-plane cleavage. The boundary between supra- and infrastructure, marked by disharmonic or complex structures, is located near the upper limit of the schists. Although it often lies at the top of the Lower Palaeozoic, where Silurian black schists provide an incompetent horizon for unsticking, this boundary is oblique to the stratigraphy, descending eastward into the Cambro-Ordovician and rising westward into the Devonian. Tectonic style is evidently related to metamorphic grade. Studies of metamorphic fabrics suggest that temperatures rose to a maximum during the late

Table 3.6. HISTORY OF THE HERCYNIAN COMPLEX OF THE PYRENEES (based on de Sitter, Zwart and others)

	Sedimentation	Deformation and Metamorphism		Granite-emplacement
Stephanian (locally late Westphalian)	post-tectonic non-marine sediments including coals			
		4	E–W folding in supra- and infrastructure; migmatisation continues in infrastructure	intrusive granodiorites
		3	conjugate folds and crenulations in infrastructure, sillimanite-bearing assemblages and migmatites	intrusive granites
		2	infrastructure only, N–S axes, new schistosity, with andalusite-cordierite assemblages	
		1	main E–W folding, with steep axial-plane cleavage in suprastructure, flat schistosity and greenschist facies assemblages in infrastructure	
		phases of deformation		
CARBONIFEROUS → Dinantian (locally includes early Westphalian)	sandy shales, often with a dark chert group at the base, usually less than 1 km			
DEVONIAN	shales and limestones, including griottes, up to 1 km			
SILURIAN	black slates			
CAMBRO-ORDOVICIAN	psammites, pelites, conglomerates, rare limestones, thickness unknown			
PRECAMBRIAN	crystalline basement			? early granites now transformed to orthogneisses

stages of deformation; the occurrence of synkinematic andalusite and cordierite in the regional schists points to the existence of a high geothermal gradient and this inference is borne out by the observed thinness of the schist zone and of its calculated overburden for which Zwart gives the figures of 1000 m and 3500–4000 m respectively.

The Hercynian complexes of the Iberian peninsula lie on strike from, and south-west of, those of the Pyrenees. The northern sector of *Cantabria* is characterised by east–west structures developed in a cover-succession ranging from late Precambrian to mid-Carboniferous. Towards the north-western tip of the peninsula, the structures turn sharply northward, and northerly or north-westerly trends dominate the pattern in *Galicia and Portugal* where the cover is associated with a reworked Precambrian basement (p. 64). In these western regions, the cover is partially metamorphosed and migmatised along a low-pressure trend and the fold-complex is invaded by huge bodies of Hercynian granite. In Cantabria the cover is non-metamorphic and only locally invaded by granites.

6 Hercynian mineralisation

The granites emplaced in late Carboniferous times in western Europe are associated with a suite of ore-deposits characterised by tin, tungsten, silver, copper, lead and zinc (Fig. 3.9). A zonal arrangement of elements is apparent in Cornwall where tin-tungsten deposits cluster in and around the granites while copper and lead-zinc deposits associated with fluorite, barytes etc. are dispersed at progressively greater distances. A remarkable feature of this mineralisation is the fact that the peripheral members of the suite extend northward far into the foreland. Small lead deposits in the Pennines, the Lake District and southern Scotland and major lead–zinc deposits in the central parts of Ireland are of Hercynian age. The distinctive tin mineralisation decreases in importance eastward, diminishing to very minor proportions in the Bohemian massif, and is not recorded in the Urals. The Appalachian belt of North America, presumably once continuous with the Hercynides, is not characterised by tin, though one or two small occurrences are recorded in Newfoundland, the region closest to the European portion of the belt.

7 Upper Carboniferous sedimentation

As a consequence of the Sudetic and earlier disturbances, virtually the whole of the Hercynian fold-belt emerged as a new land-mass. Immediately adjacent to the Hercynian front lay the *foredeep* in which Dinantian passed up without a break into Namurian and, beyond it, the *foreland* in which a sub-Namurian discordance was commonly recorded. In the narrow foredeep basins, rapid subsidence allowed the accumulation of thick non-marine sequences with marine intercalations in the *external* coal-basins of Belgium and the Ruhr. Within the elevated Hercynian belt to the south, post-tectonic sediments were laid down unconformably on folded Palaeozoics in narrow structurally controlled basins.

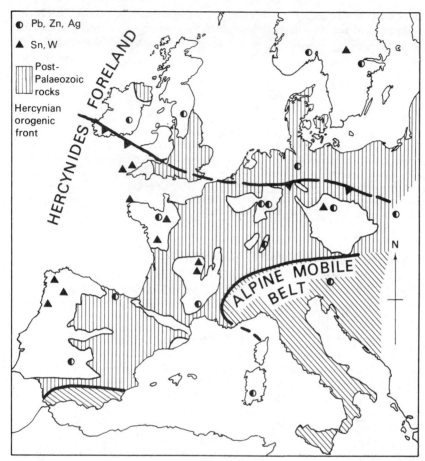

Fig. 3.9. Hercynian mineralisation in western Europe; deposits of tin and/or tungsten and of lead—zinc—silver are shown in relation to the orogenic belt

The coal measures of these *internal* coal-basins are wholly non-marine and mainly of Stephanian age.

External basins. A continuous tract of foredeep deposits, partially concealed by Mesozoic and Tertiary cover, extends from South Wales and south-west England along the skirts of the Brabant massif, supplying the coal-mining districts of northern France, the Namur basin in Belgium, the Aachen region, and the Ruhr in Westphalia. Even to name these localities is to recall more than a century of wars and political intrigues, so profound has been the importance of the coalfields in the history of industrial Europe. To the east of the Brabant massif lie concealed foreland coal-basins in the Belgian Campine, South Holland and northern Westphalia, to the north-west lie the foreland basins, exposed and concealed, of the British Isles. The foredeep basin of Upper Silesia fringes the

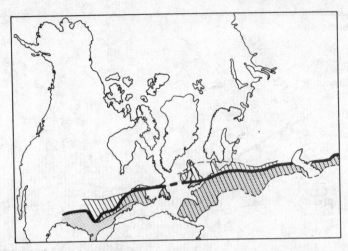

Fig. 3.10. The Pennsylvanian coal-measure facies in relation to the Hercynian—Appalachian mobile belt (essentially after Trueman, 1964). The heavy line marks the Hercynian front.

Bohemian massif and may have been open to the seas covering the Russian platform (Fig. 3.10).

In general terms, the Upper Carboniferous of the foredeep begins with marine shales and sandstones carrying goniatites. The Upper Namurian shows a change to deltaic facies containing occasional marine intercalations. Coals are thin and poor near the base; thicker and more numerous deposits in the Westphalian successions in the foredeep seldom extend into the Stephanian, since deposition was terminated by the Asturic disturbances before the end of the Carboniferous. In the foreland area of Britain, continental conditions set in rather sooner, and the Namurian consists mainly of delta-complexes of sandstones and shales. The most famous of these is the *Millstone Grit* of the Pennines in which non-marine detrital sediments with poor coals alternate with thin marine beds. The principal productive coal measures are of Westphalian age and much of the Stephanian is missing. Zoning of the paralic coal-measures has been achieved palaeontologically by the use of non-marine lamellibranchs, together with goniatites in the marine bands, and on lithostratigraphic grounds, by the tracing of distinctive marine bands, coals, palaeosoils, and ironstones.

Considerable variations in total thickness are apparent, 2000 m of strata in France are equivalent to 6000 m in Westphalia. Even in the foreland basins of Britain, 700 m of coal measures in South Staffordshire are equivalent to 1500 m in Lancashire. It is all the more surprising to find that the principal *marine bands* of the sequence are remarkably persistent, extending for many hundreds of kilometres parallel to the Hercynian front and at least 300 km perpendicular to the front. Four principal bands (the Subcrenatum, Clay Cross, Mansfield and Top of the Yorkshire coalfield) are recognised in most coalfields of England and Wales.

The regularity of the marine bands is linked with the more general phenomenon of rhythmic deposition which characterises the coal measures not

Table 3.7. COAL-MEASURE CYCLOTHEMS

A British example: The Middle Coal Measures of the Nottinghamshire coalfield, 200–250 m in thickness, consist of nine cyclothems of the following type:

> coal
> seat-earth (palaeosoil)
> sandstone
> siltstone
> grey mudstone with shells
> *black shales with marine fossils*
> (coal)

A North American example: The typical cyclothem of the Pennsylvanian Coal Measures of Illinois is of the following type (Weller, Henbest and Dunbar). Marine sediments are of decreasing importance towards the east.

approximately
15–18 m

> *Shale with ironstone concretions*
> *marine limestone*
> *black shale with limestone concretions*
> *impure marine limestone*
> *shale*
> coal
> underclay (palaeosoil)
> limestone
> sandy shale
> sandstone, locally unconformable on underlying cyclothem

(marine sediments in italics)

only of Europe but also of North America (Table 3.7). The deposition of each *cyclothem* may be taken to result from initial subsidence, which led in some instances to marine incursion and the formation of a marine band. The succeeding beds record the silting-up, afforestation and accumulation of plant litter which provided the material of the coal seam. Although tectonic forces are considered to control the intermittent subsidence, a contributary factor may have been the compaction of the diffuse plant matter to a coal-seam only a fraction of its original thickness.

Internal coal-basins. In the central and southern parts of the Hercynides, post-orogenic deposition on the eroded surface of the Sudetic fold-belt began before the end of the Carboniferous period and was continued without major breaks into the lowermost Permian. The first deposits accumulated in narrow troughs, usually fault-bounded, and are preserved as outcrops sandwiched between, or on the borders of, the crystalline complexes of the Massif Central, the Vosges and the Bohemian massif. The important *Saar basin* lies at the southern margin of the Rhenish Schiefergebirge and other basins are incorporated in the Saxo-Thuringian tract. The structural control of deposition is especially clear in the Massif Central where half a dozen small basins occur along a single fault-lineament, the north-north-easterly '*Sillon houiller*' which bisects the V-turn in tectonic trend of the older structures (Fig. 3.4).

The deposits of the internal basins are entirely of continental facies. They include polymict conglomerates and sandstones and, like other post-tectonic sequences of molasse-type, reach considerable thicknesses; the Saar basin, for example, contains up to 5 km of late Carboniferous. Coals are developed at many levels as components of cyclothems similar to those of the external basins; although their areal extent is small individual seams reach up to 10 m in thickness.

8 Terminal orogenic stages: the Asturic disturbances

In late Westphalian and early Stephanian times the precarious equilibrium of the paralic-plain environment was destroyed by the last major orogenic disturbances of the Hercynian cycle. These disturbances were most effective in the foredeep zone, where the newly deposited coal-measures and the underlying strata were thrown into overturned folds or displaced northward on low-angle thrusts. In the more southerly zones, the fold-complexes formed in the Sudetic and Bretonic phases responded mainly by block-movements along old and new dislocations. On the foreland, open warping of the cover took place in response to block movements in the Precambrian and Caledonian basement. Plutonic activity did not attain an importance comparable with that reached during the Bretonic and Sudetic phases and although a number of granites, including those of south-west England, appear to date from the Asturic phase, they bulk much less large than the granites already referred to.

Many of the structures of the foredeep zones are known in detail because of their importance in the exploitation of coal-seams. The recognition by Gosselet of overthrusting along the *Faille du Midi* of Belgium, in fact preceded and facilitated the recognition of thrust-nappes in the Alps. The main coalfields of the Ruhr and the Ardennes, and the Bristol coalfield in Britain, lie in tracts of folded and disrupted rocks. In the Ruhr, the Upper Carboniferous is tightly folded on east-north-east lines and steep north-facing fold-limbs are frequently replaced by reverse faults. The fold-style appears to be disharmonic and folding may not involve the basement. West of the Rhine, the coal measures occupy the deep asymmetrical *Namur synclinorium* whose inverted southern limb is overridden by Devonian and Lower Carboniferous along a major low-angle thrust. The displacement on this thrust (known in the east as the *Faille Efielienne* and in the west as the *Faille du Midi*) increases westward, reaching a maximum of some ten kilometres. Within the Namur synclinorium itself, the coal measures are shuffled and reduplicated by small flat thrusts.

Fault-blocks and massifs in the Hercynides and their foreland. We have seen in earlier sections that the sedimentary basins and the tectonic structures built up in the Hercynian cycle were influenced at many points by differential movements in the basement. Some fudamental lineaments probably existed in Precambrian times, while others were formed during the Palaeozoic, at various moments up till the terminal stages of the orogeny. The Devonian transcurrent fault-set, which includes the north-easterly *Great Glen Fault* in Scotland with a sinistral displacement of over 100 km, may be mentioned as an example (Fig. 2.3). Before the onset of post-orogenic deposition, at any rate, the Hercynides and much of their foreland had become traversed by an irregular grid of faults and 'uplifts' defining a host of little blocks and basins (Fig. 3.11). The

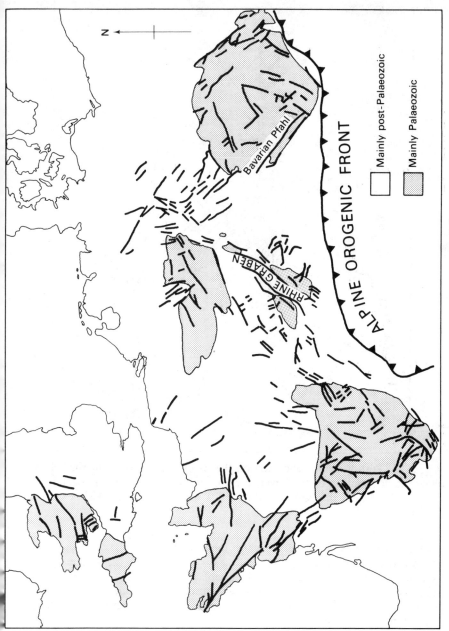

Fig. 3.11. Major lineaments and faults in central and western Europe

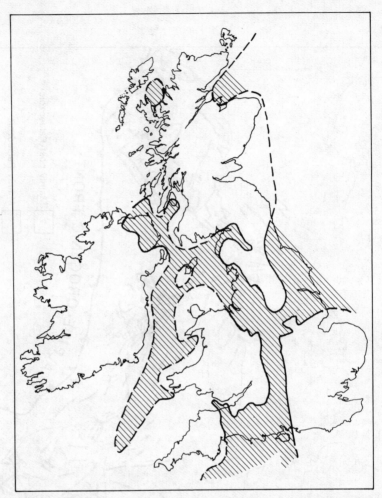

Fig. 3.12. Post-orogenic deposition on the Hercynian foreland; the shaded areas show the distribution of the lowest division of the Triassic, and the sketch-map serves to indicate the scale of the basins developed (after Audley-Charles, 1970)

outlines of the positive massifs in which the Hercynian complexes of central and western Europe are exposed today and the early parts of the Oslo—Rhine graben system (p. 213), were quite largely established before the Mesozoic cover-succession began to accumulate. No other consolidated mobile belt known to us exhibits so complex a pattern of shattering and submergence beneath the blanketing platform-sediments.

The development of uplifts and basins on the foreland is well illustrated in the British Isles where broad uplifts such as the north—south Pennine arch and narrower lineaments such as the Malvern axis (on which Precambrian basement overthrusts Lower Palaeozoic) are flanked by basins in which the major coalfields lie. The post-orogenic Permo-Triassic formations which rest unconformably on these structures are largely concentrated in small, roughly equidimensional basins beneath which the basement is depressed to depths of a few kilometres (Fig. 3.12).

IV The Ural Mountains

The remarkable lineament of the Urals, with its continuation in Novaya Zemlya, extends for well over 3000 km as a topographical and geological entity separating the Russian and West Siberian platforms (Fig. 3.1). Like the Hercynides of Europe, the Urals represent a long-lived mobile belt initiated in late Proterozoic times, deformed during several orogenic episodes and stabilised soon after the end of the Palaeozoic era. Discrepancies between the polar wandering curves for Palaeozoic periods from the cratons on either side indicate that the Uralide suture marks the site of a wide Palaeozoic ocean. A range of economic deposits including iron, chromium and aluminium is associated with the mobile belt and has contributed to the industrial development of the region.

1 Outline of structure and history

The main part of the Urals runs almost north and south as a mountainous tract of folded Precambrian and Palaeozoic rocks. Its southern continuation, masked by a post-Palaeozoic cratonic cover, probably links up with Palaeozoic fold-belts entangled in the late Phanerozoic mobile tract south of the Caspian Sea. Its northern continuation loops sharply westward into the arc of Novaya Zemlya and may have submarine connections with the Palaeozoic mobile tract of the Taimyr Peninsula to the east.

From the west to east, the following structural units can be distinguished in the Urals proper (Fig. 3.13).

(a) *The western foreland,* represented by the Russian platform. Over much of the foreland adjacent to the Urals, a few kilometres of Palaeozoic deposits rest on the Precambrian basement, but towards the south and the north, these deposits are overlain by Mesozoic, Tertiary or Quaternary sediments.

(b) *The Cisuralian foredeep*, characterised by thick basin-sequences of Carboniferous, Permian and Triassic rocks. These sequences, laid down above earlier platform-deposits, are made largely of detritus shed from the rising Uralides. They were folded during the terminal stages of the orogenic cycle but remained essentially non-metamorphic.

(c) *The miogeosynclinal zone* occupying the western and axial parts of the Ural Mountains and most of Novaya Zemlya. Much of this zone is occupied by Proterozoic and early Palaeozoic 'anticlinoria' between which lie downfolds of later Palaeozoic sequences without major volcanic intercalations. At least the lower parts of the sequence show low grades of metamorphism.

Table 3.8. THE EVOLUTION OF THE URALIDES

		Russian platform	Foredeep region
	Neogene and Quaternary	thin marine or non-marine cover	
	Palaeogene	mainly marine deposits in south, including U. Cretaceous Chalk, partly non-marine in north	
	U. Triassic ↑		
URALIDE CYCLE	L. Triassic ↑	fusilinid limestones followed by red beds and evaporites	flysch and molasse-type formations: sandstones, shales, conglomerates, evaporites, up to 8 km
	U. Carboniferous		
	M. Carboniferous ↑ U. Devonian	limestones, shales, evaporites, sandstones: oil in the Volga–Uralian oilfield occurs in late Devonian sandstones or with Devonian salt-domes	
	M. Devonian ↑ Ordovician	incomplete sequence of shallow-water and non-marine deposits; sandstones, limestones, shales	
PREURALIDE CYCLE	late Cambrian ↑ Upper Proterozoic	undeformed blue clays, sandstones, gravels of Valdai assemblage (Sinian)	(no evidence)

(d) *The main Urals deep fault zone*, marked by a broad tract of dislocation-metamorphism and by elongated massifs of serpentine, gabbro and associated rocks. This fundamental fracture-zone separating the miogeosynclinal and eugeosynclinal tracts is virtually continuous for over 2000 km and probably marks the suture on the site a former ocean.

(e) *The eugeosynclinal zone* occupied by Proterozoic and Palaeozoic sequences with abundant volcanic intercalations, folded, metamorphosed and invaded by numerous large granites. Its eastern part is masked by a Tertiary or Quaternary cover which, in the central Urals, extends almost the full width of the zone. Most of the mineral wealth of the Urals is concentrated in the

Miogeosynclinal zone	Eugeosynclinal zone	West Siberian platform
block-uplift and erosion		thin non-marine cover
marine and non-marine deposits in local basins		marine and non-marine deposits including brown coals
block-uplift and erosion, limited post-orogenic deposition in intermontane tracts		
orogenic deformation and uplift, regional metamorphism and emplacement of granite massifs in eugeosynclinal zone		
mainly greywackes and slates, up to 3 km	greywackes, limestones, coal measures, minor volcanics, up to 3 km	folded Palaeozoic equivalent to that of eugeosynclinal zone forms basement near Urals
sandstones, shales, limestones with few volcanic intercalcations, rarely more than 3 km	sandstones, shales, limestones with thick spilites, keratophyres and other volcanics total up to about 10 km	(no evidence)
	uplift and erosion	
up to 12 km of psammites, pelites and dolomites, folded and metamorphosed	detrital sediments with basic and acid volcanics, folded, metamorphosed and intruded by granites	
	Archaean basement	

eugeosynclinal tract where the industrial centres of Sverdlovsk and Magnitogorsk are situated.

(f) *The eastern foreland* represented by the west Siberian platform. The Mesozoic, Tertiary and Quaternary cover-rocks on the flanks of the Urals rest on folded Palaeozoics of the eugeosynclinal zone and obscure the eastern orogenic front. Further east, where the cover thickens to some 4 km, it may rest on a Caledonian or Precambrian basement and incorporate Palaeozoic strata.

The evolution of the Urals mobile belt is considered to have involved two major cycles, the earlier terminating in the Cambrian period with the formation of the *Preuralides* and the later terminating in the Triassic period with the *Uralides proper*. The Triassic mountain-belt was rapidly degraded by erosion and the present Urals are the result of successive block-uplifts in late Triassic to Pliocene times. The development of the entire Uralide system is given in a generalised form in Table 3.8 which is elaborated in the next few pages.

2 History of deposition

The contrasting histories of the three main depositional zones are apparent from Table 3.8. Thick late Proterozoic successions (possibly including Cambrian among the insecurely-dated formation assigned to the Sinian, see Part I) are represented in the Preuralide anticlinal massifs: an apparent contrast between a western facies, lacking volcanic intercalations, and an eastern facies carrying volcanic material, foreshadows the distinction between miogeosynclinal and eugeosynclinal zones maintained through the Uralide cycle. The corresponding deposits of the Russian platform as seen far to the west on the shores of the White Sea and around Leningrad are astonishingly little altered, indeed the famous Precambrian blue clay of the Leningrad region is still plastic and unconsolidated. A geosynclinal facies of folded rocks occupies a tract branching from the Urals system towards the mouth of the White Sea (Fig. 3.13).

The Ordovician-Devonian deposits of the Urals form two distinct assemblages: a thick eugeosynclinal sequence containing intercalations of spilites, cherts, basalts, keratophyres and (mid-Devonian) andesites; and a thinner, almost wholly sedimentary, miogeosynclinal assemblage. The late Devonian and Lower Carboniferous sequences of both zones contain little volcanic material and include poorly sorted detrital sediments. The platform-deposits of the western foreland which correspond with these early and mid-Palaeozoic sequences have non-marine intercalations at several levels and include important Devonian evaporites which give rise to salt-domes in the region east of the Volga. Major oilfields in this region are supplied by Devonian reservoir-rocks.

Towards the end of Lower Carboniferous times, orogenic movements ended widespread sedimentation in the Urals proper. Post-orogenic conglomerates occur locally in intermontane tracts but the main accumulations of syn-orogenic and post-orogenic sediments are found in the Cisuralian foredeep basins where poorly sorted marine and non-marine detrital sediments, many of them red beds with evaporites, overlie the older platform-sequences. They clearly represent wedges built out westward from the mountain-tract over the old foreland and pass westward into thinner and finer deposits. The filling of the marginal basins was followed by folding in the course of the Triassic period.

3 The main Urals deep fault

The suture which divides the miogeosynclinal zone from the eugeosynclinal zone (Fig. 3.13) dips steeply to the west and is flanked on the west by a line of elongated massifs in which Proterozoic or early Palaeozoic rocks are exposed.

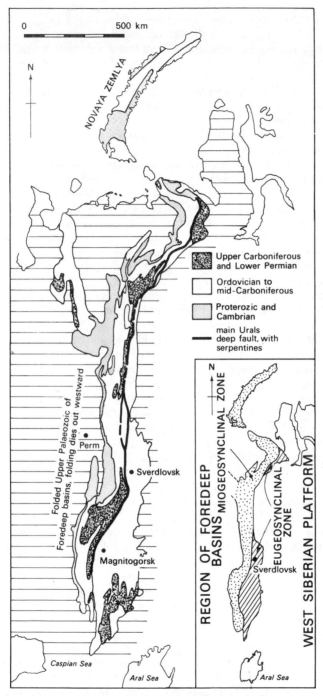

Fig. 3.13. The main geological units of the Uralides

From Ordovician till Permian times (and perhaps throughout the late Protero-zoic, see above), contrasted styles of activity characterised the regions to east and to west. The eugeosynclinal zone to the east was characterised by volcanic activity up till mid-Devonian times and by regional metamorphism accompanied by migmatisation, with voluminous granite-intrusion, in the orogenic stage. The miogeosynclinal zone received little or no volcanic material and though metamorphism of greenschist facies took place, neither migmatites nor large granites are present.

The fracture-zone and its splays are followed for long distances by igneous masses among which serpentines, dunites, peridotites, gabbros and plagioclase granites are recorded. These intrusions, which carry major deposits of chromite, nickel and platinum, are believed to be mainly of late Ordovician and early Silurian age. The lineament is marked also by a narrow, discontinuous tract in which the operation of a high-pressure type of metamorphism is recorded. Glaucophane-schists, with eclogitic bodies, are present to the east of Magnitogorsk and at various localities scattered along the length of the Urals and kynanite-gneisses are also represented; these outcrops provide the first instance encountered in this book of a high-pressure facies series developed on a significant scale (see later, p. 199).

The gross structure of the Urals, in which a geosynclinal belt characterised by the Steinmann trinity of spilites, cherts and serpentines, and invaded by granites, is flanked by a major suture marked by ultrabasic intrusions and by high-pressure metamorphism, has much in common with the structure of an active continental margin such as that of eastern Asia against the Pacific (Chapter 10). The possibility that the main deep fault marks the suture produced by the elimination of an ocean and the union of two previously separated continental plates has occurred to some authors. Such an interpretation raises difficulties, the coincidence between the geosynclinal zones of the Preuralides and the Uralides may be mentioned for one, but would be well worth testing by further palaeomagnetic comparisons of early Palaeozoic or Proterozoic rocks from the adjacent cratons.

4 Tectonics and granite-emplacement

The cover-sequences laid down during the Uralide cycle rest unconformably on rocks which had already been folded and consolidated during the Preuralide cycle. In general, the Preuralide structures follow the same trend as those of the later fold-belt and several of the basement massifs which form 'anticlinoria' in the final structure seem to have functioned as 'highs' during the Ordovician-Carboniferous period of sedimentation. The basins which flanked these massifs evolved during orogenic disturbances into the great 'synclinoria' of the fold-belt.

The larger folds and fractures are remarkably continuous along the axial direction, emphasising the linear grain of the belt. Complex folding is seen at low structural levels and within some thick incompetent pelitic groups. In the eugeosynclinal zone, migmatites appear locally at depth and granitic massifs are concentrated along certain anticlinal zones. In the miogeosyncline, the Preuralide massifs suffered little folding in Palaeozoic times, though many fractures were developed and the intervening Palaeozoics of the synclinal tracts were tightly folded. The majority of these structures were formed before the end of the Carboniferous period, but late Carboniferous and Permian strata in the clastic wedges of the foredeep were thrown into regular north–south folds which become broader and flatter westward toward the foreland.

Granite-emplacement in the Ural tract is assigned to two main periods. The Preuralide granites are concentrated mainly in the eugeosynclinal zone. In the southern Urals a nepheline-syenite is associated with rapakivi granites, and syenites south of Sverdlovsk, which may be rather younger, are associated with emeralds and other precious stones. The main granites of the Uralide cycle, emplaced in the eugeosynclinal zone, are post-Lower Carboniferous, though a variety of small Devonian or early Carboniferous intrusives also occur. Huge pyrometasomatic magnetite deposits are situated at the contacts of certain granites emplaced in limestones: Mount Magnitnaya, above Magnitogorsk, is described as a mountain made of magnetite. Small concentrations of gold, tungsten, arsenic and other metals are also widely distributed.

V Post-orogenic developments

1 Permo-Triassic deposition

By early Permian times, the uplift accompanying the terminal stages of Hercynian orogeny had led to the emergence of a land-area which included not only the Hercynian orogenic belt but also the greater part of its northern foreland. Before the end of the period, the entire Appalachian belt and much of its north-western foreland were also above sea-level, probably providing a link with the equally large continental area of Gondwanaland (Fig. 7.1). A large part of the European land-area lay within a hot-arid climatic zone. The exceptional extent of the land-masses and the climatic conditions led to the widespread deposition of continental formations of piedmont, desert and sabkha facies, which are traditionally grouped together by British geologists as the New Red Sandstone. Shallow seas invaded central and northern parts of Europe at only two stages – the Zechstein Sea was developed as a land-locked gulf from an arctic sea in late Permian times and received initial anaerobic deposits of dark bitumous shales containing up to 2 per cent of copper (Kupferschiefer). The Muschelkalk Sea was a shallow extension of the Tethyan ocean formed in mid-Triassic times. The nature of the deposits is illustrated by two classic successions detailed in Table 3.9. To the east, marine conditions lasted on throughout the Permian in the Cisuralian foredeep and adjacent parts of the Russian platform where stabilisation was delayed until the Triassic (p. 92). The ocean basin of Tethys remained in being, its shore being pushed far to the south in Permian times but advancing into the Alpine area in Triassic times.

The lowest formations may be regarded as a Hercynian molasse. They frequently include breccias, conglomerates and coarse sandstones, especially where they border on uplifts. The Verrucano, occurring in fault-bounded basins within the Hercynian complexes of the Alps and other parts of southern Europe, is of this type and is sometimes associated with andesitic volcanics. At higher

Table 3.9. PERMO-TRIASSIC SUCCESSIONS IN GERMANY

B TRIASSIC OR CENTRAL GERMANY

Keuper:	lagoonal green and red marls, dolomitic limestones, gypsum and anhydrite, sandstones with plant fragments (>450 m)
Muschelkalk:	marine series of shelly limestones, marls and anhydrite ending with lagoonal shales and thin coals (200–400 m)
Bunter:	red and mottled sandstones, conglomerates and shales, ending with dolomites and gypsum (>700 m)

A PERMIAN OF NORTH GERMANY

Zechstein:	basal conglomerate followed by: cupriferous shales (*Kupferschiefer*, 0–60 m); dolomitic limestones with impoverished fauna (5–10 m); evaporites including dolomite, gypsum, halite, salts of potassium and magnesium
Oberrothliegend:	red sandstones and conglomerates, often with discordant base (*c.* 600 m)
Unterrothliegend:	shales with occasional thin coals and volcanic intercalations (*c.* 1000 m): downward passage to Stephanian in some internal coal-basins

levels, and in broad basin-areas, red or green sandstones, shales and marls are the predominant rocks. Dune-bedded eolian sandstones are widely distributed and evaporites occur at several levels. The Permian evaporites related to the Zechstein sea are the most important of these, providing major deposits of halite and of potassic and magnesian bittern salts (including the great Stassfurt deposits) and supplying a swarm of *salt-domes* which invade the Mesozoic and Tertiary cover in parts of northern Europe and the North Sea (Fig. 3.14). Salt-tectonics related to domes are of importance in connection with the distribution of oil and natural gas in these areas.

2 Igneous activity and graben-formation

Late-orogenic volcanics including andesites and tuffs are sometimes associated with the molasse-like clastic formations such as the Verrucano of the Alps mentioned above. On the foreland, the late stages of orogeny were marked by igneous activity of other kinds. In northern Britain, Carboniferous basic lavas and small plugs of alkaline type were followed at about the end of the period by tholeiitic sills and dykes including the well-known Whin sill. Alkaline complexes were emplaced in the old alkaline province of the Kola peninsula (U.S.S.R.) and in southern Norway where they are associated with a structure which must be mentioned in a little more detail.

The *Oslo-graben* is a north-north-east rift structure some 50 km broad and 200 km in exposed length. The throw of the boundary faults is only 2–3 km in the south and decreases northward, so that the graben disappears in the vicinity

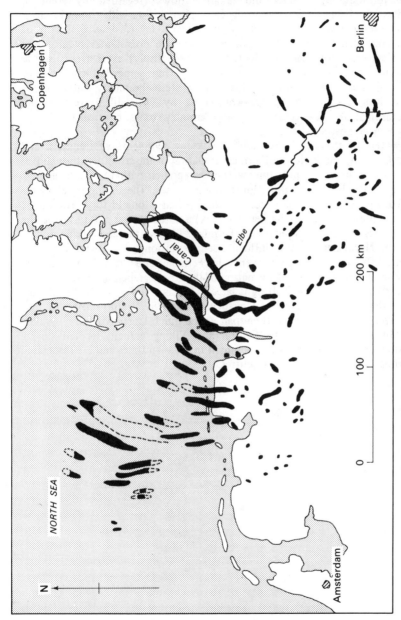

Fig. 3.14. Salt-tectonics in the post-Hercynian cover of part of the European craton, identifying subsurface diapirs of Permo-Triassic evaporites in and near the southern North Sea (based on the Tectonic Map of Europe)

of Lake Mjøsa. Within the fault-trough the classic Lower Palaeozoic succession of the Caledonian foreland is overlain with minor unconformity by Permian red-beds followed by basalts and alkaline flows (regarded by some as ignimbrites) carrying characteristic oligoclase phenocrysts with rhombic cross-section (*rhomb-porphyry*). Alkaline ring-complexes, regarded as the roots of calderas, emplaced in the graben-succession, include ornamental syenitic and monzonitic rocks (nordmarkite and larvikite respectively) extensively used as facing-stones in northern Europe. In the older but somewhat similar ring-complex of Fen, lying just outside the western boundary fault, many igneous-looking alkaline rocks are considered by Oftedahl to result from alkali-metasomatism (*fenitisation*) of the country rocks controlled by the emplacement of a carbonatite plug.

The waters of the North Sea and the young sediments of Denmark and northern Germany mask the southern continuation of the structure, though it is thought to continue at depth, linking the Oslo-graben with the *Rhine-graben* extending to the border of the Alpine mobile zones. The dating of elements connected with the rift-system indicates that it had a long history. The oldest alkaline complex in Norway (the Fen complex) is dated at over 550 m.y. and the main volcanics and ring-complexes are Permian. The Central and Viking graben beneath the North Sea began to influence sedimentation in Permo-Triassic times and were affected by igneous events in Permian and Jurassic times. The structural arch which preceded formation of the Rhine-graben in central Europe is of late Mesozoic age, the faulting is largely Tertiary and much of the alkaline igneous activity is late Tertiary or Pleistocene. The *Rhine–Oslo lineament* thus evolved over roughly the same time-span as the much larger African rift-valley system. In the *North Sea*, the Central and Viking graben, located during exploration for oil, began their development in Permian times and are associated with Permian and Jurassic volcanic centres.

4

The Appalachians and Interior Lowlands of the North American Craton

I North America in Phanerozoic Times

After the completion of the Grenville orogenic cycle, the three great units of the present North American continent were defined. The cratonic block formed by the stabilisation of successive Precambrian structural provinces assumed a roughly triangular form, with its base in the Arctic islands and its apex near the Mexican border. The *Canadian shield* in the northern part of this block remained, for the most part, a positive region receiving little or no sediment. The southern part, now forming the *interior lowlands*, was repeatedly inundated and received a cover of platform sediments.

Flanking the cratonic block, two long mobile belts came into existence before the end of the Precambrian. The *Cordilleran belt*, along its western side, retained its mobility throughout Phanerozoic time (Chapter 11). The *Appalachian belt* was stabilised at about the end of the Palaeozoic era, and thereafter the eastern part of the continent entered on an entirely new phase of development which is dealt with in later chapters. The *Appalachian cycle* responsible for the evolution of the Appalachian belt, lasted from late Precambrian to earliest Mesozoic times, with a span of more than 500 m.y.

The Appalachian belt (Fig. 4.1) extends for over 3000 km, from Newfoundland to Alabama, with a breadth of at least 600 km. North-east of Newfoundland it is truncated by the continental margin and may originally have been joined to the Caledonides of western Ireland (Fig. 2.1). In the Southern States, the arcuate *Ouachita belt* extends westward to the Mexican border and central America. The southern Appalachians and Ouachita belt are heavily masked by post-Palaeozoic sediments of the coastal plain.

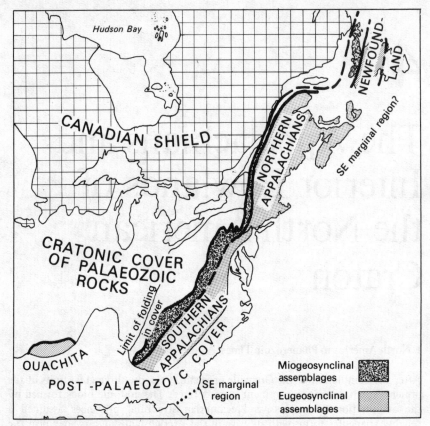

Fig. 4.1. The Appalachian and Ouachita mobile belts

II Stratigraphical and structural outline

1 The Appalachian geosyncline

The Appalachian geosyncline, the model most familiar to Hall, Dana and other early proponents of the geosynclinal thory, has passed into geological literature as a standard of reference. A distinctive feature of this geosyncline is the occurrence of two parallel zones containing rocks of contrasting sedimentary facies and tectonic style (Fig. 4.1). The *eugeosynclinal zone*, extending the full length of the belt, includes abundant detrital sediments with numerous volcanic groups and ultrabasic intrusions, and its rocks are frequently metamorphosed. The *miogeosynclinal zone*, occupying the north-western part of the mobile belt adjacent to the foreland, is broad in the southern Appalachians but narrow and poorly defined in the north. It includes abundant carbonate sediments, few volcanics or ultrabasic intrusions and is non-metamorphic. The Appalachian cycle, using this term in a broad sense, involved several orogenic events with

Table 4.1. THE APPALACHIAN GEOLOGICAL CYCLE IN OUTLINE

	Eugeosynclinal zone, northern and southern Appalachians	Orogenic disturbances	Miogeosynclinal zone of southern Appalachians
U. TRIASSIC	Newark Sandstone in fault-troughs, basic dykes and sills	Post-orogenic throughout the belt Allegheny disturbances	Formation of main folds and thrusts
MID-DEVONIAN TO PERMIAN	In *northern Appalachians*, late-orogenic, largely non-marine, clastic formations, unconformable on folded (Devonian–Pennsylvanian) In *southern Appalachians*, orogenic disturbances continued	Post-orogenic at northern end of belt, late-orogenic in central and southern Appalachians	sedimentation continued with only minor breaks. Clastic wedges built out north-westward from sourcelands within Appalachian belt. Clastic and detrital sediments of these wedges in part non-marine (red-beds, coal-measures), interfinger with shallow-water marine sediments of orthoquartzite facies
EARLY DEVONIAN	terminal orogenic folding and granite-emplacement in northern Appalachians, folding, granite-emplacement in south	Acadian disturbances	
LATE ORDOVICIAN AND SILURIAN	renewed geosynclinal deposition, with volcanic activity		
MID-ORDOVICIAN	folding, metamorphism and granite-emplacement, probably throughout eugeosynclinal tract	Taconic disturbances	
CAMBRIAN AND EARLY ORDOVICIAN	geosynclinal deposition, volcanic activity, serpentine-emplacement	early disturbances	sedimentation in platform environment, limestones abundant
LATE PRECAMBRIAN AND INFRACAMBRIAN	folding, granite-emplacement deposition of late Precambrian detrital sediments and volcanics		

which progressive changes in sedimentation are associated. The principal events are schematically summarised in Table 4.1, which is intended for reference in later sections.

The basement. Wedges, domes and thrust-sheets of basement rocks crop out in and at the north-western border of the eugeosynclinal zone (Fig. 4.2). These rocks are predominantly metamorphic or granitic and are separated by a major unconformity from the oldest members of the geosynclinal succession. They have yielded maximum ages in the region of 1100 m.y. and appear to have been part of a province of Grenville age on which the Appalachians are superimposed.

The cover-succession. The cover-succession of the Appalachians ranges in age from late Precambrian to Permian. Its upper age-limit is given by the dates of 1100–900 m.y. related to the Grenville orogeny in the basement, so that the time-span of deposition was of the order of 400–600 m.y.

Precambrian cover-rocks are largely confined to the eugeosynclinal zone where the earliest basins of subsidence were initiated. From early Cambrian to mid-Ordovician times deposition took place in both eugeosyncline and miogeosyncline though the facies laid down in the two zones were conspicuously different.

Extensive orogenic disturbances, the *Taconic* disturbances, which took place during the Ordovician period brought about considerable changes in the pattern of deposition. From this time onwards, massifs were repeatedly elevated in the central and south-eastern part of the mobile belt, interrupting deposition, restricting the width of the basins and yielding erosional detritus. Much of this detritus was laid down in *clastic wedges* which spread north-westward from the source-regions over the old miogeosynclinal zone and onto the foreland. By late Devonian times, much of the northern Appalachians had been stabilised and deposition thereafter took place mainly in late-orogenic intermontane basins. In much of the central and southern Appalachians and the Ouachita belt, stabilisation was delayed until the end of the Palaeozoic, and the post orogenic deposits are of Triassic age.

2 The tectonic zones

In the central and southern Appalachians, a number of zones characterised by contrasting tectonic and metamorphic styles can be distinguished (Fig. 4.2). In Newfoundland and the northern Appalachians there is a less well defined but broadly similar arrangement (Fig. 4.4). In general terms, a north-western zone characterised by unmetamorphosed or low-grade cover-rocks and by comparatively simple folding and thrusting can be distinguished from a south-eastern zone of higher-grade metamorphism and of more complex structures in which partially reactivated basement plays an important part.

The north-western tectonic zone is best represented in the southern Appalachians where the *Valley and Ridge province*, consisting entirely of miogeosynclinal rock, is characterised by regular folds and thrusts which die away westward

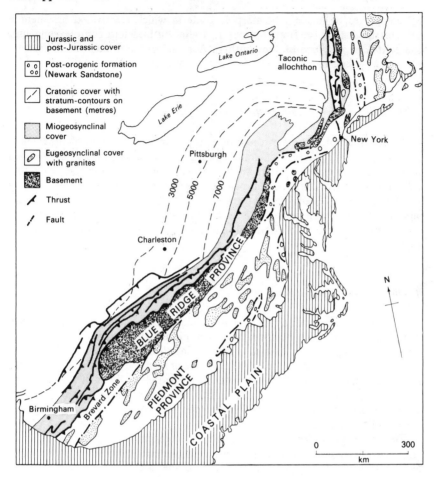

Fig. 4.2. The southern Appalachians

towards the foreland. In the northern Appalachians, the north-western zone includes rocks of both miogeosynclinal and eugeosynclinal type, the latter extending in some places to the margin of the belt. The boundary against the foreland is here usually marked by a definite thrust-front.

The junction of the north-western and south-eastern zones appears to be a line of considerable geological significance along which relative uplift of the south-eastern zone took place. It is marked in some places by a steep metamorphic gradient, in others by thrusts and almost everywhere by conspicuous gravity and magnetic gradients.

Immediately south-east of this junction lies a line of basement massifs. In the south, the *Blue Ridge province* of basement rocks interfolded with cover extends continuously for 800 km with a breadth of only about 50 km and is continued further north by a string of isolated massifs. To the south-east, they are flanked by complexes of higher metamorphic grade in which reactivated basement is difficult to distinguish from the cover. These south-eastern complexes include the *Piedmont province* of the southern Appalachians and much of the eastern parts of New England, the Maritime Provinces and central Newfoundland.

3 The south-eastern margin of the Appalachians

In the metamorphic terrain of the Piedmont province and its equivalents there is commonly some reduction of grade towards the south-east, which suggests an approach to the eastern margin of the Appalachian belt before the blanketing sediments of the Atlantic coastal plain or continental shelf are reached. Drilling in Florida has revealed little-deformed middle Palaeozoic rocks which could represent the cover on or near the eastern foreland (p. 120); possibly, therefore, a fragment of the old foreland still adheres to the American continent in this region.

4 The north-western foreland

The north-western foreland of the Appalachian belt is, of course, the North American craton which persisted as a tectonic entity throughout Phanerozoic time. It so happens that east of a line passing through Oklahoma, Kansas and Iowa, the rocks exposed are mainly Palaeozoic or Precambrian, covering much the same time-span as those of the Appalachians themselves. We shall therefore deal with this eastern part of the craton in this chapter and leave the western part, covered largely by Mesozoic and Tertiary deposits, to be considered along with the cordilleran mobile belt.

The cover-successions of the craton are generally of shallow-water or continental facies and include little or no igneous material. They show considerable lateral variations which are related to the occurrence of a number of cratonic basins separated by *swells* (Fig. 4.3). These basins, a few hundred kilometres in their longest diameter, contain Palaeozoic sequences up to some 5 km in total thickness. The intervening swells are characterised by basement outcrops or by thinner and less complete successions with many unconformities. The *Canadian shield* which flanks the most northerly parts of the Appalachians, is comparable with the uplifts of the interior lowlands in that it was, on balance, positive and retained only local remnants of a cover.

III Early Stages of Deposition in the Appalachians

In this section we shall carry the history of deposition from the inception of the first geosynclinal basins up to the first widely recorded tectonic and plutonic events, that is, from the late Precambrian to the mid-Ordovician. During this period the classic contrasts between miogeosynclinal and eugeosynclinal deposition were most clearly expressed. The American subdivisions of the Palaeozoic

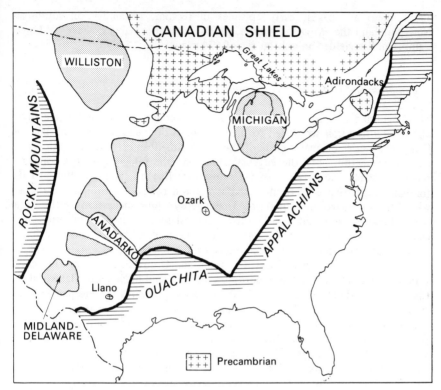

Fig. 4.3. The North American craton, showing the larger basins and uplifts (based on Ham and Wilson, 1967)

systems mentioned in the succeeding pages are compared with the British subdivisions in Tables 2.5 and 3.4.

1 Precambrian cover-successions

Late Precambrian successions with maximum thicknesses of 6—8 km occur in south-east Newfoundland (the *Avalonian*) and south-east Nova Scotia, in Maryland (the *Glenarm* group) and in the Great Smoky Mountains of Tennessee and North Carolina (the *Ocoee* and associated groups) and probably also in the Piedmont province, where the stratigraphy is obscured by metamorphism and deformation. All these groups include poorly sorted detrital sediments such as greywackes, arkoses and conglomerates. Basaltic, andesitic and rhyolitic lavas and acid pyroclastics are widely distributed. Glacigene sediments have been reported only from Newfoundland. All the thick Precambrian groups lie in or close to the eugeosynclinal zone which, it appears, had already been delineated in late Precambrian times. The thicknesses of the groups, the abundance of clastic material and of volcanics and the local presence of unconformities within or above them all suggest early episodes of tectonic instability. In south-east Newfoundland the *Holyrood granite*, which is unconformably covered by Lower Cambrian, has yielded an ischron age of 574 m.y.

2 Cambrian and Ordovician successions

The beginning of the Lower Palaeozoic era was marked, as in Europe, by a *marine transgression* which led to the deposition of transgressive Cambrian and Ordovician series on many parts of the North American craton. Deposition now took place within both the eugeosynclinal and the miogeosynclinal zones. In terms of facies, the deposits of the miogeosyncline are similar to those of the foreland, consisting of carbonate sediments and rocks of orthoquartzite facies with little or no volcanic material. The eugeosynclinal succession on the other hand, includes abundant poorly sorted detrital sediments with volcanic inter-calations. Both types of succession in the mobile belt reach maximum thicknesses many times greater than that of the foreland succession.

In the *northern Appalachians*, Cambrian and Ordovician of miogeosynclinal type crop out in narrow tracts bordering the foreland in Newfoundland and the Champlain lowlands (Fig. 4.4). Cambrian orthoquartzites of feldspathic sand-stones are followed by thicker limestones and dolomites extending from early Cambrian to early Ordovician, but often lacking Middle Cambrian. They are followed, sometimes unconformably, by predominantly shaly rocks of mid-Ordovician (Trenton) ages. The succession, some 2500 m in thickness, resembles the Durness succession and its equivalents in Britain and Greenland (p. 22) and like these successions, its fauna is of 'Pacific' facies.

Cambrian and Ordovician of eugeosynclinal type crop out in a number of broadly anticlinal tracts to the east of the regions mentioned. Rocks of the same type also form a group of sheets with a total thickness of 600–700 m resting on the quartzites and limestones in the region south of Lake Champlain. These exotic sheets, together constitute the *Taconic allochthon* and are regarded as remnants of a cover-series which has been transported westward by gravity, sliding from initial positions above, or to the east of the Green Mountain basement massif (Fig. 4.4).

The more westerly parts of the eugeosynclinal zones are occupied largely by argillaceous sediments, sometimes graptolitic, which interfinger with the carbonate sediments to the west. Some remarkable Ordovician breccias containing immense blocks of quartzite and limestone are interbedded with these rocks in western Newfoundland, and near Quebec. Their presence suggests that sub-marine faulting or slumping brought material eastward from a shelf-sea environment. Still further east, the Cambrian and Ordovician are almost wholly detrital, including both graptolitic pelites and greywackes and reaching thick-nesses estimated at up to 16 km. Oolitic, hematitic, and chamositic iron ores are of importance in the Lower Ordovician of south-east Newfoundland. Pillow-lavas, acid volcanics and pyroclastics, occurring principally within the Ordovician are prominent. Finally, in the most easterly tract of the Avalon peninsula in Newfoundland, Cambrian and Ordovician of platform type reappear, suggesting that the south-eastern border of the mobile belt was near.

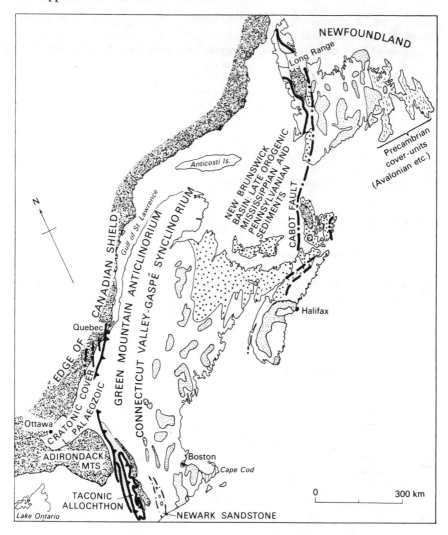

Fig 4.4. The northern Appalachians

The faunas of these rocks compare more closely with those of central Britain than with those on the north-west border of the Appalachians.

In the Valley and Ridge province of the *southern Appalachians*, miogeosynclinal sediments form an almost unbroken succession extending from Lower Cambrian to Pennsylvanian. The Cambrian and early Ordovician parts of this succession total more than 3 km and are of orthoquartzite facies. Basal sandstones are followed by dolomites and limestones, with minor intercalations of sandstone and shale. As in the more northerly region, the environment of the miogeosynclinal zone appears to have been one of shallow seas receiving little detritus and stable enough to allow the accumulation of chemical and organic sediments over long periods of time.

No satisfatory comparisons can be made with corresponding sediments in the adjacent eugeosynclinal zone, for although cover-rocks are extensively exposed in the Piedmont province their stratigraphy is little known. These rocks are seen at relatively low grades of metamorphism in the Carolina slate belt towards the east where they include acid, intermediate and basic lavas, serpentines, pyroclastics and pelites. Elsewhere, they are represented mainly by schists, amphibolites and gneisses. Local finds of fossils have confirmed the presence of Cambrian and Ordovician in addition to Devonian and possibly Carboniferous strata. Isotopic dating of certain volcanic and pelitic rocks indicates that metamorphism had begun by about 450 m.y.

IV Mid-Ordovician Disturbances

1 Tectonic and plutonic episodes

The early phases of deposition in the Appalachian belt were marked by many local indications of geological instability. Stronger disturbances during the later part of the Ordovician period brought about permanent changes in the pattern of deposition and caused widespread folding and metamorphism. They mark the *Taconic* 'orogeny', named from the Taconic Mountains in the northern Appalachians.

The Taconic episodes are indicated by unconformities within the Ordovician, or between Upper Ordovician, or Silurian, and older rocks. Such unconformities are widely recorded over the north-western part of the northern Appalachians but tend to become less conspicuous in the interior of the belt. In the Valley and Ridge province, the only part of the southern Appalachians in which the Lower Palaeozoic succession is well known, there are no major angular unconformities at these levels. A second stratigraphical feature generally attributed to the Taconic orogeny is the appearance in the Middle or Upper Ordovician of considerable thicknesses of rather coarse sediments, apparently derived from newly elevated lands in the interior of the mobile belt. These sediments make the earliest of the *clastic wedges* which are recurring features of the later phases of deposition in the Appalachian belt. In northern regions, the 'Queenstown delta' began to form in New England with the deposition of Caradocian sandstones passing westward into shales. Ashgillian red-beds above the sandstones and shales suggest that the delta was built up above sea-level before being swamped by subsidence in early Silurian times. A Middle Ordovician clastic wedge of sandstones and shales overlies the carbonate sequences of the Valley and Ridge province in eastern Tennessee, reaching a maximum thickness of over 2 km (Fig. 4.6A).

A still closer link between the Taconic disturbances and Appalachian sedimentation is apparent in the Taconic Mountains of New York and the adjacent states. Here, the early Palaeozoic of the Taconic allochthon (p. 106) rests on an autochthonous Cambrian–Ordovician succession of orthoquartzite facies. The autochthonous sediments immediately below the displaced sheets are usually shales of Caradocian age, and at some localities these shales carry jumbles of sub-angular fragments of many rock-types including not only quartzites and dolomites but also greywackes and shales. Some of the included blocks appear to be derived from the allochthonous succession and it is inferred that the boulder-beds were laid down ahead of, and subsequently overridden by, the allochthon.

Evidence of Ordovician folding, metamorphism and granite-intrusion is fairly widespread in the northern Appalachians. In the north-western part of the belt a considerable angular unconformity often separates the first post-Taconic deposits from the underlying rocks. A number of granitic bodies from Newfoundland to New England have yielded radiometric ages of from 500–450 m.y. and some are covered unconformably by Silurian strata; similar ages have been obtained from the Piedmont province.

2 The Appalachian serpentines

The presence of hundreds of pods of serpentinised ultrabasic igneous rocks is a feature of the Appalachian belt which has aroused much interest. These serpentine bodies are almost restricted to the eugeosynclinal zone and are distributed over a distance of 2500 km along it; many of them are ranged in a few (often two) tracts parallel to the structural grain. Individual bodies, often less than a kilometre in length, are made up partly or wholly of serpentinised olivine with enstatite, augite, chromite and such secondary minerals as talc. They usually show signs of post-crystalline stress and in some localities contain workable asbestos deposits.

The Appalachian serpentines are of *alpine type*, to use the nomenclature of Benson (1926). They occur as strings of small steep bodies, commonly in association with sedimentary rocks of greywacke type. Most writers appear to favour the Ordovician period as the main time of emplacement, connecting this event with the onset of widespread orogenic compression. Hess suggested in 1938 that the serpentine belt was developed as a result of partial fusion in a deep peridotitic region tapped by down-buckling of the granitic crustal layer into this deep layer. More recent authors have envisaged an origin controlled by the 'consumption' of oceanic crust at an island arc-oceanic trench system whose contents are represented by the eugeosynclinal filling (compare Bird and Dewey, 1970).

V Later Evolution: the Northern Appalachians

1 Silurian and Devonian deposition

In the northern region, deposition was resumed in the interior of the mobile belt after the Taconic disturbances in late Ordovician or early Silurian times and was continued in many places without major breaks until early or middle Devonian. The main Silurian-Devonian outcrops occupy an irregular tract extending from Gaspé to the coast of Connecticut (Fig. 4.4). This tract, referred to as the *Connecticut Valley–Gaspé 'synclinorium'*, occupies the site of a relatively narrow basin and contains a succession thicker than those laid down over the anticlinal belts on either side.

The Silurian-Devonian succession of the Connecticut Valley—Gaspé synclinorium includes much material of greywacke facies and reaches thicknesses of some 5 km. It ends, in some localities, with estuarine or continental clastic sediments of Middle or Upper Devonian age. Andesitic, rhyolitic and basaltic lavas and pyroclastics are interbedded with the sediments, indicating that vulcanicity was resumed after the Taconic episodes. In central Newfoundland, the Silurian contains coarse polymict conglomerates carrying not only volcanic and sedimentary material but also granitic boulders probably derived from unroofed Taconic intrusions.

These Silurian-Devonian outcrops lie well within the mobile belt and appear to represent deposits of marine basins and shelf-seas essentially similar to those formed during earlier stages of evolution. They may be contrasted with two clastic wedges which were built out towards the foreland from land-sources in the southern part of the northern Appalachians. These wedges are largely deltaic and fluviatile and contain little or no volcanic material. Their thick ends lie close to the western boundary of the metamorphic tract, while their thin ends fan out over the foreland. The earlier and smaller wedges of fine clastic sediments, with red-beds and evaporites, form the Ludlovian *Salina group* of New York State. The great *Catskill delta* began to accumulate in early Devonian times and extended in the course of the period as far north-west as Lake Erie and as far south-west as Virginia. In its thickest, proximal part, the delta-wedge reaches over 3 km and consists largely of sandstones with red-beds which appear to represent sub-aerial flood-plain deposits; towards the outer limit of the delta, dark shales predominate in a zone which apparently represents its submarine part. The Catskill delta is regarded as a product of Devonian tectonic disturbances, the *Acadian* episodes in the interior of the mobile belt.

2 The Acadian disturbances and the structural and plutonic history

The Acadian disturbances marked the last important stages of tectonic and metamorphic activity in the northern Appalachians and were responsible for the major structures of the region. Stratigraphical and radiometric evidence suggests that the principal Acadian episodes were mid-Devonian. Lower Devonian sediments follow the Silurian without a major break in many places and are themselves folded and intruded by granites. Mississippian and locally late Devonian strata are unconformable on underlying rocks and show relatively open folds. Granites have yielded ages in the range 400—360 m.y.

The structural pattern of the northern Appalachians is a result of the superposition of Acadian on Taconic structures. In this composite pattern we may recognise, from north-west to south-east, a marginal thrust-zone, a zone of low-grade cover-rocks, a zone of basement massifs and a metamorphic zone of cover and basement (Fig. 4.4).

At the *marginal thrust-zone* which marks the orogenic front, a number of low-angle thrusts with easterly and south-easterly dips carry folded cover-rocks, usually Cambrian and Ordovician, over the flat-lying Cambrian and Ordovician of the foreland. This thrust-zone, sometimes referred to as Logan's Line, can be traced from Gaspé to northern New York State. To the north in Newfoundland and to the south in New York, folding dies out more gradually. Contrasts in the thickness or facies of the Cambrian and Ordovician above and below the thrusts indicate that the tectonic boundary of the fold-belt was close to the margin of the earlier Appalachian basins of deposition.

The western *zone of low-grade cover-rocks* includes miogeosynclinal successions as well as the more widespread eugeosynclinal rocks. Most of this zone is made up of Ordovician and Cambrian showing strong, but not recumbent, folding and commonly exhibiting a slaty cleavage. In New York State and New Hampshire, the *Taconic allochthon*, representing slide-sheets of detrital Cambrian-Ordovician emplaced during Taconic episodes, is folded along with the Cambrian-Ordovician carbonate sediments on which it rests.

The *zone of basement massifs* enters the northern Appalachians in the region of New York as an extension of the Blue Ridge province of the southern Appalachians. In Vermont a string of elongated massifs is flanked by Lower Palaeozoic strata. To the north of the Green Mountains massif the basement plunges down below the cover and the zone is continued into Gaspé by an anticlinal tract of Cambrian flanked by Ordovician.

Although the basement massifs are bounded on the west by steep or overturned junctions which are often marked by faults, they are regarded as essentially autochthonous masses rooted in their present position. Relics of an autochthonous cover are sometimes little disturbed or are nipped down in disrupted folds between upthrust wedges of crystalline rocks. The basement itself is made up of gneisses, granites and metasedimentary schists containing relics of marbles. In the western parts of the massifs, its characters are essentially those acquired during the Grenville cycle; radiometric dating from the massif west of the Hudson River has given both whole-rock and K–Ar mineral ages older than 800 m.y. In the eastern parts there is more evidence of Appalachian reworking; on the east side of the Hudson River, for example, newly developed shear-zones traverse the basement and dating of minerals gives apparent ages ranging from 800–320 m.y.

The *metamorphic zone* of the northern Appalachians extends from the zone of basement massifs to the Atlantic coast with a maximum breadth, in the Maritime Provinces, of some 500 kilometres. In Newfoundland, the metamorphic zone east of the Long Range is evidently out of alignment with respect to the corresponding structure to the south. South of Cape Cod, the zone is cut back by the Atlantic coast to a narrow remnant in the vicinity of New York, but it broadens southward to unite with the Piedmont province of the southern Appalachians. The essential continuity of the belt is attested by its constant relationship to the arcuate zone of basement massifs and is a subject to which we shall return later.

The country-rocks of the metamorphic belt consist of the contents of the old eugeosynclinal basin, together with a few reworked bodies of basement. The cover-series ranges in age from late Precambrian to mid-Devonian and includes, besides the characteristic detrital sediments, numerous thick volcanic groups, and the serpentines of the Appalachian serpentine belt. Carboniferous, and locally even Upper Devonian sediments rest in the Maritime Provinces on rocks previously metamorphosed and folded. Numerous granitic bodies of several types and ages invade the country-rocks.

The distribution of Cambrian and Ordovician on the one hand, and of Silurian and Devonian on the other, define a number of broadly anticlinal and synclinal tracts which can be traced as sub-parallel zones for considerable distances and which appear in some areas to refold earlier isoclines (Fig. 4.5). The axial planes of these major folds are usually steep and follow the trend of the fold-belt as a whole. Numerous domes of gneissose granitic rocks, the *Oliverian magma-series*, distort the linearity of the grain in a tract extending north-north-east through Vermont and New Hampshire towards the western side

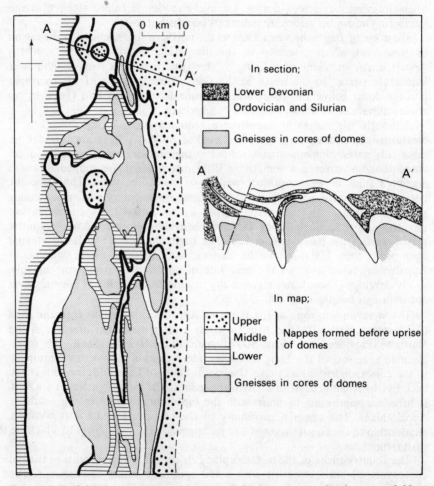

Fig. 4.5 Refolded nappes in the Appalachian eugeosynclinal zone of New England; recumbent folds in Ordovician-to-Devonian cover-rocks with westward vergence are distorted by gneiss domes derived from Ordovician volcanics or from older rocks (based on Thompson and others, 1968)

of the belt. Most of the domes lie at low levels in the succession; they are tentatively regarded as intrusions of Ordovician age but could perhaps have originated as basement domes.

Metamorphism is mainly of Barrovian type marked by the classic index minerals chlorite, biotite, garnet, staurolite (kyanite is usually absent) and silli-manite. At the western margin of the belt, remarkably steep metamorphic gradients are displayed locally. East of the Hudson River in Dutchess County, for example, the grade increases from biotite- to sillimanite-zone over a distance of less than 15 km. The transition is seen only where the cover-series laps around the northern end of the Hudson Highlands basement massif; the recognition of this transition in a classic study by Balk and Barth (1936) provided the first clear demonstration that high-grade rocks such as the Manhattan schists of New York could be equivalent to the little altered cover-rocks of the western tract.

In the interior of the metamorphic belt, considerable variations of grade are recorded. In general, the grade is low in the north-east, higher towards the south. Local variations appear often to be related to the distribution of the domes already mentioned, or of tongues of gneissic granites, and the isograds may be steep or overturned near the margins of these bodies. The domes and tongues are referred to by many names – the Oliverian magma series, the Bethlehem Gneiss and several others; they are concordant with the country-rocks, share the regional foliation and are regarded by some authors as being at least partly of replacement origin. Regional cooling over most of the northern Appalachians appears to have taken place by Devonian times, for metamorphic minerals yield dates in the range 360–380 m.y. In eastern Maine, however, much younger apparent ages of metamorphic rocks, in the region of 250 m.y., suggest late Palaeozoic activity. These apparent ages may (Faul et al., 1963) reflect no more than a mild reheating without actual recrystallisation; they mark the most northerly extent of late-Palaeozoic plutonic activity in the Appalachians (p. 120).

Almost all granitic bodies of the northern Appalachians occur within the metamorphic zone. The oldest dated granites (p. 105) are late Precambrian. A widely distributed set (the 'Highlandcroft magma-series') is Taconic as is, most likely, the Oliverian magma-series. Most of the large intrusive bodies, however, are related to the Acadian disturbances and have given radiometric ages of 400–360 m.y. Discordant granites (the *New Hampshire magma-series*) include bodies with well-defined contact aureoles.

3 The subsequent history

The Acadian disturbances marked the terminal orogenic stage in the northern Appalachians. The succeeding strata, mainly Mississippian and Pennsylvanian, rest unconformably on the eroded surface of folded rocks and granites. They are largely of continental facies and occupy narrow basins separated by highland massifs which rose repeatedly during the period of deposition. In the centres of the basins, many kilometres of sediment accumulated while, at their margins, the successions are thin, are interrupted by many non-sequences and include some volcanic components.

The principal intermontane deposits of the northern Appalachians border the Gulf of St Lawrence in the Maritime Provinces and Newfoundland, while smaller outcrops occur around Rhode Island, The Mississippian includes continental sandstones and shales, followed by a marine limestone (the Windsor formation) associated with red-beds and evaporites; the Pennsylvanian is entirely continental and includes sandstones, shales and thin coals. These deposits are extensively faulted and towards the south-west of the basins the flexures induced by differential subsidence are accentuated by folding powerful enough to produce a slaty cleavage. Following these late-orogenic Upper Palaeozoic sediments, and unconformable or all older rocks, is the *Newark Sandstone* of Upper Triassic age. This formation, which extends intermittently the length of the Appalachians, is made up of continental sandstones, shales and conglomerates deposited in fault-bounded troughs. Basic lavas, dykes and sills are associated with the sediments, one of the most conspicuous being the gravity-stratified *Palisade sill* of New Jersey.

VI Later Evolution: the Southern Appalachians

In the miogeosyncline of the southern Appalachians, the Taconic disturbances had little direct impact, deposition continuing with only local interruptions throughout the Ordovician period. Indirectly, however, these disturbances brought about a lasting change in the distribution of facies, for the elevation of new lands within the mobile belt provided source-regions from which detrital sediment spread over the earlier carbonate-rocks. The maximum thicknesses of later Palaeozoic deposits near the eastern border of the miogeosyncline are of the order of 7 km; with early Palaeozoic deposits, the total thickness of sediment in the miogeosyncline rises to at least 9 km, making a great prism almost wholly free from volcanic material which thins westward to about 2 km over a horizontal distance estimated at about 400 km.

The first of the clastic wedges in the southern Appalachians was the mid-Ordovician *Tellico-Sevier wedge* which, reaches a maximum thickness of over 2 km, thinning both distally and laterally away from the source (Fig. 4.6A). In its thickest part there is seen an alternation of shales and sandstones, with local conglomerates near the base and red, possibly terrestrial, sandstones near the top. These detrital deposits interfinger with carbonate sediments which continued to accumulate near the western border of the miogeosyncline throughout the middle Ordovician, finally passing up into shales and sandstones in late Ordovician and Silurian times.

The Silurian period was marked by a lull in orogenic activity through most of the southern Appalachians and, consequently, Silurian deposits in the Valley and Ridge province are generally thin and are shaly or calcareous. Only in the north does the Silurian thicken towards the *Salina Group*, which includes oil-bearing sandstones (p. 110). The dominant unit of the Devonian system is the *Catskill delta*. In the southernmost part of the Appalachians, Devonian is absent or represented by very thin deposits, sometimes unconformable on the underlying strata. The Middle and Upper Devonian thicken north-eastward, reaching a total of over 3 km in central Pennsylvania. As might be expected,

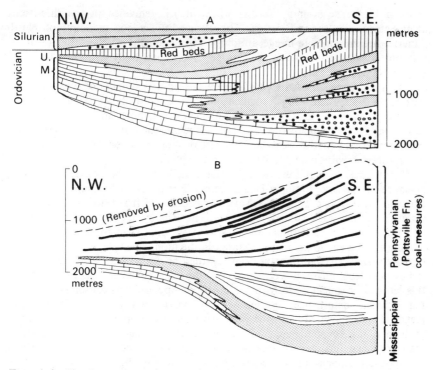

Fig. 4.6. Clastic wedges deposited near the north-western border of the Appalachians: (A) the Tellico-Sevier wedge in eastern Tennessee (based on King, 1950); (B) the Pottsville Formation in Alabama (based on King, 1954)

since the source-region lay well to the north, the sediments related to the Catskill delta appear progressively earlier in the succession as one goes north-eastward up the miogeosynclinal zone. In Virginia, at the southern edge of the delta, red-beds appeared in the Givetian; in Gaspé the coarse clastics of the proximal parts are Lower Devonian.

The Mississippian period began in the southern Appalachians in relative tranquility, with carbonate deposits advancing again into the north-western part of the miogeosyncline and deposition of clastic sediments continuing in the south-eastern part of the zone. In late Mississippian times a new delta began to build out in the south across the miogeosynclinal zone and onto the foreland beyond. The growth of this delta dominated the pattern of sedimentation during the building up of the *Pottsville Formation* with maximum thicknesses of about 3 km in Alabama (Fig. 4.6B).

During the build-up of this wedge, a regional geographical change of great importance was taking place. This was the emergence on the site of the old miogeosyncline, and the adjacent foreland, of a broad plain spilling the sea back into the central parts of the North American craton. This *paralic plain*, standing only just above sea-level, was backed by rising land-masses in the interior of the Appalachian and Ouachita mobile belts. It was, for much of its existence, covered by dense swampy forests whose debris remained in the thickening sedimentary pile as coal seams. The Pennsylvanian coal-measures of North America are comparable in age, lithology and geological setting with the Upper Carboniferous coal-measures of Europe which are dealt with in Chapter 3 (Fig. 3.10, Table 3.7).

The Pennsylvanian of the paralic plain is seen in a number of synclinal cores within the Valley and Ridge province and more extensively on the less disturbed *Allegheny plateau* to the north-west. The sediments are almost entirely terrestrial (marine incursions are less frequent than in the European coal-measures except in the west) and consist of feldspathic sandstones, shales and coals, often showing a distinctive rhythmic sequence.

The youngest strata exposed in the synclines of the Valley and Ridge province are generally Middle Pennsylvanian but occasionally range up to early Permian. The Upper Triassic Newark Sandstone, the first formation laid down when deposition was resumed, is post-tectonic. The interval between these rocks, late Pennsylvanian or Permian and Triassic, saw the last major phase of folding, metamorphism and granite-emplacement in the mobile belt.

This period has traditionally been regarded as *the* 'Appalachian orogeny' and it has been customary to contrast the southern Appalachians consolidated by this late-Palaeozoic episode with the northern Appalachians consolidated during the Acadian disturbances. Recent studies, particularly those involving radiometric dating, have led to a widespread revision of this opinion among American geologists. It is becoming apparent that although little deformation took place in the miogeosynclinal zone before late Palaeozoic times, the metamorphic tract of the southern Appalachians had suffered orogenic disturbances from Ordovician times onward and was largely stabilised before the end of the era. For these reasons, Woodward (1957) proposed the name *Alleghenian* for the late Palaeozoic disturbances in place of the more comprehensive older term 'Appalachian orogeny'.

VII Structure of the Southern Appalachians

The three principal tectonic zones of the southern Appalachians (p. 102 Fig. 4.2) recall the zones of the northern part of the belt. The Piedmont and Blue Ridge provinces are essentially continuous with the metamorphic zone and zone of basement massifs which have already been discussed. The Valley and Ridge province, on the other hand, does not extend much further north than Pennsylvania. It is a tectonic zone peculiar to the southern Appalachians which does not appear to have shared to the full extent in the earlier deformational history of the orogenic belt.

1 The Valley and Ridge province

The thick prism of non-metamorphic cover-sediments of the Valley and Ridge province includes many competent psammitic and calcareous divisions which thin gradually north-westward towards the foreland. On the south-east side, the present thickness of the cover falls abruptly to nothing against the Blue Ridge province where the basement emerges at the surface. The combined evidence of geological structure, gravity (Fig. 4.7) and magnetic surveys suggests that in West Virginia, Maryland and Pennsylvania, the basement stands as a gigantic wall, overturned westward and plunging down to depths of 6 or 7 km beneath the thickest part of the prism. Further south, the basement extends westward in flat thrust-sheets over the sedimentary prism. It follows that the miogeosynclinal prism, over 1200 km in length and today little more than 200 km in breadth, occupies an asymmetrical trough in the basement with a steep wall on the east and a floor rising gradually towards the west.

The outer (north-western) margin of the Valley and Ridge province is marked by an abrupt decrease in the intensity of folding. The Upper Palaeozoic strata of the foreland in the Allegheny plateau show only open folds and associated faults. The tectonic front itself is defined by a tract of low-angle thrusts and of

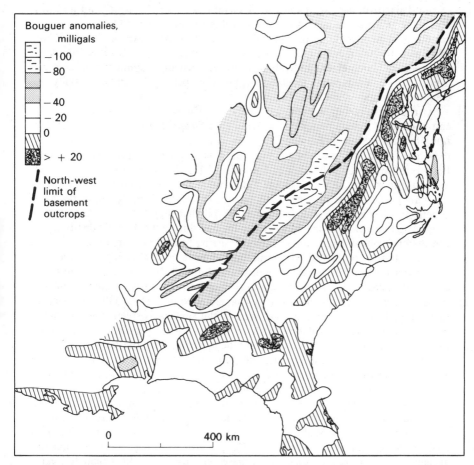

Fig. 4.7. Bouguer gravity-anomalies in relation to the Blue Ridge province (after P. B. King, In: Lowry *et al.*, 1964)

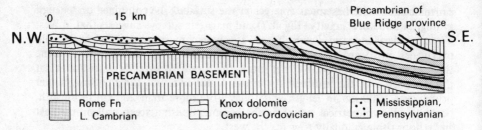

Fig. 4.8. Sketch-map of the Valley and Ridge province (based on Rodgers, 1970)

steeply-dipping, sometimes overturned, rocks which produce a great westward-facing step in the cover: to the east, older and more tightly-folded rocks mostly of mid-Palaeozoic age are exposed. A second 'step' of the same kind mid-way across the province brings up isoclinally folded Cambrian and Ordovician which are the dominant rocks in the south-eastern part of the province. The dominant folds and faults trace out a simple series of open arcs extending for many hundreds of kilometres. Individual folds are usually asymmetrical and are overturned towards the foreland while the principal faults partially or completely replace their steep limbs.

At the north-eastern end of the Valley and Ridge province, the folds plunge east-north-east from a major culmination (the *Susquehanna* or *Juniata culmination*) to bring down Pennsylvanian in the anthracite basin of Pennsylvania. As they rise again on the eastern side of this depression, the folds begin to flatten out and over a distance of less than 25 km the entire fold-set fades away into an almost flat sheet of Upper Palaeozoic. The characteristic structures of the Valley and Ridge province thus terminate rather abruptly along the axial direction; it has been suggested (Rodgers, 1964) that they end against the western margin of the region which had been deformed during the Taconic disturbances and thereby consolidated into a buttress resisting further deformation.

Towards the south-west, the structures show a gradual change from a style dominated by beautifully regular concentric folds in the competent sandstones and limestones — the classic structures featured in early accounts of the Appalachians — to a style increasingly characterised by eastward-dipping dislocations (Fig. 4.8). Many of these are seen at the surface as steep reversed faults, but appear to flatten off downward into thrusts almost parallel to the bedding, whose effect is to strip off and reduplicate the upper layers of the cover.

The character of these thrusts, the concentric forms of the folds and the absence of basement outcrops suggest that the Valley and Ridge structures do not persist in depth. Rodgers among others has compared these structures with those of the Jura Mountains, suggesting that the cover was deformed independently, parting from the basement along *décollements* located in incompetent Cambrian shales which in some localities contain evaporites. This analogy is not universally accepted — Cooper, for example, considers that the anticlinal folds were developed over basement uplifts which originated early enough to influence the thickness and facies of the cover sediments; but it appears to be supported by seismic and gravity data. Estimates for the shortening of the Palaeozoic cover achieved by sliding over the basement range up to about 80 km (Gwinn, 1970).

2 The Blue Ridge province

The junction between the Valley and Ridge and Blue Ridge provinces, at which the basement rises from depths of several kilometres to positions at or near the surface, is marked by a steep gravity gradient linking negative anomalies over the Valley and Ridge province with positive anomalies over the metamorphic terrain (Fig. 4.7). This gravity feature continues far to the north along the zone of basement massifs to Gaspé, confirming the essential continuity of the zone with the Blue Ridge province.

The country-rocks of the province include a crystalline basement and a late Precambrian and early Palaeozoic cover. In these country-rocks are emplaced trains of small serpentine bodies, continuing the serpentine belt of the northern Appalachians, and a few Palaeozoic granites similar to those described in the next section. The basal unconformity is usually still recognisable as a structural and metamorphic discontinuity, but the cover is commonly metamorphosed at least to slate grade and the basement is partially reworked. In the Valley and Ridge province, cleavage is developed only locally near the south-eastern margin.

The basement of the Blue Ridge province consists largely of gneisses and amphibolites, with areas of charnockite and numerous bodies of more homogeneous granitic gneiss. Complex radiometric age-patterns have been recorded: zircons tend to give dates of about 1000 m.y., appropriate to rocks of the Grenville province, while micas give dates down to about 250 m.y., indicating reheating during several Palaeozoic episodes. The cover succession includes late Precambrian units such as the Ocoee Group which are not represented further west, as well as Cambrian and Ordovician similar in facies to those of the Valley and Ridge province. The pelites and volcanics commonly appear as slates, greenschists and phyllites and even the more competent units may be strongly deformed. A classic study by Cloos (1947) in the South Mountain anticline of Maryland showed that oolites in limestones were distorted to ellipses with axes giving ratios of up to at least 5:1.

In the northern part of the Blue Ridge province the basement outcrops, like those of the northern Appalachians, appear to represent more or less autochthonous anticlinal wedges overturned towards the foreland and carrying remnants of a folded and dislocated cover. Where the basement lies below ground-level, as in the South Mountain anticline, this cover is directly continuous with that of the Valley and Ridge province. From central Virginia southward, however, the basement rocks of the Blue Ridge are largely allochthonous, and are seen to overlie the Valley and Ridge cover-sequence on low-angle thrusts (Fig. 4.7). The overridden cover is revealed in windows exposed well toward the south-east side of the province, suggesting tangential displacements of the order of 60 km. King (in Lowry et al., 1964) notes that the gravity gradient which coincides with the western margin of the autochthonous basement-outcrops falls back to the south towards the south-east beneath the allochthonous basement-outcrops, following the line of the autochthonous basement rising from depth.

3 The Piedmont province

The Piedmont province is occupied by repeatedly deformed metamorphic rocks associated with an almost equal bulk of granitic material emplaced during one or more stages of Appalachian plutonic activity. In the southern part of the region, its junction with the Blue Ridge province is a remarkably straight feature known as the *Brevard zone* which is traceable for over 500 km. There are consistent lithological differences between the rocks on either side of this lineament which is marked by a broad zone of phyllonites and blastomylonites. It is usually regarded as the site of a transcurrent fault on which considerable dextral horizontal displacement took place and may also represent the 'root-zone' of the basement slices forming the southern part of the Blue Ridge province. Further north, the boundary of the province is marked by rather sharp increases of metamorphic grade, and in some instances by early dislocations such as the 'Martic overthrust' of Maryland. The structural patterns are complex and irregular when compared with those of the unmetamorphosed zone to the north-west. Elongated dome-patterns are seen in many places, sometimes associated with interference-structures. In Pennsylvania and Maryland, regenerated basement rocks enter into the structure as cores of isoclinal folds and mantled gneiss domes (the *Baltimore gneiss* of Maryland). Uranium–lead determinations of zircons have yielded ages of about 1000 m.y., relating to pre-Appalachian (Grenville) events, but K–Ar and Rb–Sr dates for minerals from the same rocks give values of 280–380 m.y. reflecting the effects of reworking.

The main areas of high-grade metamorphism are concentrated in the central and western parts of the province. Here, the principal rocks are schists, amphibolites and migmatitic gneisses, showing a Barrovian type of metamorphism ranging up to sillimanite grade. The *Carolina slate belt* on the eastern flank reveals little-altered slates, greywackes, tuffs and lavas of uncertain age, while undeformed Ordovician, Silurian and Devonian occur in a belt known as the *Suwanee basin* hidden beneath the coastal-plain sediments of Florida. Igneous bodies emplaced in the Piedmont zone include early gabbros and diorites and an array of granites among which foliated concordant types and large discordant plutons are distinguished.

The dating of tectonic and metamorphic events in the Piedmont province is still uncertain. Radiometric dates scatter from approximately 550 m.y. down to about 230 m.y. The youngest dates, falling in the Permian period, have been derived from a rather narrow tract in the west of the province. One or two late-tectonic granitic intrusions in Tennessee have yielded mineral ages of up to 380 m.y. (Davis, Tilton, Wetherill, 1962), while dates tentatively associated with the younger of two episodes of syntectonic crystallisation in Pennsylvania (Lapham and Bassett) fall round around 330 m.y. In this latter region, structures formed during the same episode fold the Martic thrust which evidently belongs to an early phase of deformation.

From this kind of evidence, from the appearance of an Ordovician clastic wedge in the Valley and Ridge province, and from the local recognition of a mid-Ordovician unconformity, it has been inferred by King, Rodgers and others that folding and metamorphism began during the Taconic disturbances and were largely ended by mid-Carboniferous times. Many of the vast late-tectonic granites may be late Devonian in age. In this respect, the metamorphic zone of the south has more in common with the corresponding zone of the northern Appalachians than with the non-metamorphic Valley and Ridge province in which major folding did not begin until late Pennsylvanian times.

VIII Mineralisation in the Appalachian Cycle

It is convenient to digress at this point to mention the ore-mineralisation in the Appalachians as a whole. The subject can be dealt with rapidly for, in fact, the cycle in North America appears to have been a rather barren one. In the Canadian sector, small sulphide deposits with copper, lead, zinc and gold in the eugeosynclinal cover recall similar deposits of the European Caledonides. The Appalachian serpentines are not associated with important metalliferous ores, though they yield asbestos and talc where metamorphosed. The immense mid-Palaeozoic granites and related rocks have little associated mineralisation and in this respect contrast with the late-Palaeozoic granites of the Hercynian belt in Europe (p. 82). There are, however, small deposits carrying Au and Ag in the Blue Ridge province and hydrothermal deposits with Zn, Pb, Cd and Ag in the Valley and Ridge province.

On the other hand, the sedimentary deposits of the margin of the Appalachian belt and its foreland are of great economic value, providing the *Pennsylvanian coal-measures*, the original basis of industrial development in the eastern United States. Considerable oil reserves occur in the gently-folded Allegheny plateau, the principal reservoir rocks being Middle Devonian, Mississippian and Pennsylvanian sandstones.

IX Post-Orogenic Deposits

The terminal stages of orogenic activity in the Appalachians were followed, as in many mobile belts, by the accumulation of post-orogenic clastic sediments in intermontane basins. These deposits form the *Newark Sandstone* laid down in fault-bounded troughs running parallel to the orogenic grain. They are of Upper Triassic age and were therefore separated by an interval of some 50 m.y. from the youngest (early Permian) cover-deposits which preceded stabilisation. Newark Sandstone outcrops of essentially similar types occur, mainly within the metamorphic zone, up almost the entire length of the Appalachians, indicating a remarkable uniformity of conditions during the final stage of the cycle. Basic minor intrusions are constant associates of the sediments, though lavas such as are found in the northern Appalachians are lacking in the south.

The Triassic deposits are of molasse type, consisting of red sandstones and shales of continental facies, which rest with profound unconformity on older rocks. The faults bounding the Triassic outcrops appear to have been active during the period of deposition, for conglomeratic rocks near the borders interfinger with the finer clastic deposits in the interior. The development of the rift-structures containing the Newark Sandstone marked a turning-point in the history of eastern North America for the later evolution of this region was dominated by processes related to fracturing and ocean-development.

X Palaeozoic Evolution of the North American Craton

The history of the Canadian shield and interior lowlands during the Palaeozoic eras provides one of the best-documented examples of the development of a cratonic block of continental dimensions and, as such, is worth considering from a number of angles. The contrast between the enormous *shield area* of exposed Precambrian in Canada and Greenland on the one hand and the *platform area* of the interior lowlands on the other, reflects a long-term tendency for the

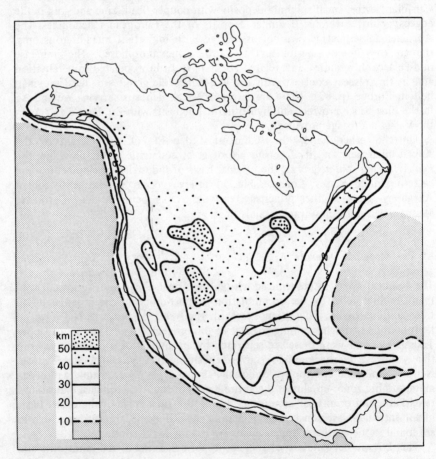

Fig. 4.9. Preliminary data showing variations in crustal thickness in North America; the contours show the depth of the Moho (based on Kanasewich)

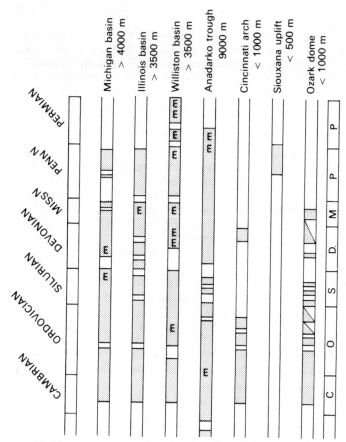

Fig. 4.10. Palaeozoic successions of the North American craton. The Cambrian to Mississippian deposits are predominantly carbonate, the Pennsylvanian and Permian predominantly detrital; (E) = evaporites; blank area = no deposits; the figures above each column give maximum thicknesses (based on Ham and Wilson, 1967)

southern part of the craton to sag, relative to the northern part. The old-established *palaeoslope* is reflected even today by the southward course of the great rivers of the interior. The characteristic ovoid plan of the principal basins of subsidence in the interior lowlands and within the shield (Fig. 4.3) differs conspicuously from that of the long narrow geosynclinal and subsidiary troughs developed in the Appalachian belt. When the outlines of these basins are superimposed on the map of tectonic provinces in the underlying basement little correspondence is apparent, but the intervening 'uplifts' appear to parallel old lineaments in the basement (see Part 1, p. 89). The basins persisted for several hundred million years and were in some instances (Fig. 4.9) associated with notable thickening of the crust as a whole, indicating that they represent fundamental structures.

The Palaeozoic deposits of the craton (Fig. 4.10) are overwhelmingly those of shallow seas and low-lying lands, principally sandstones and carbonate rocks in the older systems and principally terrestrial sandstones, shales, coals and evaporites in the Pennsylvanian and Permian. Repeated non-sequences or minor unconformities interrupt the successions in and on the flanks of the shield and uplifts, while shorter but still numerous breaks are recorded in the basins. The thicknesses of individual systems are usually only a fraction (generally no more than a tenth) of those of the same systems in the Appalachian and Cordilleran mobile belts. Maximum thicknesses of Lower and Upper Palaeozoic together are of the order of 5 km in the Michigan, Illinois and Delaware basins, less than 3 km in the other basins, and usually less than 2 km elsewhere. Volcanic components are almost entirely absent.

The strata which follow the basal unconformity in the eastern parts of the craton are nowhere older than Middle Cambrian and are more frequently Upper Cambrian or Ordovician. Around the margins of the Canadian shield and along the Sioux uplift, which projects from the south-western corner of the shield, Silurian or Upper Palaeozoic strata commonly rest directly on Precambrian. The first deposits, whatever their age, are generally thin basal conglomerates or sandstones. They are followed by dolomites and limestones interbedded with subordinate shales and sandstones. Until mid-Palaeozoic times, the detritus supplied to the craton was limited and the coarser sediments were therefore restricted. Only a few sandy formations such as the Chazyan St Peter Sandstone extend over a substantial part of the craton.

In the absence of abundant detritus, *chemical-organic* sediments accumulated widely in Lower Palaeozoic times. For the most part, these sediments were limestones and dolomites, but from time to time the shallow and possibly restricted seas attained such high salinities that evaporites were deposited in formations up to several hundred metres thick.

The carbonate sediments of Cambrian and Lower Ordovician ages carry faunas not unlike those of the Appalachian miogeosynclinal zone and of the regions along the western border of the Caledonides in Greenland and Europe. The Beekmantown limestone (Canadian), for example, has long provided a standard for comparison with the Durness Limestone of north-west Scotland. Silurian limestones occur widely in Arctic Canada, around Hudson Bay and in the interior lowlands, many of them being of reef facies. Devonian reef-complexes fringe the western border of the Canadian shield, where they are associated with major oil-fields.

The earliest evaporites occur in the Late Ordovician of the Williston basin. More widespread deposits of gypsum, anhydrite and rock-salt occur in the Upper Silurian in the Williston basin, in the Michigan basin (up to 600 m) and thence eastward in the Salina Group to the margin of the Appalachians (p. 110). Evaporites are interbedded with the predominantly detrital Devonian sediments in the same regions and with the continental Permian over a considerable area further south, suggesting that the craton moved through a warm-arid climatic belt during the Upper Palaeozoic era (p. 226).

From late Silurian to Permian times, the deposits of the craton were predominently *detrital*. The change in lithology suggests an increase in the supply of land-derived detritus related to the elevation of new land-masses in the eastern and southern mobile belts. Furthermore, a spread of sediments of continental facies indicates that much of the craton itself rose intermittently above sea level.

The effects of Acadian disturbances are recorded by unconformities within the Devonian. Early Devonian sediments, when present, include a good deal of marine carbonate-rocks. Late Devonian sediments, usually unconformable on older rocks, are characterised by widespread black or grey slates, by red-beds in the Catskill delta and by evaporites in the Williston and Michigan basins. The Lower and Middle Mississippian include the last extensive Palaeozoic limestones of the craton. They are followed by variable sandstones and shales which spread more widely than clastic formations of any earlier episode. The Pennsylvanian, generally unconformable on the older rocks, is dominantly clastic and partly continental. The Permian is missing altogether from much of the northern and eastern parts of the craton; in the south and west it is largely clastic.

The characteristic features of the Pennsylvanian strata are the cyclic repetition of a few lithological types, the occurrence of coals and the interleaving of marine and non-marine beds (compare Table 3.7). The deposits of the low-lying forested plains extending north-westward from the newly-elevated lands of the Appalachian and Ouachita belts merged westward with those of the remaining shallow seas and repeated fluctuations in the relative levels of land and sea led to the interfingering of marine and non-marine facies across a wide central zone. Limestones of the Palaeozoic cover are host-rocks of the important *Mississippi Valley* type of mineral-deposits. Sulphides of lead, zinc and copper, with barytes and fluorite, are concentrated in porous host-rocks around the cratonic basins; they are thought to have been derived from black pelites accumulated near the centre of an anaerobic basin and distributed by connate waters migrating up-dip to the basin-rim.

XI The Ouachita Belt and Associated Units

The narrow southern apex of the North American craton is bordered by an arcuate Palaeozoic orogenic belt which runs parallel to the north coast of the Gulf of Mexico (Fig. 3.1). The deposits of the Mississippi Valley and coastal plain obscure not only the southern portion of this belt but also its junction with the Appalachian system. Its exposed portions, in Arkansas, Oklahoma, Texas and New Mexico, reveal remnants of a number of troughs whose Palaeozoic infillings are deformed and metamorphosed to varying extents. Numerous borehole-records allow these structures to be traced for some distance eastward beneath the coastal-plain sediments. The strongly-folded strata of the belt range from Cambrian to Pennsylvanian while the oldest post-orogenic sediments are Permian. Ouachita and Appalachian structures approach each other almost at right angles and must meet beneath the coastal plain. Although the structures of the Valley and Ridge province may terminate against the Ouachita belt, drillhole data suggests that the Ouachita structures turn sharply north-east to unite with the interior zones of the Appalachian system. At its western end the Ouachita belt approaches the Cordilleran belt of Mexico almost at right angles and appears to be truncated by the latter.

The basins of deposition. Within and along the border of the Ouachita system, a number of troughs with differing depositional and tectonic histories are distinguished. They fall into three main categories: (a) basins of the fold-belt proper, exposed in the Ouachita Mountains (Oklahoma and Arkansas) and the Marathon area (Texas); (b) marginal basins of late development; and (c) the Anadarko trough which runs obliquely through the cratonic block.

(a) The *basins of the fold-belt proper* came into existence at or before the beginning of the Palaeozoic, though the bulk of their in-filling was late Palaeozoic. Cambrian to Devonian deposits total at most a couple of kilometres and are of strange facies, consisting mainly of black shales and cherts. These sediments suggest deposition in a starved basin of moderately deep water. They are followed by Mississippian and early Pennsylvanian turbidites with slide-breccias, other sandstones and shales and occasional layers of tuff. These late Palaeozoic flysch-like deposits mount up to over 10 km. The main period of deformation, apparently late Pennsylvanian, led to the development of an arcuate fold-system associated with complex faults in style not unlike those of the Valley and Ridge province. Marginal thrusts show northward displacements of as much as 80 km. Low-grade metamorphism affected at least the older parts of the cover in the Mississippi area. Post-orogenic Permian sandstones and red-beds are unconformable on the basin-fillings.

(b) The *marginal basins* along the northern border of the Ouachita fold-belt carry great thicknesses of late Palaeozoic clastic sediments which, it is presumed, were derived from the erosion of lands elevated within the developing orogenic belt. In the angle between the Ouachita and Appalachian tracts, the *Black Warrior Basin* contains 3 km of sediment all of Pottsville Formation (early Pennsylvanian) age. Further west, the *Arkoma basin* has up to 8 km of Pennsylvanian and the *Val Verde* and *Delaware* basins of Texas up to 6 km of Permian red-beds, sandstones and evaporites. The Pennsylvanian basin-fillings are essentially syntectonic sediments laid down rapidly and subsequently folded and fractured. The Permian basin-fillings are of molasse type and have suffered little subsequent deformation.

(c) The *Anadarko trough* (or Wichita geosyncline) projects north-westward into the craton from the Ouachita arc (Fig. 4.3). Unlike the marginal basins already referred to, it was initiated in Cambrian or even Precambrian times, received volcanic contributions at an early stage, and appears to have evolved as an independent unit flanked on either side by fault-bounded 'uplifts'. The basin, some 500 km in length and 120 km in width, received up to 12 km of Palaeozoic supracrustals of which nearly half were Pennsylvanian and Permian. Considerable thicknesses of lavas, tuffs and greywackes at the base of the succession are early Cambrian or Precambrian in age. They are followed by Cambrian to Devonian platform-deposits, mainly limestones, and then by late Devonian and Mississippian shales, sandstones and limestones which culminate in thick Pennsylvanian sandstones, shales, limestones and conglomerates. The Permian, which follows, includes sandstones, red-beds, marine limestones and evaporites. In the centre of the basin these strata are conformable on those below and there seems to have been no interruption of deposition. Towards the margins of the basin, however, a considerable unconformity separates the Permian from the older rocks. Folding and faulting, strongly influenced by the form of the basin and the irregularities of its floor, took place during two main tectonic episodes; the earlier or *Wichita* episode is regarded as early Pennsylvanian and the later or *Arbuckle* episode as late Pennsylvanian.

XII Retrospect: the Appalachians in Relation to the Caledonides

This chapter and Chapter 2 have dealt largely with the portions of what can now be viewed as an originally continuous mobile belt, extending from Spitsbergen to Mexico (Fig. 2.1). The total length of this belt was well over 6000 km and as it is perhaps the longest portion of a deeply eroded mobile belt for which adequate documentation is available, it may be worth enquiring into the nature of the variations along the length of the structure. It will at once be apparent that although the early histories of the various parts of the belt had much in common, considerable differences developed towards the closing stages (Fig. 4.11).

Along the whole length of the system, early basins of subsidence were differentiated from the adjacent cratons well before the beginning of the Lower Palaeozoic era. Thick Precambrian successions accumulated in these basins before the oldest cover-rocks were laid down on the cratons. In all regions, with the possible exception of the Ouachita belt, early disturbances involving folding and often accompanied by metamorphism or granite-emplacement occurred in late Precambrian or early Palaeozoic times, say prior to 460 m.y. From this time onward, divergencies began to appear. In the northern regions, notably in Spitsbergen and Greenland, the accumulation of thick basin-deposits ended in mid-Ordovician times. In parts of Scandinavia, Britain and the northern Appalachians, accumulation of this type of deposit continued until late Silurian or Devonian times. In parts of the southern Appalachians and the Ouachita belt, it continued into the Pennsylvanian, ending at about 300 m.y. Folding, metamorphism and granite-emplacement which were widespread up till the end of the Lower Palaeozoic era became more or less restricted to the southern regions during the Upper Palaeozoic.

The diagrammatic colums in Fig. 4.11 show that the southward extension of stabilisation was irregular. Over the entire northern tract, as far south-west as Newfoundland, the terminal stages of widespread granite-formation, uplift and onset of accumulation of molasse-type Old Red Sandstone took place at about 410–390 m.y., although in Scandinavia and Britain these stages followed almost immediately on the last phases of marine basin-filling, whereas in Greenland and Spitsbergen they followed a pause of nearly 100 m.y. In the southern tract, the terminal stages were reached at later times and followed immediately on the final phases of marine basin-filling; indeed in the Ouachita belt the bulk of the fill in the initial basins was not deposited until Pennsylvanian times.

The Appalachian belt of North America thus reveals a lengthwise variation in crustal mobility which contrasts with the variations over the corresponding time-span in Europe, where an entire branch of the mobile belt system – the Caledonides of the north-west – was immobilised in Devonian times but a second branch – the Hercynides – remained active. These contrasted records imply (see Fig. 4.12) that an initial arrangement of three cratonic plates separated by two mobile belts was converted in Devonian times to a simpler arrangement of a single mobile belt flanked on the south by an African craton and on the north by a Laurasian craton incorporating both the North American–Greenlandic and the Baltic cratons, welded together by the stabilised Caledonides. The Caledonian branch of the mobile system which was immobilised at an early stage is flanked on either side by Precambrian

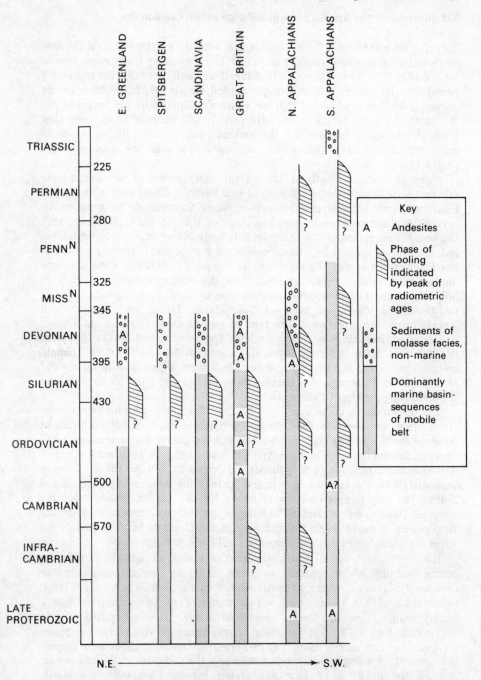

Fig. 4.11. Comparative columns showing the timing of critical events in several sectors of the Caledonian-Appalachian mobile belt

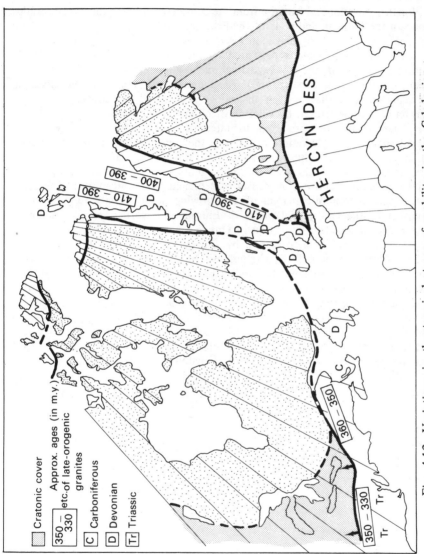

Fig. 4.12. Variations in the terminal stages of mobility in the Caledonian-Appalachian belt; arrowheads mark clastic wedges overlapping onto adjacent craton; lettered formations are of molasse facies

HERCYNIDES

Cratonic cover

350 — 330 Approx. ages (in m.y.) etc.of late-orogenic granites

C Carboniferous

D Devonian

Tr Triassic

410 – 390

400 – 390

410 – 390

360 – 350

350 – 330

shields – that is, by regions with a long-term positive tendency – whereas the sectors which retained their mobility until the end of the Palaeozoic are flanked by platform-regions carrying Phanerozoic cover-sequences of varying thicknesses. The regional differences of crustal regime indicated by these contrasts must, we suggest, have been controlled by a mechanism operating on a global scale.

Further questions which arise from the consideration of regional relationships concern the location and scale of the Palaeozoic ocean basins. The Appalachian miogeosynclinal and eugeosynclinal sequences have been compared with the post-Palaeozoic successions deposited at the Atlantic margin of the North American continent, the eugeosynclinal pile resembling the turbidite-prism at or near the continental slope (p. 266). The abundant volcanics in the eugeosynclinal sequence indicate that, if such an analogy is accepted, the continental margin was an active one fringed by volcanic island arcs. Once this model is adopted, it has been found an easy step to interpret the structural history of the Appalachians and Caledonides in terms of the consumption of an oceanic plate at an active margin. Bird and Dewey (1970) envisage an early phase (late Precambrian–early Ordovician) of sedimentation at the inactive margin of a widening ocean, followed by phases of sedimentation, vulcanism and tectonic activity at the mobile margin of a narrowing ocean, finally terminated by collision of the continental plates of Europe and North America. A similar history, involving the elimination in Ordovician and Silurian times of a broad oceanic basin, has been proposed by Dewey and Pankhurst (1970) for the Caledonides of Scotland. This hypothesis in its simplest form is difficult to reconcile with the apparent continuity in tectonic and metamorphic–plutonic evolution over some 400 m.y. It necessitates recognition of a 'suture' within each belt representing the site of the eliminated ocean basin (pp. 49–50), or the identification of portions of the belt floored by crust of oceanic type for which the evidence seems doubtful. Further geophysical and geological studies may well clarify these and other essential points. Present palaeomagnetic data suggest that the 'Proto-Atlantic' on the line of the Caledonides was not more than 1000 km in width.

5

Gondwanaland in Late Proterozoic and Early Palaeozoic Times

I The Supercontinent of Gondwanaland

The object of this chapter is to outline the geological cycle which began in late Precambrian times in the continents of the southern hemisphere. It reached its terminal phases in early Palaeozoic times and was followed by very extensive stabilisation of the crust. It is referred to by many names: Katangan, Mozambiquian, Pan-African, Indian Ocean, Adelaidean are all in common use for what are, in essentials, equivalent cycles.

The ending of the late Precambrian—early Palaeozoic orogenic cycle seems to have marked a decisive stage in the development of the southern continents. At this time, these continents entered on a distinctive geological evolution which is reflected in many aspects of their later Phanerozoic history. The initial concept of *Gondwanaland* as a geological entity embracing the continental masses of Africa, peninsular India, Australia, South America and Antarctica, owed much of its force to this evidence of a common geological history. Its special significance arises from its connection with the hypothesis of continental drift which envisages a physically continuous Palaeozoic 'supercontinent' of Gondwanaland made up of the continental masses mentioned (Fig. 5.1).

II The Mobile Belts and Cratons

The mobile belts which came into existence in Gondwanaland during late Precambrian times form a network enclosing a number of relatively small

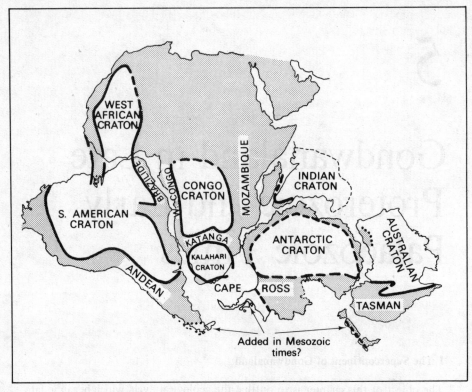

Fig. 5.1. The early Phanerozoic mobile belts of Gondwanaland

cratons. They are schematically represented on the continental reconstruction of Fig. 5.1 and may be summarised as follows:

Africa. Orogenic events of late Precambrian and early Palaeozoic age in Africa, it has been estimated, affected nearly half the crystalline rocks of the continent. Along the eastern side of the continent, portions of a wide belt of north—south grain and of high metamorphic grade crop out at intervals from Mozambique and Madagascar in the south, to Egypt and Arabia in the north. This belt has received many names; we shall use Holmes' original term, the *Mozambique belt.* Along the western side of the continent a little-known zone includes sectors in *Dahomey, West Congo* and *South-West Africa.* Linking these north—south zones, and separating the Precambrian cratons of central and southern Africa is an arcuate orogenic tract, made up from east to west of the *Zambesi, Katangan* and *Lufilian* and *Damaran* portions. To the south of the Kalahari craton a second linking belt is now incorporated in the younger Palaeozoic *Cape fold-belt.*

India. The main part of peninsular India constituted a craton in late Precambrian times within which sporadic activity continued in the Satpura and Delhi belts until 700—800 m.y. Further activity took place during the *Indian*

Ocean orogeny (500–600 m.y.) in Ceylon and the adjoining parts of southern India, extending for some distance up the east coast.

Australia. The Precambrian craton of Australia is flanked to the east by the *Adelaidean orogenic belt* developed on the site of the *Adelaide geosyncline.*

South America and Antarctica. In South America, a conspicuous zone of reworking (the *Brazilides*) follows the eastern margin of the Brazilian shield while other zones active in late Precambrian or early Palaeozoic times appear to be incorporated in the Andean belts. In Antarctica, extensive reworking during the period 400–600 m.y. has been indicated by the dating of samples from many coastal localities and a mobile belt incorporating the *Ross Supergroup* follows the Transantarctic Mountains.

Figure 5.1 reveals a striking fact: the network of mobile belts is closely followed by the margins of the present-day continental fragments. The coasts of central and southern Africa, for example, follow the two great north–south belts of the Mozambiquian system, and the coasts of Brazil, Antarctica and south-east India coincide for long distances with belts of the equivalent systems. Small continental fragments such as Madagascar, Ceylon and the Seychelles consists largely of rocks of this age. Conversely, older Precambrian rocks predominate in the interiors of the present continents; only the Katangan belt of Africa extends through the centre of a continental mass. This aspect of the orogenic pattern suggests that when disruption of Gondwanaland began, fracturing followed the lines of the recently consolidated mobile belts.

III Africa

1 *The Katanga, West Congo and Damara belts and the adjacent cratons*

The broad arcuate belt of late Precambrian orogenic activity which separates the ancient cratons of the Kalahari and East Africa (see Part I, Fig. 6.2) passes through the *Copper Belt of Zambia* and the mining district of Katanga in the former Belgian Congo. These regions constitute the *Katanga belt* proper. The south-eastern portion of the arc is made by the *Zambesi belt* which links up eastwards with the Mozambique belt. The south-western portion of the arc, the *Damara belt*, may be linked in a similar way with the north–south mobile belt of south-western Africa. The border of this arcuate system against the stable craton to the south is exposed in the south-east where the Rhodesian massif projects into the angle between the Zambesi and Mozambique belts. The corresponding junction with the craton to the north is seen in Congo and Tanzania, where folded Katangan cover-rocks pass into little disturbed cover-formations resting on a crystalline basement. The rich copper and uranium mineralisation of the Katanga belt is concentrated mainly in Zambia and Katanga from which most details of the stratigraphy and structure are derived (Fig. 5.2).

The Pre-Katangan basement is revealed in domes and anticlines in the Copper Belt and more widely as a regenerated complex in the Zambesi belt. The basement in the Copper Belt includes a number of pre-Katangan granites to

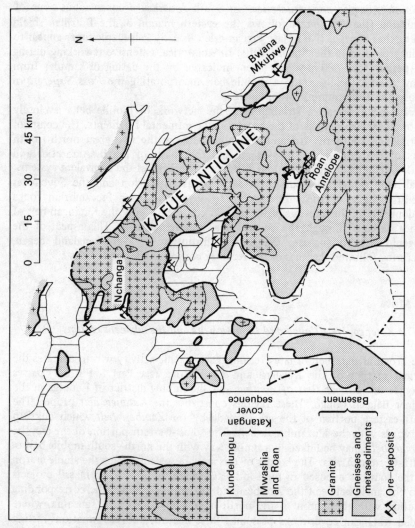

Fig. 5.2. The Copper Belt of Zambia (basd on a compilation by Mendelsohn, 1961)

Bwana
Mkubwa

KAFUE ANTICLINE

Roan
Antelope

Nchanga

N

0   15   30   45 km

Katangan
cover
sequence

Kundelungu

Mwashia
and Roan

Basement

Granite

Gneisses and
metasediments

Ore-deposits

which the copper mineralisation was at one time erroniously attributed. Towards the south-east, the basement incorporates the *Umkondo Group* of psammites, pelites and limestones which rests undeformed on older crystalline rocks in the Rhodesian shield. The dating of basic intrusions has shown that this group was deposited before 1500 m.y. and it is therefore not equivalent to the Katangan cover.

The Katanga 'System' (for parts of which the name *Muva* has sometimes been used) forms a thick series of predominantly shallow-water psammites and pelites with occasional dolomites (Table 5.1). Volcanic rocks are notably rare, though basalts are important in the equivalent Bukoban of the East African craton (Part I). Two horizons marked by boulder-beds and usually regarded as tillites have been recognised and similar, probably equivalent, beds occur over enormous distances in the Damaran and West Congo belts.

Towards the south and west of the Copper Belt, the Katangan reaches maximum thicknesses of up to 10 km. It thins northwards towards the craton, certain divisions wedging out altogether and others being reduced in thickness. Continental sandstones form a considerable part of the equivalent foreland deposits (the Bushimay of Katanga and the Abercorn and Bukoban of Tanzania); it has been suggested that the first of these deposits represented a molasse related to the Karagwe—Ankolean belts of central Africa (Part I).

The Lower Roan of the Copper Belt begins with terrestrial deposits, including anhydrite-bearing evaporites and dune-bedded psammites. They rest on an irregular erosion surface and are followed by a transgressive marine sequence of current-bedded sandstones, shales and dolomites, which provides the main copper-bearing horizons. The ore-deposits lie principally within shales, less commonly in dolomites, and consist mainly of sulphides: pyrite,chalcopyrite, bornite and chalcocite occurring in decreasing order of abundance. The sulphides are moulded on the grains of the host-sediment and it was origially thought that the copper was entirely epigenetic. The discovery that the dome-like granite bodies over which the ore-bearing horizons are draped belong to the basement, with evidence that mineralisation preceded folding and metamorphism, has led to a hypothesis of syngenetic origin. Garlick, for example, has compared the deposits with those of the Mansfeld *Kupferschiefer* (p. 95) and envisages deposition in an anaerobic (Black Sea) environment.

The upper division of the Katanga 'System', the *Kundelungu*, spreads northward beyond the limit of the fold-belt in Katanga and is composed mainly of shallow-water detrital sediments with two glacigene intercalations. The lower boulder-bed – *Grand Conglomérat* of Katanga – lies at or near the base and reaches thicknesses of several hundred metres: it includes finer-grained bedded sediments interpreted as fluvioglacial deposits. The upper boulder-bed – *Petit Conglomérat* of Katanga – is generally thinner, lies up to 3000 m below the top of the succession, and is followed by sandstones, shales, calc-shales, cherts and limestones.

In the West Congo belt, a sedimentary sequence broadly similar to the Katangan can be traced for 1000 km from Angola through Congo to Gabon. This sequence consists mainly of mudstones with interbedded sandstones and limestones with two principal boulder-beds. Towards the east, it is little deformed and shows several minor breaks due to intermittent uplift. Towards

Table 5.1. THE KATANGA 'SYSTEM'

	Katanga		Zambia (Copper Belt)	
Kundelungu	Kalule Group: sandstones, shales, limestones, dolomites	} U. Kundelungu c. 3500 m	Upper Kundelungu: shales, quartzites	} Kundelungu 'Series'
	Kyubo Group: sandstones, shales, calc-shales		Middle Kundelungu: shales, tillite	
	Plateau Group: shales, sandstones		Lower Kundelungu: shales, dolomites and shales, tillites	
	'Petit Conglomérat' tillites, c. 50 m	} L. Kundelungu 0–2500 m		
	sandstones, limestones, shales			
	limestones, dolomites, conglomerate			
	'Grand Conglomérat' tillites and periglacial beds 300 m			
Mwashya 800 m	black shales, sandstones, dolomitic shales		Mwashia: black shales, shales dolomites, quartzites 600 m	} Mine Series
Roan 1500+ m	Mofya Group: limestones, shales		Upper Roan: dolomites, shales, quartzites 600 m	
	Dipeta Group: phyllites, sandstones, dolomites			
	'Mine Series' Group: dolomites, sandstones, shales dolomitic sandstones		Lower Roan 0–1000 m	'hanging wall' quartzites, shales
				Ore, shales, dolomites
				'footwall' sandstones
(no base)			(basement of granites, gneisses)	

the west, more intense folding is seen and the sequence thickens to over 5 kilometres.

The Kundelunguan boulder-beds contain, in an almost unbedded argillaceous matrix, scattered angular fragments of granite, shale, quartz, quartzite and dolomite. They occur conformably in a shallow-water marine succession and have been regarded by Cahen and many others as glacigene deposits. The boulder-beds of the West Congo belt which are probably stratigraphically equivalent to them have, on the other hand, been interpreted as slide-rocks or associates of turbidites by Schermerhorn and Stanton who refer to them by the non-commital term *tilloids*.

From Angola, supracrustal rocks extend southward into Damaraland and the northern part of South-west Africa where they curve eastward, probably linking up beneath the younger cover with the Katangan outcrop. Two tracts of differing metamorphic grade occur. The high-grade rocks which lie to the south were at one time classified as Archaean, and were distinguished from the supposedly younger ('Algonkian') low-grade series. Both are now assigned to the *Damara 'System'*, the high-grade rocks being the *Swakop facies* and the low-grade rocks the *Outjo facies* (Table 5.2). The Outjo facies is regarded as being of miogeosynclinal type. It has a maximum thickness of some 7 km, contains little volcanic material and is composed dominantly of chemical or well worked detrital sediments. The Swakop succession has a high proportion of detrital sediments and contains basic and ultrabasic intrusives which give it a eugeosynclinal aspect. It is bounded on the south by a basement-ridge marking the front of the mobile belt and passes by a rapid transition across this front into an almost unmetamorphosed shelf-facies.

Table 5.2. THE DAMARA 'SYSTEM'

OUTJO FACIES (low-grade)	SWAKOP FACIES (high-grade)
3 MULDEN SERIES quartzite, sandstones, conglomerates, greywackes: maximum thickness about 2000 m	3 KHOMAS SERIES biotite-sericite-schists with micaceous quartzites and calc-rocks, ortho-amphibolites and serpentines, estimated thickness 3000–12 000 m
←——— *(local unconformity)* ———→	
2 OTAVI SERIES a three-part series, starting with dolomites, shales and greywackes followed with discordance by a tillite-clastic group and then by thick dolomites; maximum thickness about 6000 m	2 HAKOS SERIES a three-part series with a central tillite group associated with iron ores, overlain and underlain by mixed sequences of schists, quartzites, greywackes and marbles. Unconformity at base of tillite or transition at base of Hakos Series. Thickness up to 2800 m
←——— *(unconformity)* ———→	
1 NOSIB FORMATION quartzite, phyllites, greywackes, basic lavas and agglomerates. Maximum thickness 5000 m	1 NOSIB FORMATION quartzites, phyllites, schists, calc-rocks, and ortho-amphibolites. Maximum thickness 4000 m
←——————————— *(major unconformity)* ———————————→	

In Namaland, the southern part of South-west Africa, this shelf-facies is probably represented by the flat-lying *Nama 'System'* which extends southward for some thousand kilometres. The Nama, consisting of quartzites followed by limestones, dolomites and shales, is seldom more than 200 m in thickness. It is underlain in some places by several hundred metres of tillites and other glacigene sediments (*the Numees Formation*) and capped by the *Bushmannsklippe Formation* which contains more tillites and has yielded fossils akin to the Infracambrian Ediacara fauna (p. 151) as well as Lower Cambrian Archaeo-cyathids.

Structure and metamorphism. The dominant structures of the Copper Belt are broad anticlines and domes of basement rocks mantled by the ore-bearing Lower Roan. The great *Kafue Anticline* can be followed north-westward parallel to the tectonic trend for more than 150 km and passes into a string of coalescing domes. On its south-western flank a richly mineralised zone includes the districts of Broken Hill, Roan Antelope and Bancroft; on its north-eastern flank are the mines of Bwana Mkubwa. As in many other fold-belts, the mantled gneiss domes of the Copper Belt were formed in association with rejuvenation of granitic material. The *Nchanga granite*, for example, is a basement granite and contributed detrital feldspars to the Lower Roan; it is, however, extensively sheared, possibly metasomatised, and has given an isochron age of 570 m.y. Tight folds and dislocations in the cover are earlier than the rise of the domes.

Metamorphism is usually only of greenschist facies in the copper-mining areas; indeed much of the Upper Katangan is almost unaltered. The grade increases southward towards the interior of the Katangan belt where rocks of kyanite or sillimanite grade are widespread. Migmatites are of relatively little importance, though gneisses of migmatitic aspect are among the reworked rocks of the basement. Towards the southern margins of the belt, the grade of metamorphism falls off again. A southward transition in reworked pre-Umkonda basement from kyanite to chlorite grade over a distance of about 25 km is recorded by Vail from the Zambesi belt (1965).

The dating of events in the Katangan cycle is clear within broad limits, but remains controversial in detail. The presence of molasse-like deposits near the base of the Katangan suggests that it began to accumulate soon after the ending of Kibaran orogeny at 1200–1000 m.y. The occurrence of the (presumably Infracambrian) Kundelunguan tillites suggests that deposition continued until some time near the beginning of the Palaeozoic. The highest beds of the Upper Kundelungu are thought by some to post-date much of the folding in Katanga. A date for the uprise of mantled gneiss domes which completed the evolution of the main tectonic pattern is given by isochrons for the Nchanga granite at 570 m.y. Uranium minerals associated with pegmatites and epigenetic veins have given dates suggesting episodes at about 720 m.y., at 620 m.y. (associated with transverse faulting) and 520 m.y. In addition, many mineral ages from pegmatites, granites and sulphide veins fall in the range 500–420 m.y. These ages are ascribed by Cahen and Snelling (1968) to relatively unimportant thermal episodes, but are linked by Clifford, Vail and others with a major *Damaran* metamorphic event.

2 *The Mozambique Belt*

The Mozambique belt can be traced for almost the full length of the African continent, with a north and south extension of not less than 4000 km (Fig. 5.3). Detached portions in the oceanic basin to the east make much of Madagascar and the Seychelles. The western front of the belt truncates a number of very old tectonic patterns in the cratons of Rhodesia and Tanzania. High grades of metamorphism are characteristic of the belt, which contains many rocks of granulite facies. Prior to 1948, the gneisses of the Mozambique belt in East Africa had been regarded as among the oldest rocks of the region and it was the demonstration that the belt truncated older structures and yielded young radiometric dates which, more than any other evidence, won gradual acceptance for the criteria employed by Holmes in his classic study of African geochronology (1951).

The apparent ages of rocks in the Mozambique belt fall mainly in the range 650–400 m.y. and the final phases of orogeny are assigned by most authors to

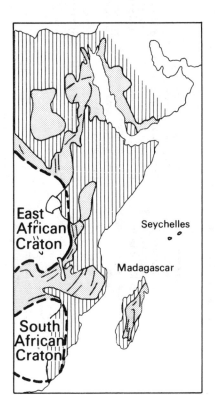
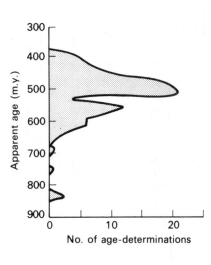

Fig. 5.3. The Mozambique belt (sketch-map), with a histogram of age-determinations of metamorphic rocks and granites from the belt; a few apparent ages > 900 m.y. are omitted (based mainly on Cahen and Snelling)

Table 5.3. THE MOZAMBIQUE BELT

	SOMALIA, EGYPT, SUDAN	WEST NILE KENYA	TANZANIA, MALAWI	MOZAMBIQUE, MADAGASCAR
Apparent ages of pegmatites or granites (m.y.)	Granites 650–470	Pegmatites c. 835, 650–600, 480	Pegmatites c. 600, 480	Pegmatites c. 600
Range of apparent ages (m.y.)	650–450	660–430	650–400	630–470
Supracrustal series, where separated	*Inda Ad* group in Somalia, sometimes greenschist facies, includes mudstones, sandstones, limestones. Low-grade metasediments also in Ethiopia, elsewhere metasediments are of high grades and uncertain relationships	*Turoka Series* (Kenya) = mantle series?, quartzites, pelites, calc-schists, graphitic marbles, usually of high metamorphic grades, amphibolite facies	*Mafingi Series* of psammites (Malawi) possibly other groups of metasediments or metavolcanics	*Umkondo group* pelites, psammites, limestones, unmetamorphosed on foreland but metamorphosed up to kyanite grade in Mozambique belt: dating of basic intrusion on foreland indicates that group was deposited before 1500 m.y.
Characters of metamorphic products	Migmatitic gneisses with relics of supracrustal rocks, sometimes evidence of polymetamorphism with retrogression of early granulite facies to amphibolite facies	Gneisses predominate (= *Basement System* of Kenya), widely distributed evidence of retrogression from granulite facies to amphibolite facies in gneiss complexes. Migmatites form large areas	Gneisses with metasedimentary relics (= Usugaran) evidence of retrogression from granulite to amphibolite facies, sometimes associated with migmatisation	Gneisses predominate, sometimes of granulite facies, with marbles, scapolite-gneisses
Other features	Layered gabbros intruded between two metamorphic episodes in Somalia. Granites not abundant, late synorogenic granodiorites, post-tectonic granites	Discrete intrusive granites not abundant	Syenites c. 550–600 m.y.	Anorthosites

periods within this range. Much of the belt, however, is undoubtedly polycyclic. Although metamorphosed supracrustal sequences are separable from an older basement in several regions, it is by no means certain that all, or indeed any, of them represent basin-deposits laid down during the cycle which led up to the Katangan orogeny. The Umkondo group already mentioned (p. 135), has been shown in some places to be of middle Precambrian age; Lower Palaeozoic strata or Infracambrian glacigene groups have not so far been identified. Hence, although Holmes (1965) considered that the development of the Mozambique belt was preceded by the formation of a geosyncline, this interpretation has still to be confirmed. The real unity of the belt is shown by its structural and metamorphic characters, as is illustrated by the details summarised in Table 5.3.

Marginal zones of the Mozambique Belt. The crystalline rocks of the western foreland are mostly over 2000 m.y. in age and there is commonly a discordance between their structural patterns and those developed within the Mozambique belt. There is often also a conspicuous metamorphic contrast, for whereas the rocks of the Mozambique belt are usually of high grades, those of the forelands incorporate little-altered supracrustal series.

East of Lake Victoria, east—west belts of Nyanzian—Kavirondian on the foreland are deformed, sliced by movement-zones and metamorphosed to high grades as they are traced across the orogenic front. Sanders (1965) shows that in the vicinity of the Nandi escarpment of Kenya the reworked basement disappears eastwards beneath a *mantle series* of pelites, quartzites and calc-schists representing a younger cover. In central Tanganyika, migmatitic Dodoman gneisses are separated from high-grade gneisses in the interior of the Mozambique belt by a broad steep zone in which the gneisses are intensely foliated and sometimes granulated.

Towards the south, in Malawi, the tectonic front is obscured by later sediments. On the Rhodesian—Mozambique border, however, the front is well defined against the granitic rocks and intervening supracrustal belts of the Rhodesian massif. Here, the tectonic and metamorphic zones at the southern border of the Zambesi belt (p. 138) turn through a broad arc and continue southward along the border of the Mozambique belt. The eastward increase of metamorphic grade is associated with changes in the style of deformation in the basement, an outer zone marked by the development of widely spaced zones of shearing flanks a zone of penetrative shearing which is succeeded eastward by a region of plastic deformation. Across the marginal tract the Umkondo group, which lies almost horizontally on crystalline rocks in the Rhodesian massif, becomes strongly folded and is metamorphosed up to kyanite grade. Johnson (1968) and Vail record that the Umkondo psammites, pelites and limestones show an eastward change from inshore to offshore facies approximately along the line bounding the area of folding. This observation, if it proves to be of wider application, might suggest that the Mozambique belt was developed on the site of a much older sedimentary basin.

Interior of the Mozambique belt. Gneisses (often including migmatites with metasedimentary host-rocks) are the predominant components of the interior, while homogeneous granitic rocks on the one hand and lower-grade supracrustal series on the other, are of restricted distribution. A separation of supposedly older and younger components, the latter including altered supracrustal rocks and possibly representing cover-series of unknown ages, has been attempted in some areas (Table 5.3).

The characteristic high grades of metamorphism are expressed not only by the dominance of gneisses but also by the occurrence of charnockites and other rocks of granulite facies. These rocks not infrequently exhibit evidence of retrogression to amphibolite facies associated with refolding and migmatisation. Pegmatite veins are widely distributed, but although late-tectonic granites occur in some regions, for example in Somalia and the Seychelles, they are almost absent from many parts of the belt.

Radiometric dating has yielded surprisingly uniform results over the vast length of the Mozambique belt (Fig. 5.3). Although much of the belt appears to be polycyclic, dates older than about 900 m.y. are remarkably uncommon, a fact which may be connected with the consistently high grades of metamorphism. Cahen and Snelling (1968) attach particular importance to concordant U—Pb dates obtained from minerals from cross-cutting pegmatites, because such pegmatites record episodes of injection which can be dated relative to the fabrics which they cross. From scattered determinations, they suggest that in Kenya, Tanzania and Malawi, pegmatites were emplaced at about 835 m.y. 650—600 m.y. and 480 m.y. In Somalia, late-tectonic granites are dated at 500 m.y. Cahen and Snelling conclude that effective regional metamorphism had ended by about 600 m.y. and that the spread of K—Ar and Rb—Sr dates down to about 400 m.y. reflects variations in the time of post-orogenic uplift and consequent setting of the radiometric clock. On the other hand, Clifford (for example 1969) and his co-workers attribute the preponderance of apparent ages younger than 600 m.y. to a major *Damaran* episode of metamorphism at about 500 m.y.

IV India and Sri Lanka

The greater part of peninsular India had, as we saw in Part I, been stabilised at or before 900 m.y. On the platform so produced, cratonic sediments began to accumulate here and there, the oldest of which, the *Vindhyan*, have been described in Part I. The scanty radiometric evidence suggests that a belt of tectonic and plutonic activity of very late Precambrian or early Palaeozoic age crossed the south-eastern tip of India and continued into Ceylon. The activity recorded in this belt is referred to an Indian Ocean orogeny by Aswathanatayana and others.

In the extreme south of India, and in coastal regions at least as far north as Madras, a complex age-pattern of dates ranging from about 1700 m.y. down to about 400 m.y. appears to have been produced by the partial reworking of the Dharwar and Eastern Ghats belts. The structure of the region is little known and the position of any late Precambrian orogenic front and the presence or absence of a cover-succession remain to be determined.

In Sri Lanka, the superposition of successive tectonic and metamorphic events is more clearly displayed (Fig. 5.4). The central mountains of the island are made up of charnockitic gneisses which contain numerous relics of quartzites, sillimanite-quartzites, sillimanite-garnet-gneisses, marbles and calc-granulites. This assemblage, the *Highland Series*, has a rather regular layering with a consistent north-north-west or northerly trend. Flanking the main massif of the Highland Series and enclosing one or two smaller massifs of similar type are two NNE belts composed mainly of biotite-gneiss, hornblende-gneiss and granite-gneiss. These belts, known as the *Vijayan Series*, are predominantly migmatitic; their structural pattern is irregular both on a small scale and on a regional scale.

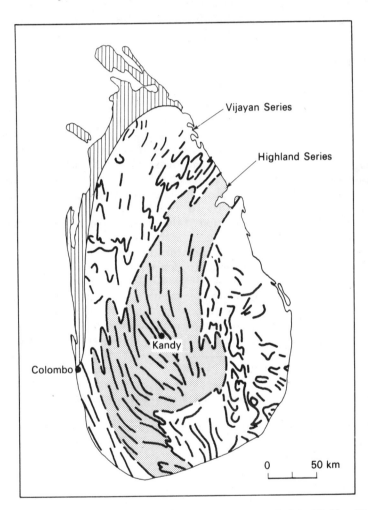

Fig. 5.4 The polycyclic complexes of Sri Lanka. The massif of the Highland Series is composed of gneisses and metasediments of granulite facies yielding whole-rock Rb–Sr ages > 2000 m.y. and mineral ages scattering down to 450 m.y. The Vijayan Series is migmatitic and consists largely of reworked derivatives of the Highland Series yielding whole-rock Rb–Sr ages of about 1250 m.y. and mineral ages scattering down to 450 m.y. (Based on compilations by Cooray, 1962, and geochronological studies by Crawford and Oliver.)

Until recently, the Highland Series, with its relatively high proportion of metasediments, was regarded as a younger group resting unconformably on a granitic basement. Field studies such as those of Cooray (1962) show, however, that the two assemblages are connected by a transition-zone in which the granulite facies gneisses and metasediments are down-graded to amphibolite facies, migmatised, crumpled and disrupted until they take on the aspect of Vijayan gneisses. Pegmatites associated with migmatisation of the Vijayan are dated at 500–600 m.y. A few age-determinations suggest that the Highland Series dates back to the Eastern Ghats or even the Dharwar cycle.

V Australia

1 The Adelaide mobile belt

The Adelaide mobile belt which wraps round the south-eastern margin of the Precambrian craton of Australia, is exposed mainly within the State of South Australia and thanks to its proximity to the state capital has been studied in detail. The history of the *Adelaide geosyncline* with its 16 km pile of Infracambrian and Cambrian rocks, together with that of the adjacent cratonic regions, occupies a very special place in Gondwanaland geology. Such important features as the Infracambrian glacial deposits, the occurrence of very late Precambrian faunas and the relation of the Infracambrian to the Cambrian are unusually well documented here.

The Adelaidean mobile belt, with the fringing deposits of its forelands, occupies an arc stretching for over 1100 km from Kangaroo Island in the south to the latitude of Lake Eyre in the north (Fig. 5.5). Its possible northward continuation is obscured by younger sediments, as are portions of its eastern and western boundaries; an important extension of the belt runs north-eastward into New South Wales, reaching the famous mining district of Broken Hill before it, too, is lost beneath the younger sediments of the Great Artesian Basin. The Adelaide geosyncline had as its *western foreland* the presumed Archaean rocks of the Gawler block (Part I, p. 142) with a cover-sequence laid down during the Adelaidean cycle. The *north-eastern foreland,* mostly concealed beneath younger deposits, is considered to be a second Archaean block.

The pre-Adelaidean basement. Near the eastern side of the mobile belt the basement rocks of the Olary–Broken Hill area, form the Willyama complex of gneisses and metasediments carrying the important silver–lead–zinc mineralisation of *Broken Hill.* Although the main phases of mineralisation were Proterozoic, there are indications that regeneration took place during the Adelaidean orogeny (ores of Thakaringa type).

At the southern end of the belt a string of basement inliers appears in an arc of axial culminations in the *Mount Lofty Ranges.* They are made up largely of mica-schists (sometimes carrying kyanite or sillimanite), augen-gneisses and other migmatites, granitic gneisses, lenses of granite and amphibole-pyroxene granulites ('diorites') of uncertain origin. At the north-eastern end of the Flinders Ranges, metasediments and granites crop out in anticlinal cores between Mount

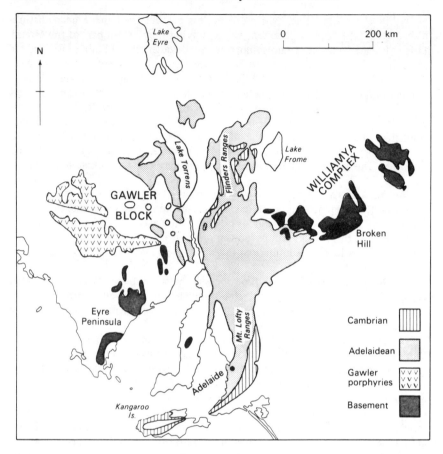

Fig. 5.5. The Adelaidean mobile belt

Painter and Prospect Hill. Finally in the Peake and Denison Ranges west of Lake Eyre, the Peake Series of metasediments, migmatites, augen-gneisses and rapakivi gneisses, regarded as components of the basement, are seen in mechanical contact with the Adelaide system.

The Adelaide geosyncline and its filling. In the Adelaide geosyncline an immense thickness of dominantly shallow-water sediments, the *Adelaide System,* followed by the *Cambrian,* was accumulated, amounting in aggregate possibly to some 16 km. The initiation of deposition is recorded by the 'Grand Unconformity' of early South Australian geologists at which the little-altered geosynclinal filling rests on the metamorphic basement. Deposition was interrupted by crustal movements and was terminated by the Adelaidean orogeny. Thinner and less complete sequences rest on the Precambrian craton and, like those of the geosyncline, these are characterised by Infracambrian glacial deposits.

Within the geosyncline, two more or less parallel tracts have been distinguished (Fig. 5.5). The more extensive, occupying the western part of the Mount
Lofty Ranges and continuing northward through the Flinders Ranges, is
characterised by a rather constant succession. It suffered little orogenic
disturbance until at least Middle Cambrian times and remained largely
unmetamorphosed. This *non-metamorphic tract* provides the most detailed
record of deposition. In the eastern parts of the Mount Lofty Ranges and their
extension southwards into Kangaroo Island, on the other hand, unstable
conditions set in at least as early as the Lower Cambrian, variations of thickness
and lithology were common and metamorphism reached high grades in some
localities. This unstable metamorphic tract we may call the *Kanmantoo trough*.

The Adelaide 'System'. In the western Mount Lofty Ranges, the type area of
the non-metamorphic facies, it has been customary to recognise three sub-
divisions of the Adelaide System. Elsewhere, however, and even locally in the
type area, further thick groups appear conformably beneath these divisions.
Selected successions are summarised in Table 5.4 and certain aspects of them are
elaborated below.

The *Willouran Series* in the type locality between Lakes Eyre and Torrens
consists of thick quartzites and slates which are followed conformably by
equivalents of the Torrensian. Similar shallow-water deposits occur locally
elsewhere.

The *Torrensian Series* is a transgressive deposit of very variable thickness. It is
thin or absent south of Adelaide and in the Olary, Broken Hill and Mount Fitton
areas, but reaches some 3000 m in the Adelaide region and the northern Flinders
Ranges. Still further north in the inliers of the Peake—Denison Ranges,
equivalent beds are estimated to reach over 6000 metres.

The *Sturtian Series* that follows includes glacigene deposits confidently
interpreted as moraines, boulder-clays, massive tills, varved clays and fluvioglacial sands and gravels. These deposits, magnificently exposed, especially in the
Flinders Ranges, sometimes reach thicknesses of over 5500 m with their
associated interglacials, an astounding thickness for sequences of glacial facies.
The Sturtian Series transgresses widely over the Australian craton where tillites
may rest directly on the basement.

From variations in the thickness and character of the deposits, it has been
inferred that a mountainous Willyamia was the source of the ice-supply. On the
margin of Willyamia some deposits lying directly on the basement have been
interpreted as terrestrial moraines. The vast majority of the boulder-beds,
however, are regarded as products of marine deposition from ice-masses drifting
down the length of the geosynclinal trough; these deposits show a general
thickening towards the north-east and north. Mawson, uniquely familiar with
both regions, has compared conditions in the Adelaide geosyncline with those of
the present Antarctic where the land-ice rides out into deep water as the Ross
Ice-shelf and gives rise to immense icebergs which finally deposit their load of
debris over the ocean floor.

The *Marinoan Series*, the uppermost division of the Adelaidean, follows
conformably on the Sturtian within the geosyncline and reaches some 1400 m in
the Flinders Ranges. It is made up of red, purple or chocolate detrital sediments,

together with dolomitic limestones and current-bedded quartzites; the majority opinion seems to be that these sediments were deposited largely in lagoonal or lacustrine environments of paralic or even continental type; some beds have been interpreted as loess.

The Pound quartzite. Between strata which are universally accepted as Marinoan, and the lowest indubitably Cambrian *Archaeocyatha Limestone* there intervenes, in conformable sequence, a thick quartzite especially well seen in the central and northern Flinders Ranges. Of late years the Pound Quartzite has yielded a diverse fauna of soft-bodied creatures quite unlike faunas characteristic of lowest Cambrian strata elsewhere; this is the *Ediacara fauna*, named from the locality at which it was found. In view of the distinctive character of the fauna and of the lithological contrasts between the Pound Quartzite and the overlying Cambrian, it is convenient to place the dividing line between the Adelaide and Cambrian systems at the top of the Pound Quartzite. An effectively continuous stratigraphical record links the fossiliferous Infracambrian with the Lower Palaeozoic (Table 5.5).

The Cambrian. The normal non-metamorphic Cambrian of the Adelaide geosyncline follows evenly on the Pound Quartzite, where this has been recognised, and reaches a maximum thickness of around 5000 m in the Lake Frome region of the northern Flinders Ranges.

In the Kanmantoo trough, unstable conditions set in early in the Cambrian, leading to the accumulation of a thick greywacke assemblage, the *Kanmantoo Group.* In the northern part of Kangaroo Island the Cambrian sequence begins with shales and sandstones, followed by a thick conglomerate containing boulders of Archaeocyatha Limestone, as well as of granite and quartzite. Above the conglomerate is a shale horizon yielding a trilobite fauna characteristic of the upper zone of the Lower Cambrian and above this, thick arkoses, pelites, quartzites and conglomerates.

2 The foreland and craton

At the western margin of the mobile belt, there are conspicuous changes of thickness and facies in the Adelaidean sediments. In the Hummocks Range at the head of Gulf St Vincent, a marginal thrust-zone cuts out part of the transition. Here, Adelaidean of the Mount Lofty geosynclinal type, together with Pound Quartzite and Cambrian limestones, rests with a thrust junction on an autochthonous foreland cover-succession. The upper glacial horizon of the Sturtian and the succeeding Marinoan in the thrust mass are much thinner than their equivalents in the interior of the geosyncline and these groups are lacking from the foreland succession in Yorke Peninsula. The Pound Quartzite is represented by some 200 m of brownish current-bedded sandstones topped by shales in the thrust-mass and on the foreland by red conglomerates, grits and arkosic sandstones that thin rapidly westward. The Lower and Middle Cambrian of Yorke Peninsula are made up largely of limestones some 600 m in thickness and yielding brachiopods and trilobites. Calcareous conglomerates, following the

Table 5.4.

	1 ADELAIDE REGION	2 OLARY REGION
MARINOAN	Quartzites and slates Chocolate slates and quartzites Slates Quartzites Arkoses, quartzites Chocolate slates	
STURTIAN	Brighton Limestone: dolomites and limestones Tapley Hill Slates: laminated slates and siliceous limestones, 3000 m Sturt Tillite: tillites, quartzites, 300 m	Fluvioglacial quartzites Upper Tillite, 65–500 m Interglacial Beds: laminated calcareous slates, minor quartzites, dolomites, 5000 m Lower Tillite: 7 major tillites with detrital sediments, up to 2300 m
TORRENSIAN	Mitcham quartzites and arkoses Glenosmond Slates and dolomites Beaumont Dolomite Upper Phyllites and dolomites Stonyfell Quartzite Lower Phyllites and dolomites Aldgate Sandstone and conglomerate	Slates Calcareous dolomitic beds Basal conglomerate BASEMENT
WILLOURAN	Wakefield River Group; phyllites, quartzites, 400 m BASEMENT	

ADELAIDEAN SUCCESSIONS

3 CENTRAL FLINDERS RANGES, BIBLIANDO	4 MYRTLE-SERLE AREA	5 MOUNT FITTON
	Archaeocyatha Limestone (CAMBRIAN)	
———————	Grey slates Pound Quartzite	———————
	Purple and green shales with dolomites Dolomitic limestones	Calcereous siltstones and dolomites
Calcareous Beds lacustrine Lacustrine shales with greywackes, 3000 m Upper Glacial Series 1300 m tillites interbedded with detrital sediments interglacial arkoses Lower Glacial Series 300 m tillites, interbedded with detrital sediments	Upper Tillite 260 m Laminated shale and dolomite Massive dolomite Laminated shale and dolomite Boulder Tillite 230 m	Upper Boulder Tillite Laminated shales with dolomite Boulder Tillite with siltstone layers, 300 m Laminated hornfels with minor boulder beds Tillite Varved beds Basal Boulder Tillite
	Magnesite Beds and sandstone Massive quartzite with basal conglomerates BASEMENT	BASEMENT

Table 5.5. THE INFRACAMBRIAN–CAMBRIAN SUCCESSION OF LAKE FROME

(top)	*Lake Frome Group* 290 m *Red-beds*
	chocolate-coloured shales and sandstones, with subordinate current-bedded sandstones and quartzites, ending the geosynclinal sedimentation
	No fossils recorded but regarded as Middle Cambrian
MIDDLE CAMBRIAN	*Wirrealpa Limestone* (*'Obolella' Limestone*) 115 m
	massive and mottled limestone with *Redlichia, Helcionella, Girranella, 'Obolella'; Archaeocyatha* abundant.
	Base of Middle Cambrian
LOWER CAMBRIAN	*Belly Creek Formation* 770 m *Red-beds*
	chocolate-coloured micaceous shales, sandstone and siltstone
	Oraparinna Shale 270 m
	shales with rubbly limestone atop; poor trilobites and *Hyolithus*
	Bunkers Sandstone 270 m
	Archaeocyatha Limestone 830 m
	with abundant *Archaeocyatha* and also horizons yielding abundant trilobites, brachiopods and hyolithids of Lowest Cambrian age.
POUND QUARTZITE	up to 300 m, feldspathic quartzites, often current-bedded or ripple-marked, with shale layers and fine conglomerates: the *Ediacara fauna*.

limestones, contain pebbles of fossiliferous Lower Cambrian, again revealing the effects of early Cambrian movement. In the foreland area of the Eyre Peninsula the flat-lying *Moonabie Grit*, equated by some with the Adelaidean, is followed with slight unconformity by the *Corunna Formation* which bears some resemblance both to the Marinoan Series and to the Pound Quartzite and both are intruded by the *Gawler Porphyries*.

The *Amadeus trough* is considered by some to constitute to the north an east–west continuation of the Adelaide geosyncline (Fig. 6.2). Its evolution, however, followed a different course: the very thick sedimentation, continued from Adelaidean until mid-Palaeozoic times, was controlled by sporadic faulting, folding was largely related to block-movements and there was no metamorphism and no invasion by granites.

The trough appears to have been defined initially by fracturing of the crystalline basement to produce a complex graben. Repeated vertical movements made room for the accumulation, without serious interruptions, of shallow-water Adelaidean, Cambrian and Ordovician sediments. The *Adelaidean* begins with thick sandstones followed by algal dolomites and pyritic shales. After a minor break, a 1600 m sequence of psammitic and carbonate rocks with several glacial and fluvioglacial intercalations was laid down and, after a second break, a thick sandy and dolomitic series passing up into greywackes and thence into Lower Cambrian strata. The *Lower Palaeozoic* extends through Cambrian and Orodvician as a sequence of pelites, psammites and carbonate rocks. It is followed, with marked unconformity, by three to four kilometres of unfossiliferous clastic sediments, possibly of Lower Devonian age, laid down during a

Table 5.6. CRATONIC SUCCESSION OF HALLS CREEK AREA, EAST KIMBERLEYS
(based on Dow (1965), dates from Compston and Arriens (1968))

CAMBRIAN	Lower-Middle	*Negri-Group* (*c.* 600 m) shale and limestone, minor sandstone
	Lower	*Antrim Plateau Volcanics* (*c.* 1000 m) tholeiitic basalt, minor agglomerate

←——————————— *angular unconformity* ——————————→

	Albert Edward Group (*c.* 1900 m) shale, sandstone (dolomite)

←——————————— *angular unconformity* ——————————→

Ranforth Formation (*c.* 600 m) sandstone, siltstone, shale, ferruginous sandstone, tillite: 685 ± 70 m.y.

LATE UPPER PROTEROZOIC

Moonlight Valley Tillite (0–150 m) tillite followed by thin pink-cream dolomite: 740 ± 30 m.y.

←——————————— *? angular unconformity ?* ——————————→

Frank River Sandstone (250 m) dolomite sandstone, rare tillite lenses

Fargoo Tillite (0–150 m)

←——————————— *angular unconformity* ——————————→

Stillwater Sediments (4000 + m)

---

period of tectonic unrest (Chapter 6 p. 162). The total thickness of the fill of the Amadeus trough is about 11 km.

Far removed from the Adelaide mobile belt, a cratonic cover of Adelaidean and early Palaeozoic is widely developed from the Kimberleys in Western Australia eastwards through the Northern Territories into Queensland and the foreland of the Tasman belt. The cover-rocks lie in broad basins now separated by faulting and their distribution and lithology indicate that, as in the Amadeus trough, differential vertical movements controlled the pattern of sedimentation. Dow's (1965) sequence for the Halls Creek area of the East Kimberleys provides an example of this cratonic succession (Table 5.6).

3 The Ediacara fauna

The remarkable fauna of soft-bodied organisms found in the Pound Quartzite at Ediacara in the Flinders Ranges was recorded by Sprigg in 1947 and has been discussed mainly by Glaessner and his associates (Glaessner, 1951, 1960, 1961, Glaessner and Daily, 1959). An excellent summary has been given by Rhodes (1966).

The fauna of nearly a thousand specimens (preserved mainly as casts in the bases of sandstone beds overlying shales) includes representatives of several phyla. There are about six genera of medusoids, together with sea-pens or pennatulate anthozoa, some remarkable annelids and two invertebrates of new types which may possibly be echinoderms. The sandstone of the matrix shows current-bedding, ripple-marking and sun-cracks and is evidently of shallow-water origin. The fossils appear to have included bottom-dwellers – there are innumerable burrows and the pennatulids may, like the modern sea-pens, have grown with their base buried in sand – which presumably represent the natural population of the environment; the pelagic medusoids and possibly also the annelids may have drifted in from deeper waters and been stranded, as jellyfish often are on modern beaches.

The Ediacara fauna occurs some 35–60 m below the top of the Pound Quartzite and is separated by some 200–230 m of sediment from the lowest fossiliferous horizon in the overlying Ajax (*Archaeocyatha*) limestone (Table 5.5). This limestone carries typical Cambrian fossils including archaeocyathids, sponge spicules, hyolithids and brachiopods and contains no representatives of the Ediacara fauna. Elements of the Ediacara fauna, or of very similar types, have been recorded elsewhere from strata known or suspected to lie at no great distance below the base of the Cambrian, for example, in northern and western Australia, South-west Africa, Charnwood Forest, Engand, North America and eastern Siberia.

A conspicuous feature of the Ediacara fauna, and one in which it contrasts with the succeeding Cambrian faunas, is the absence of skeletons or shells. Only three phyla and only a few species of each appear to be represented, whereas Lower Cambrian faunas have nine phyla and over nine hundred known species. Ruganov has concluded that the initial appearance of fossils with hard parts was synchronous in certain groups throughout the world and has proposed to use this event to define the base of the Cambrian. Successive Cambrian faunas, however, show evidence of a notable evolution. In the Ajax Limestone of southern Australia, the archaeocyathids have calcareous skeletons, the sponge spicules are siliceous and the brachiopods and hyolithids are phosphatic. Trilobites do not appear until higher in the Lower Cambrian and when they do appear have tests which are less calcareous than those of later forms. In general, early Cambrian forms were small, primitive, thin-shelled and with shells made mainly of chitin, silica or phosphate. Typical 'shelly' forms appeared in variety later in the Cambrian.

The transition from thin non-calcareous to thick calcareous shells and skeletons took place over different periods in different groups. Furthermore, this transition was spread over a period estimated by Rhodes (1966) to have been as long as 170 m.y. in certain groups. The early faunas are restricted and cosmopolitan faunas do not seem to have come into existence until the later Cambrian and Ordovician. It is thought by some palaeontologists (for example Glaessner, 1962, Rhodes, 1966) that the acquisition of hard parts was not a consequence of a change in habitat but is more likely to have been connected with changes in physiology, perhaps in the excretory processes. Other authors have seen this event as a consequence of the build-up of atmospheric oxygen to percentages not far from those of the present atmosphere: but such a build-up is assigned to a much earlier time by most geologists.

4 The Adelaidean orogeny

Australian opinion (for example, Campana, 1955, 1957, *Explanation of Tectonic Map of Australia*, 1962) has attributed the deformation of the contents of the Adelaide geosyncline to 'jostling and warping of the Archaean basement of the geosyncline with consequent complex folding and faulting of the overlying Upper Proterozoic and Cambrian sediment, and frequent development of *en echelon* fold patterns' (*Tectonic Map,* p. 29). The stable Gawler block on the western flank and the Willyama complex with its hidden southward continuation on the eastern flank provided the framework within which disturbances took place. As we have seen, movements began early in the Cambrian at the southern end of the fold-belt, but the more widespread phases were somewhat later, possibly at the close of Middle Cambrian times. The overall picture is of a belt of miogeosynclinal type affected by fairly mild orogenic disturbances. There are no major ophiolite suites and metamorphism of high grades or regional migmatisation is confined to the small Kanmantoo trough. In all these respects the Adelaidean tract contrasts with the Tasman belt to the east, whose early evolutionary stages overlapped with the Adelaidean cycle (Chapter 6).

The grain of the fold-belt is sinuous (Fig. 5.5). Variations in axial plunge define a few broad plunge depressions, such as that at the southern end of the belt in which full cover-sequences are preserved, and intervening culminations such as those in the Mount Lofty and Olary regions where basement rocks appear in domes and anticlinal fold-cores. At high levels in the cover-series, folding is relatively simple. Individual folds of Appalachian style can be followed for great distances along their axial direction. At lower structural levels, folding tends to be tighter and more complex, though metamorphism seldom rises much beyond slate grade. In the Broken Hill area (p. 144) the basement shows some retrogressive metamorphism and the 'Grand Unconformity' between basement and cover is obscured by metamorphic convergence. Elsewhere, basement-masses or low members of the cover-series form diapiric structures and piercement-wedges, appearing well above their proper positions in the sequences; such diapiric wedges of granitic basement have sometimes been taken to be intrusive magmatic granites. The western border of the fold-belt is marked by folds overturned to the west and locally by recumbent folds whose middle limbs are replaced by low-angle thrusts. On the foreland, the cover-succession shows only very broad warps.

In the Kanmantoo trough (p. 146) where much higher metamorphic grades are attained, mica-schists with andalusite, kyanite and sillimanite are developed, together with high-grade marbles and calc-silicate rocks. Some limited injection and feldspathisation are recorded and there are many bosses of granite which were intruded towards the close of the late Cambrian folding.

Mineralisation was on a relatively minor scale and was characterised by many small hydrothermal vein-deposits and local replacement-deposits of sulphides, especially auriferous pyrite, chalcopyrite and bornite with rarer galena and blende. Rejuvenation of basement deposits of lead is indicated in the Broken Hill area (p. 144) and rejuvenation of basement uranium may be responsible for uranium-bearing veins in the Adelaidean of the Mount Painter region. Syngenetic sedimentary deposits of copper and other sulphides are interbedded in the

Table 5.7. THE LATE PRECAMBRIAN-LOWER

	DAMARA BELT	AFRICA KATANGA BELT	MOZAMBIQUE BELT
GRANITES AND ASSOCIATED ROCKS	Late-orogenic granites in eugeosynclinal zone	Late-orogenic granites few, pegmatites dated at 720–500 m.y. Uranium and copper mineralisation	Late-orogenic granites few, pegmatites dated at 650–480 m.y. Little mineralisation.
METAMORPHISM AND TECTONISM	Folding and thrusting with little metamorphism in miogeosynclinal zone; high-grade metamorphism with migmatisation in eugeosynclinal zone	Folding and repeated fracturing, with greenschist-facies metamorphism in Copper Belt. Higher-grade metamorphism towards southern side of belt.	Reworking of basement imposes north–south tectonic grain. High-grade metamorphism, local charnockites; mineral dates 650–500 m.y.
COVER-SERIES	See Table 5.2 *Damara 'System'* locally over 10 000 m miogeosynclinal facies mainly shallow-water, eugeosynclinal facies includes greywackes and basic lavas. Infracambrian–early Cambrian: tillites	See Table 5.1 *Katanga 'System'* up to about 8000 m mainly shallow-water sediments, little or no volcanics: Infracambrian predominates, Lower Palaeozoic not identified: tillites	See Table 5.3 Cover-series only of local occurrence, not stratigraphically dated No tillites recorded
BASEMENT		Regenerated basement granites give mantled gneiss domes in Copper Belt	Extensive regeneration giving high-grade rocks including charnockites

Cambrian Kanmantoo rocks of the Adelaide district and sedimentary magnesite occurs in Adelaidean dolomites throughout the geosynclinal region.

VI South America and Antarctica

Some details of late Precambrian and earliest Palaeozoic events in South America and Antarctica were given in Part I (pp. 164–8). We may recall that the *Brazilides*, constituting a tract of tectonic and thermal activity, border the Brazilian shield on the east and are probably linked with similar belts partially

PALAEOZOIC CYCLE IN GONDWANALAND

S. INDIA AND CEYLON	AUSTRALIA ADELAIDE BELT	SOUTH AMERICA BRAZILIDES	ANTARCTICA ROSS BELTS
Late-orogenic granites few and small. Little mineralisation	Late-orogenic granites few and small. Little mineralisation	Possibly late granites in equivalent of Lavras Series	Syntectonic and late-tectonic granites; contact-migmatites. Granites dated 550–400 m.y.
In Ceylon, complex folding, metamorphism and migmatisation of amphibolite facies. In South India, high-grade metamorphism, mineral dates *c.* 450 m.y.	Complex folding, mainly non-metamorphic or low-grade, except in Kanmantoo trough where sillimanite-grade reached. Few migmatites 450–500 m.y. dates from basement biotites	Unfolded and non-metamorphic Lavras Series thought to pass SW into much deformed and metamorphosed Ribeira and Itajai Series, the latter invaded by granites. Rejuvenated basement rocks give 500–450 m.y.	Strongly folded and metamorphosed to grenschist facies; sillimanite grade in granite-rich areas. Reactivated basement dates 550–400 m.y.
No cover-series identified	See Tables 5.4, 5.5 *Infracambrian Adelaidean 'System'* and Lower and Middle Cambrian; possible 19 km shallow-water sediments, no volcanics	Possibly *Lavras Series* (? Infracambrian) clastics, with rocks of glacial facies	*Ross Supergroup* (Infracambrian and Cambrian): thick greywacke-pelites carbonate assemblage with acid pyroclastics
	Tillites and interglacial beds up to 5 km thick	Tillites	No tillites yet recorded
Extensive regeneration giving migmatitic gneisses of Vijayan Series in Ceylon	Little regeneration of basement rocks	Extensive regeneration giving granulite facies, charnockitic in places	Extensive regeneration of basement rocks, giving granulite facies and intrusive charnockites

embracing the shield; and that a zone of tectonic and thermal activity bordering the craton of East Antarctica is linked to a more complete orogenic belt forming the *Trans-Antarctic Mountains*. Supracrustal groups whose distribution is related to these belts include the *Lavras Series* and its equivalents of eastern Brazil, which contains tillites regarded as Infracambrian, and the thick *Ross Supergroup* of the Trans-Antarctic Mountains which, though lacking glacigene horizons, is considered to include both Infracambrian and Cambrian rocks. The timing of orogenic events in South America and Antarctica indicates that the main phases of activity fell in earliest Palaeozoic or latest Precambrian times; the spread of mineral ages down to about 450 m.y. may reflect subsequent episodes or, more likely, post-orogenic phases of uplift and cooling.

VII Generalisations

Our continent-by-continent review of the effects of the late Precambrian — early Palaeozoic geological cycle in Gondwanaland provides a basis for the summary Table 5.7. Comparison with the data concerning the Caledonian cycle in Laurasia given in Chapters 2, 3 and 4 reveals some illuminating contrasts:

(a) Most portions of the Caledonian network incorporate supracrustal cover-successions of geosynclinal thickness, whereas many parts of the Gondwanaland network consist almost wholly of polycyclic rocks.

(b) Late-orogenic molasse-formations (e.g. Old Red Sandstone) are widely distributed and reach considerable thicknesses in Caledonian belts, whereas analogous formations are almost absent from the Gondwanaland belts.

(c) Charnockites and other rocks of granulite facies are common in polycyclic Gondwanaland belts but very rare in the Caledonian belts.

(d) While there seems to have been little difference in the timing of the initiation of mobility and of the early orogenic phases, the timing of later phases appears to have been out of step. In Gondwanaland, geosynclinal successions, when present, usually contain no rocks younger than Cambrian; in the Caledonian belts, geosynclinal successions often range up to Silurian strata. The principal phases of late-orogenic granite-emplacement in the Gondwanaland belts seem to date back in many places to 600—500 m.y.; the corresponding phases in most Caledonian belts took place at about 400 m.y., though timing in the Caledonian belt of France and adjacent regions had more in common with the Gondwanaland scheme.

The scarcity of molasse-deposits formed in the immediate aftermath of orogeny poses problems in connection with the processes of stabilisation. In the Caledonian belts of Eurasia, as in most Phanerozoic mobile belts, stabilisation was associated with widespread granite emplacement and with regional uplift leading to the development of mountain tracts. Erosion-debris from these tracts forms the bulk of the molasse-formations. Does the scarcity of molasse in the Gondwanaland belts indicate that stabilisation was not accompanied either by large-scale granite-emplacement or by mountain-building? The high metamorphic grade of polycyclic complexes such as those of the Mozambique belt suggests deep erosion, but if erosion-debris was formed in the late-orogenic period, it cannot easily be accounted for: in peninsular India, and in eastern, central and western Africa the oldest sediments resting on rocks of the Indian Ocean—Mozambique cycle are Karroo (p. 161). These anomalies remain to be explained.

VIII Infracambrian Ice-ages

Evidence for a very widespread glaciation (or for two or more glaciations) towards the end of Precambrian time has been given in this chapter and in Chapter 2. The distribution of apparently reputable glacial sediments of late Precambrian age, based on the compilations of Harland (1963) and Cahen (1963), is shown in Fig. 5.6 in which the continents are restored to their possible

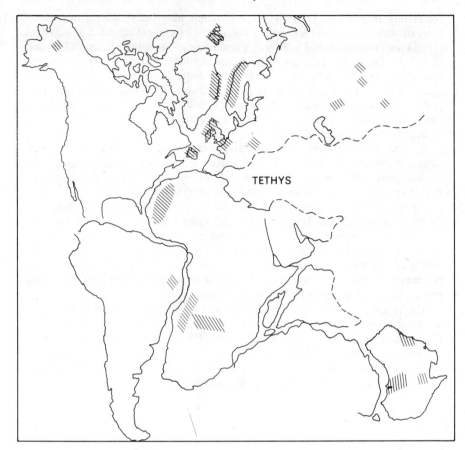

Fig. 5.6. The Infracambrian glaciation; the distribution of glacigene deposits is plotted on a reconstruction which allows for post-Palaeozoic continental drift

pre-drift positions. For the most part, the tillites and associated rocks of glacial facies appear to have been deposited in water: most likely in sea-water since they appear as intercalations in neritic sequences, and are quite often associated with stromatolitic limestones. It will be recalled, however, that terrestrial ice-deposits resting on glacial pavements have been demonstrated in several continents. The glaciation of which evidence is preserved appears to have been largely akin to that in the vicinity of the present Antarctic continent where great expanses of shelf-ice, move out from high land-masses.

The remarkably wide distribution of glacigene deposits both in Gondwanaland and in Laurasia is not significantly reduced by any probable restoration of continental masses to pre-drift positions. With the advance of knowledge it may eventually be possible to delimit separate ice-sheets and ice-shelves, but here the awkward question of contemporaneity arises. Were there many late Precambrian glaciations and, if so, what are the chances of correlating them? Two or more tillite horizons have been identified in several parts of Australia. In the Kimberley region, for example, Compston and Arriens (1968) quote dates of 740 ± 30 m.y. for samples of Moonlight Valley tillites and of 685 ± 70 m.y. for the Egan tillite of the Ranforth Formation, both of which appear to be separated by a slight unconformity from the Fargoo tillite (Table 5.6). In the Caledonides of Greenland and Spitsbergen an upper tillite group a few hundred metres conformably below the Cambrian (the Morkebjerg and Polarisbreen tillites, respectively) is separated by up to 12 km of non-glacial sediments from a lower tillite (Tables 2.5, 2.6).

It seems to us evident that a specific glacial horizon is represented in both supercontinents not far below the base of the Cambrian. This is the Infracambrian tillite which Harland used to define the base of his Infracambrian Period (p. 32). It is very widely distributed in almost every continent (Fig. 5.6) and must record a glaciation of stupendous extent, which could have had far-reaching side-effects. The possibility that there was a connection between the retreat of ice-sheets and the great Cambrian transgressions onto the cratons cannot be entirely dismissed. The comparatively brief interval between the ending of glaciation and the appearance of the diversified Cambrian fauna, too, requires to be borne in mind, since the retreat of ice-sheets must have opened up new areas for colonisation by marine organisms.

The extent, number and dating of tillites at lower stratigraphical levels remain doubtful. Little can be established at present beyond the bare fact that late Precambrian glaciations took place. Harland (1963, p. 122) favours a 'very small number of discrete, simultaneous, severe and prolonged ice-ages' in the period, for which he proposed the term *Infra-Palaeozoic era* defined by specific successions such as the Dalradian, Sinian and the like.

6

Gondwanaland in Late Palaeozoic Times

1 Gondwanaland at the Close of the Mozambique Cycle

During the earliest parts of the Phanerozoic era, the supercontinent of Gondwanaland was traversed by a network of mobile belts in whose interstices lay relatively small stable cratons. At the termination of orogenic activity in many parts of this network at or soon after 500 m.y., several previously separated cratons were welded together to produce a single, very much larger, stable unit. Orogenic activity continued in peripheral mobile zones which formed a nearly continuous girdle enveloping the newly consolidated craton (Fig. 6.1).

The enormous raft of stable continental crust which we may call *the Gondwanaland craton* constituted a crustal unit rather different from any discussed in Part I. It was larger than the cratons which had existed during the late Precambrian and early Palaeozoic cycle and although cratons of similar size may well have existed towards the end of the Svecofennide chelogenic cycle, their characters were not necessarily the same. The cratonic regions of Gondwanaland have, with a few exceptions, remained free from orogenic activity up to the present day, though they no longer form a crustal entity. The history of the Gondwanaland craton as a single entity covers a time-span of at most some 350 m.y., falling largely with the Palaeozoic, and was terminated by the disruption of the supercontinent.

Of the peripheral mobile belts (Fig. 6.1), the segments embracing western South America, west Antarctica, North Africa, the Middle East and northern India remained active long after the end of the Palaeozoic era, some even to the present day. These segments will be dealt with in later chapters, where their evolution can be viewed in perspective. The *Mauretanide belt* of West Africa, the *Cape fold-belt* of South Africa and the *Tasman belt* of eastern Australia were stabilised at about the end of the Upper Palaeozoic era and will therefore be

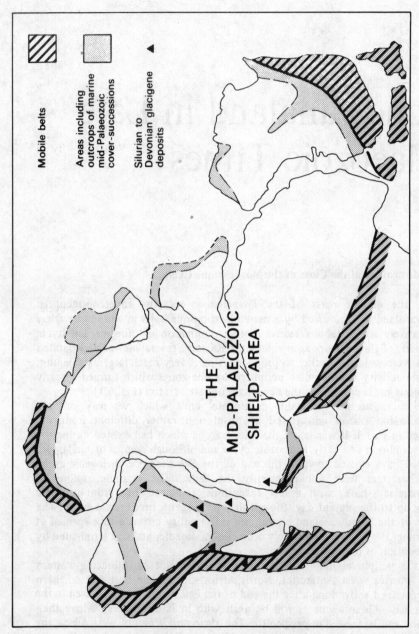

Fig. 6.1. Gondwanaland in mid-Palaeozoic times, showing the inferred emergent shield, the areas in which marine Silurian, Devonian or Mississippian sediments are preserved, and the peripheral mobile belts

Mobile belts

Areas including outcrops of marine mid-Palaeozoic cover-successions

Silurian to Devonian glacigene deposits

THE MID-PALAEOZOIC SHIELD AREA

dealt with here. Our main theme, however, will be the behaviour of the great central craton prior to its fragmentation.

II The Mid-Palaeozoic History of the Craton

1 Shield and platform areas

Stratigraphical dating of the terminal stages of the orogenic cycle which preceded unification of the craton is hampered by the general scarcity of fossiliferous cover-rocks in mobile belts of this system. As we saw in Chapter 5, pre-orogenic basin-successions seldom include strata younger than Cambrian, and radiometric dating of late-orogenic intrusions suggests that the main phases of orogenic activity came to an end in early Palaeozoic times. We next deal with the history of the craton during the stages of its evolution which occupied the periods from late Ordovician to early Carboniferous. This mid-Palaeozoic phase, lasting perhaps 200–250 m.y., was brought to an end in late Carboniferous times by radical changes in the pattern of sedimentation, which heralded an important glaciation and a period of continental deposition lasting well into the Mesozoic era.

Mid-Palaeozoic supracrustal rocks have a rather restricted distribution in the craton of Gondwanaland (Fig. 6.1). They occur mainly in South America (where they extend eastward in a number of great embayments from the Andean mobile zone), in North Africa, Arabia, northern India and northern Australia. Over most of Africa, peninsular India and southern Australia, no mid-Palaeozoic strata are exposed. Late Palaeozoic or younger cover-rocks usually rest directly on the basement, suggesting either that no mid-Palaeozoic cover was deposited or that deposition has been more than balanced by erosion. We may regard these positive regions as constituting one or more *mid-Palaeozoic shield-areas* in contrast to the peripheral cratonic *platforms* which received and retained mid-Palaeozoic cover-formations.

2 The mid-Palaeozoic shield

The mid-Palaeozoic shield-fragments in Africa and India are now partially covered by Mesozoic and Tertiary sediments. Remnants of a mid-Palaeozoic cover may be concealed in some of the younger basins of deposition but, if so, they have been overlapped by younger deposits which rest directly on crystalline rocks at the exposed margins of these basins. The mid-Palaeozoic shield areas include portions of the recently stabilised late Precambrian–early Palaeozoic orogenic system, notably the Katanga and Damara belts, much of the Mozambique belt and the belt of south India and Ceylon. As we have noted (p. 156) no molasse-type sediments laid down during and immediately after the terminal stages of the cycle have been preserved. The oldest formations resting on the Katanga or Mozambique belts are Karroo sediments, laid down more than 200 m.y. after stabilisation had taken place. If, therefore, uplift and erosion accompanied stabilisation, the erosion-products must have been subsequently removed from the mid-Palaeozoic shield area.

3 The mid-Palaeozoic cratons

An intermittent cover of Lower Palaeozoic, Devonian and early Carboniferous sediments of shallow-water marine or continental facies accumulated on the platforms fringing the Gondwanaland shield (Fig. 6.1). These sediments are preserved mainly in a number of cratonic basins initiated during deposition and emphasised by later warping. On the 'swells' which separate neighbouring basins, mid-Palaeozoic successions are thin, punctuated by unconformities, or absent altogether. A basal unconformity almost everywhere marks the resumption of deposition after the termination of the Mozambique cycle. Ordovician and Silurian platform-deposits are restricted in distribution and Devonian rocks often rest directly on the basement. It would appear that in Devonian times, platform-deposits began to encroach on areas previously subject to erosion and a much greater extension of cover-deposits marked the beginning of the late Palaeozoic stage dealt with in Chapter 8.

Australia. Mid-Palaeozoic cratonic cover-rocks are seen mainly in the central, western and northern parts of Australia, and more locally on the foreland of the Tasman belt in the east (Fig. 6.2). Much of southern Australia appears to have constituted a shield-area in mid-Palaeozoic times, for Permian or Mesozoic cover-rocks rest directly on the Precambrian or Adelaidean. The mid-Palaeozoic strata are predominantly marine sandstones and shales associated with highly fossiliferous limestones. Devonian and Lower Carboniferous rocks are widely distributed; Ordovician, Silurian and late Carboniferous rather less so.

The *Amadeus trough* (pp. 150–1), was already receiving thick deposits even before the end of the Precambrian. The younger parts of the trough-filling, often following unconformably on Ordovician or older rocks, consist largely of unfossiliferous clastic sediments – psammites and conglomerates – reaching 7 km in maximum thickness and possibly representing piedmont and alluvial fan deposits. They are thought to be largely Devonian in age and, with the underlying deposits, bring the total thickness of supracrustal rocks in the trough up to over 11 km. Although the rocks of this immense pile are much faulted and are thrown into an arcuate fold-system, they show no metamorphism and are not invaded by granites; we regard the Amadeus trough as a cratonic basin.

On the border between Northern Territories and Western Australia, marine sediments of Devonian and Carboniferous age rest unconformably on Precambrian or Cambrian in the *Bonaparte Gulf basin.* The succession reaches a maximum thickness of some 4 km and is made up of a shallow-water sandstone–limestone assemblage including reef-limestones of Upper Devonian age. Several minor unconformities interrupt the Carboniferous sequence. Further west, several kilometres of Palaeozoic sediments are seen in the *Fitzroy basin* on the north-eastern flank of the larger Canning basin which is filled mainly with Permian and Mesozoic deposits. In the Fitzroy basin the fill ranges from Ordovician to Permian and contains potentially valuable oil reserves (Table 6.1). The Ordovician, which carries glauconite and limonite after pyrite, is regarded as the source-rock.

The *Carnarvon basin* along the west coast of the continent is, like the Canning basin, occupied largely by late Palaeozoic or younger strata. Marine sandstones

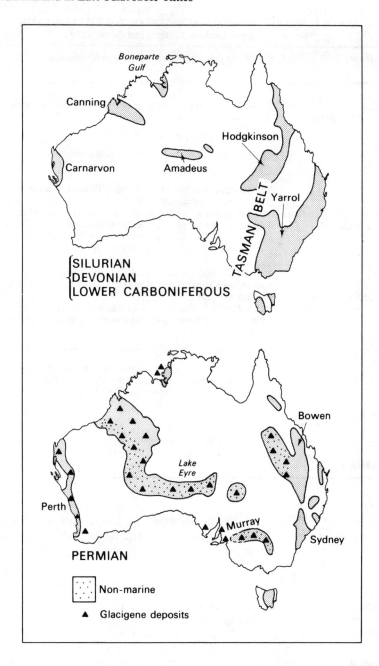

Fig. 6.2. Australia: the distribution of middle and late Palaeozoic strata — from surface and subsurface data — and the principal basins of deposition (based on Brown and others, 1968)

Table 6.1. MID-PALAEOZOIC SUCCESSION OF THE FITZROY BASIN
(Based on Guppy, Lindner, Rattigan and Casey, 1958)

PERMIAN	(Glacigene deposits followed by detrital sediments)

←———————————————— (*unconformity*) ————————————————→

LOWER CARBONIFEROUS	*Laurel Beds* (500 m) clastic limestones and siltstones with abundant brachiopods
	?disconformity
	Fairfield Beds (200 m) limestones, sandy limestones and sandstones
UPPER DEVONIAN	Famennian and Frasnian — *Limestones and conglomerates* (max. 1000 m) a complex series, predominantly calcareous, with strong lateral variation. Marginal facies include polymict fanglomerates, sometimes resting directly on the basement; elsewhere, the dominant rocks are clastic limestones, reef-limestones and oolites with abundant faunas of brachiopods, nautiloids and corals
MIDDLE DEVONIAN	Frasnian — *Sadler and Gogo Formations* (400 m) principally limestones with minor shales and conglomerates, small bioherms, corals, stromatoporoids Givetian — *Pillara Formation* (600 m) basal arkoses followed by limestones with stromatoporoid masses, abundant brachiopods and local coral reefs

←———————————————— (*unconformity*) ————————————————→

LOWER ORDOVICIAN	*Prices Creek Group* (800 m) limestones and shales followed by dolomites and limestones with intercalated sandstones: rich marine fauna with abundant trilobites (source-rocks of oil)
? CAMBRIAN	Hart Basalt

←———————————————— (*unconformity*) ————————————————→

BASEMENT	

and limestones ranging from Middle Devonian to Lower Carboniferous are, however, locally recorded both at the surface and in boreholes. These early cover-rocks are followed unconformably by glacial deposits probably of late Carboniferous age.

Antarctica. The stable region of East Antarctica carries a mid-Palaeozoic cover only in an area bordering the Beardmore Glacier and Ross Ice-shelf. Elsewhere, the late Palaeozoic-Mesozoic Beacon Group rests directly on the basement. The only members of the mid-Palaeozoic sequence which are dated palaeontologically are Devonian: marine faunas rich in Lower Devonian brachiopods have been recorded from sandstones and shales in the central Horlick Mountains, near the head of the Beardmore Glacier, but in the Alexandra Range flanking the glacier, pale-coloured Devonian sandstones and shales with fish and lycopod remains appear to be of continental facies. In west Antarctica, which formed part of the peripheral Andean mobile belt, marine Carboniferous is recorded.

South America. Mid-Palaeozoic cratonic sediments appear to have covered much of South America and were laid down mainly in shallow seas extending eastward from the Andean troughs. The Brazilian shield and the Pampean and Patagonian massifs remained for much of the time above the seas, while a number of basin-regions were outlined between them, notably the *Paraná basin* of southern Brazil, the *Piany basin* of north-east Brazil and the *Amazon basin.* In general, deposition appears to have begun early in the west. Marine Cambro-Ordovician sediments are seen near the border of the Pampean massif and on the western border of the Brazilian shield in Bolivia. In the *Paraná basin,* which was to receive thick sequences of late Palaeozoic and younger rocks, the earliest deposits appear to be Silurian or Devonian. In eastern Paraguay, graptolitic Lower Silurian lies on crystalline rocks, while in southern Brazil the Devonian rests on a peneplained surface of Precambrian.

The cover-succession of Brazil and Uruguay (Table 6.2) begins in a few localities with tillites (the *Iapo tillite*) which are followed conformably by uniform well-sorted sandstones (the *Furnas Sandstone*) and by the bituminous *Ponta Grossa Shales* carrying a rich Lower Devonian marine fauna. Somewhat similar successions, perhaps locally extending down into the Silurian, are seen further south in Argentina and the Falkland Islands where they are followed by terrestrial sediments thought to be of Lower Carboniferous age (Table 6.2). These sequences bear a close resemblance to the Cape System of South Africa (p. 173) and it has been suggested (for example Martin, 1961) that all were laid down by arms of the sea advancing northward or north-eastward from the southernmost parts of the Andean trough. Martin points out that neither the distribution of marine sediments nor the pattern of palaeocurrents responsible for deposition of the Furnas Sandstone offers any indication of the existence of a Devonian sea on the site of the present Atlantic.

An important feature shared by the mid-Palaeozoic of South America and that of South Africa is the occurrence of *glacigene deposits* (Fig. 6.1). The Iapo tillite of southern Brazil just mentioned is covered by fossiliferous Lower Devonian, as are the tillites of the Table Mountain Sandstone in South Africa (p. 173). In north-eastern Brazil both Lower and Upper Devonian glacigene deposits have been recorded in a succession made predominantly of marine sandstones. Middle Devonian tillites are known from the Precordillera of Argentina. Although information about some of these occurrences is limited, it seems reasonable to regard them as deposits associated with true ice-sheets rather than with mountain-glaciers: some are marine, while others rest on surfaces of low relief. It will be recalled that Devonian reef-limestones are conspicuous in Australia, suggesting deposition in warm seas and, consequently, indicating a climatic zoning within Gondwanaland. Palaeomagnetic studies suggest that throughout the Devonian and Carboniferous periods the eastern parts of South America remained at high latitudes as the continent migrated through the southern polar region. Such a picture would be consistent with indications of repeated glaciation.

Table 6.2. MID-PALAEOZOIC SUCCESSIONS OF THE CAPE BELT,
(Martin, 1961; Adie, 1962)

SOUTH AFRICA	SOUTH

PERMO-CARBONIFEROUS

Western part of Cape Belt

(Dwyka tillites, often unconformable on mid-Palaeozoic)

Eastern Argentina

(tillites unconformable on underlying rocks)

←——————————————————————————————————— *unconformity*

M. AND U. DEVONIAN

Witteberg Series: shales with interbedded quartzites *c.* 800 m

L. DEVONIAN

Bokkeveld Series: interbedded shales and sandstones, plant remains in upper part, marine L. Devonian fauna in lower part *c.* 750 m

SILURIAN

ORDOVICIAN

Table Mountain Series: false-bedded sandstones with scattered pebbles, purple siltstones: local tillite horizon with shales in upper part of series: max. 1600 m

(*unconformity* is a peneplain cut in basement, last granites of basement are end-Precambrian, 550 m.y.)

CAPE 'SYSTEM'

Shales and sandstones carrying a L. Devonian marine fauna 400 m

White and pink sandstones with shale intercalations, minor unconformity within series, basal conglomerates *c.* 2000 m

(*unconformable on basement*)

North Africa and Arabia. The whole of southern, central and eastern Africa, with the exception of the Cape belt and its foreland in the extreme south, lacks exposures of mid-Palaeozoic cover-rocks and forms the core of the Gondwanaland shield-area. In the north of the continent and in Arabia, however, an extensive cover, partly marine, partly terrestrial, fringes the Mediterranean border of Gondwanaland. In the extreme west, the cover enters the *Mauretanide mobile belt* which was not stabilised until the end of the Palaeozoic. Elsewhere,

SOUTH-EASTERN SOUTH AMERICA AND THE FALKLAND ISLANDS

AMERICA		

		PERMO-CARBONIFEROUS
Paraná basin	Falkland Islands	
(tillites unconformable on underlying rocks)	(tillites unconformable on underlying rocks)	

→

	Port Stanley Beds ⎫	terrestrial	DEVONIAN
Ponta Grossa Shales: shales, sometimes bituminous with L. Devonian marine fauna *c.* 500 m	Port Philomel Beds ⎬		
Furnas Sandstone: false-bedded pebbly sandstones *c.* 300 m			SILURIAN
Iapo tillite (locally developed)	Fox Bay Beds ⎫	marine	
	Port Stephen Beds ⎭		ORDOVICIAN
(*unconformity* a peneplain cut in basement			

it is almost undisturbed, though it is penetrated by a few Palaeozoic granites. The underlying basement, especially in Egypt and Sudan, includes considerable tracts which were consolidated during the Mozambique cycle.

In the westernmost parts of the craton, on the foreland of the Mauretanides, deposition of the cover-succession began early and continued in predominantly marine conditions. *Infracambrian* clastics and tillites are followed by shallow-water sandstones, limestones and stromatolitic dolomites of *Cambrian and*

Ordovician age, in a sequence generally somewhat thinner than that of the adjacent mobile belt (p. 174). This sequence is followed, often with minor unconformity, by graptolitic *Silurian* shales, black shales and limestones and by *Devonian* sandstones and limestones. Carboniferous rocks are generally absent. As the cover is traced eastward from Guinée, Mali and Morocco, the lower members become reduced, and at the same time, the facies of the remaining members changes from marine to predominantly terrestrial. In Algeria, the sequence is still dominantly marine and ranges from Cambro-Ordovician to Lower Carboniferous; both here and to the west it is penetrated by dolerite sills believed to be late Palaeozoic. In northern Libya, continental sandstones alternate with marine strata in a sequence extending from Cambrian to Carboniferous. In the *Kufri basin* of south-east Libya, a predominantly non-marine succession contains marine intercalations within the Silurian, Devonian and Carboniferous. In Egypt, there are usually no cover-rocks older than Carboniferous, but in Arabia there is a cover ranging with little variation of facies, though with big stratigraphical gaps, from Cambrian to Mesozoic and consisting mainly of continental sandstones. Evaporites of Cambrian or Proterozoic age are important in Iran.

This seems an appropriate place to mention the term *Nubian sandstone* which has been applied to many continental sandstones of North Africa and Arabia. Rocks which range in age from Cambrian to Tertiary have, at one time or another, received this name. The term Nubian sandstone, used in this very broad sense, is simply a facies term, of interest mainly for its indication of the persistence of similar environments of deposition over a very long period in North Africa. In a stratigraphical sense, the term is now usually restricted to late Mesozoic sandstones.

Northern India. Although peninsular India lacks mid-Palaeozoic cover-rocks and appears to have formed part of the shield-area of Gondwanaland, there is evidence to suggest that a platform-area carrying a cratonic cover originally fringed its northern border, intervening, as in other regions, between the shield and the peripheral mobile belt. This northern platform-zone has been swallowed up in the Himalayan mountain system and is now revealed mainly in the thrust-nappes and folded tracts of the Lower Himalayas (Chapter 7). In the *Salt Range*, a well-documented succession ranging from Cambrian evaporites through Palaeozoic and into Mesozoic formations is largely marine and is interrupted by many minor non-sequences. Further east, the stratigraphy of the cratonic cover is often scarcely known but a widespread boulder-bed, the *Blaini Beds*, commonly regarded as equivalent to the Permo-Carboniferous tillites of Gondwanaland, provides a useful time-marker. In the *Simla Hills*, two groups of unfossiliferous shallow-water detrital sediments, the Simla Slates and the Jaunsar Series, underlie the Blaini Beds and it seems reasonable to accept them as members of the Palaeozoic cover.

III Palaeozoic Mobile Belts

1 The Tasman Belt

The Tasman belt of eastern Australia extends for almost 3000 km from Tasmania to the Cape York Peninsula of Queensland. Throughout the belt the tectonic grain, emphasised by the alignment of elongated granite bodies, is roughly north and south. The eastern margin is hidden by the sea. The western margin is generally blanketed by the sediments of the Great Artesian and Murray basins, but in northern Queensland a clearly defined front – the *Tasman line* – is seen separating the rocks of the fold-belt from a foreland tract where the Precambrian basement underlies a relatively thin Palaeozoic cover. In the extreme south, the relationships of the Tasman belt are more obscure, for it is flanked by the Adelaidean mobile belt which itself remained active into early Palaeozoic times; it is possible that the Tasman and Adelaidean belts were continuous structures prior to the stabilisation of the Adelaidean belt.

The Tasman belt appears to have been initiated at, or perhaps before, the beginning of the Palaeozoic era and remained active up to, and locally even beyond, the end of the Upper Palaeozoic. During its life-span of some 400 m.y., a number of elongated north–south troughs were filled with thick sequences of sedimentary and volcanic material (Fig. 6.3). The Lower Palaeozoic sediments were almost entirely marine and were laid down very widely. Tectonic disturbances began in the Ordovician and thereafter recurred at intervals, notably in the late Silurian-Early Devonian (*Bowning orogeny*), the mid-Carboniferous (*Kanimblan orogeny*) and the Permian (*Hunter–Bowen orogeny*). The middle and the later Palaeozoic sequences, which were broadly syn-orogenic, include considerable thicknesses of deltaic and flood-plain sediments. In the south, post-Devonian sediments were largely non-marine and suffered relatively little disturbance. Towards the north-east they included more material of marine origin and were more severely folded. Basin-formation and folding continued into Mesozoic times in the extreme east where the Maryborough basin incorporates folded Jurassic.

The Tasman belt was characterised through most of its history by profuse and varied *igneous activity*. A great range of lavas and pyroclastics is associated with the sedimentary sequences in certain tracts. Serpentine bodies of Alpine type, individually small but very numerous, are grouped in zones following the tectonic trend for hundreds of kilometres. Those of Victoria and New South Wales are regarded as possibly Ordovician, those of a northern zone extending for nearly 2000 km north-north-west from the vicinity of Brisbane are regarded as Devonian. Granites are unusually abundant, especially in the southern part of the mobile belt. They range in age from Silurian to early Mesozoic. High-level granites are commonly associated with volcanic centres at which are found thick piles of acid-intermediate pyroclastics and lavas; the abundance of acid volcanics is a distinctive feature of the Tasman belt.

Early Palaeozoic supracrustals form thick sequences in two principal tracts on the southern part of the belt (Fig. 6.3). In these tracts, one extending from Tasmania northward into Victoria, and the other occupying the coastal region south of Brisbane, detrital sediments including poorly sorted greywackes are

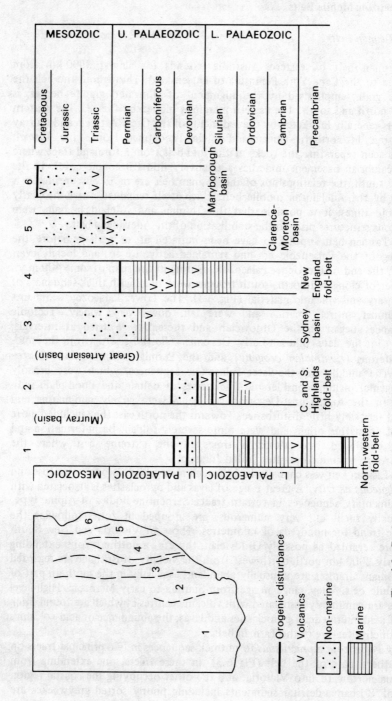

Fig. 6.3. Basin-successions in the southern part of the Tasman belt (modified after Crook, 1969)

associated with basic and andesitic lavas and pyroclastics. The sequences have few breaks and reach considerable thicknesses; in Victoria, the Cambrian locally exceeds 3 km and the Ordovician and Silurian each exceed 6 km. The two basin-tracts of Victoria are separated by a zone passing through the Canberra region and into central New South Wales which has been likened to an island arc. In this zone, the Lower Palaeozoic is relatively thin, the succession is broken by angular unconformities and its Silurian and younger components are largely volcanic; andesites, ignimbrites and acid lavas are abundantly developed while the serpentines already referred to and a number of large granites, contemporaneous with the volcanics, are emplaced in the supracrustals.

Silurian rocks are the oldest cover-strata of the marginal orogenic tract in the north of the Tasman belt, and rest directly on the basement. In the Broken River area, an embayment in the Tasman line, clastic sediments mostly of Silurian age attain thicknesses of well over 10 km. In eastern Queensland, in the interior of the fold-belt, somewhat metamorphosed Cambrian and Ordovician are again represented.

Towards the end of the Silurian the *Bowning disturbances* led to widespread folding, low-grade metamorphism and granite-emplacement. Mid-Palaeozoic basins received abundant clastic mineral — sandstones, conglomerates and shales — resulting from these disturbances. In the south and west the sediments were largely non-marine, though horizons with marine shells record occasional incursions of the sea. In the *Yarrol and Drummond basins* of eastern Queensland, marine conditions predominated. Almost everywhere, some tuffaceous material is mixed with the detritus and at certain localities there are thick groups of acid or intermediate tuffs and lavas. The maximum total thicknesses of Devonian and Lower Carboniferous are of the order of 7 km.

By late Carboniferous times, the southern parts of the mobile belt were largely stabilised and *late Carboniferous and Permian* successions in these parts lie almost undisturbed on folded cover-rocks. Glacigene deposits (the *Kuttung Series*) of Permo-Carboniferous age are represented, and the associated non-marine detrital sediments include a number of coal-seams. Acid volcanics are again associated with the sediments. In the *Bowen basin* of eastern Queensland, where orogenic activity still continued, the late-Carboniferous and Permian comprise some 12 km of alternating marine and non-marine sediments associated with andesitic volcanics and followed by a further 3 km or so of Triassic. Much of this late-orogenic basin-fill consists of fine sandstones, siltstones and coal measures. Elsewhere, thicknesses were not so great, but, nevertheless, an immense bulk of detrital late-Palaeozoic sediment was distributed within the mobile belt and spread out westward over the foreland. The *Maryborough basin* of the coastal zone north of Brisbane contains thick Triassic and Jurassic sediments of similar types associated, as ever, with andesitic volcanics.

Metamorphism in the Tasman belt is usually of low grade and affects mainly the older portions of the cover. Schists and gneisses associated with granites are seen in a number of 'structural highs' bordering the later basins of Queensland and some of them are tentatively regarded as derivatives of the basement. In the southern parts of the belt, the regional slate- or phyllite-grade of metamorphism increases abruptly around a number of small 'nodes' occupied by high-grade rocks and migmatites. Metamorphism in these nodes is of the low-pressure facies series characterised by the development of andalusite, cordierite and sillimanite in pelitic rocks. The style of metamorphism, the narrowness of the metamorphic zones and the close association with migmatites suggest unusually steep geothermal gradients leading to the development of 'hot-spots' at high levels in the crust.

The characteristic production of *granites and acid volcanics*, which continued for some 200 m.y., provides a further indication of high heat-flow in the Tasman belt. The granites are of several types, ranging from foliated or gneissose varieties to massive varieties surrounded by hornfels aureoles and to ring-complexes associated with volcanic centres. Radiometric dating has shown that there was an irregular but significant eastward shifting of the main zones of granite-emplacement from Devonian to Permian times. The earliest dated granites (410–430 m.y.) lie in south-east Victoria. The youngest, emplaced in older granites east of the Bowen basin, has given ages of 122–128 m.y. The data suggest (Fig. 6.4) that plutonic activity migrated eastwards towards the continental margin and towards the region in which basin-formation continued

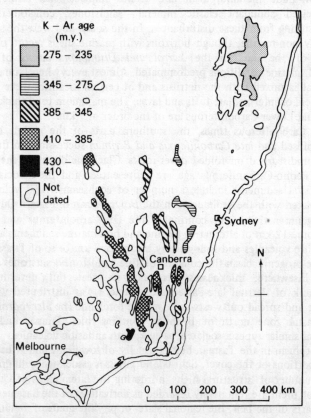

Fig. 6.4. Migration of sites of granite formation in the Tasman belt (after Evernden and Richards, 1962)

longest. *Mineralisation* of varied kinds was associated with certain granites and volcanics. Gold mineralisation in Victoria is associated with the Devonian Wood's Point dyke swarm and provides quartz saddle-reefs such as those of Bendigo. Lead—zinc and copper deposits of Devonian (and possibly also of Cambrian) age occur in western Tasmania, while minor occurrences of similar types are fairly widespread. Certain of the younger granites are associated with tin deposits.

2 The Cape belt

The Cape fold-belt crosses the southern tip of Africa from east to west, turning northward to follow the west coast for some distance before it is truncated by the continental margin. It is built on a Precambrian basement much of which was folded, intruded by granites and stabilised in very late Precambrian times (p. 132). A Palaeozoic trough on the site of the Cape belt received terrestrial and marine sediments with a maximum thickness of about 3 km, while a thinner cover of equivalent rocks extended northward over the foreland.

The *Cape 'System'* which constitutes the trough-filling and foreland succession is entirely sedimentary and detrital (Table 6.2). The *Table Mountain Sandstone* rests on a surface of low relief and appears, from the predominant southerly dip of the foreset beds, to have been derived from source-regions in the foreland. The sandstones appear to be of shallow-water origin and have been regarded as largely fluviatile or deltaic; the recent discovery of a late-Ordovician brachiopod fauna, together with trace-fossils interpreted as trilobite-trails, indicate, however, that it is at least partly marine. A tillite group consisting of up to 100 m of unlaminated sandstone and mudstone carrying a variety of angular and occasionally striated pebbles is intercalated in the Table Mountain Sandstone and has been compared with the Iapo tillite of Brazil (Fig. 6.1).

The *Bokkeveld Series* is dominantly argillaceous and frequently carbonaceous. Its lower beds contain Lower Devonian spirifers, trilobites and lamellibranchs resembling those of the Ponta Grossa shales of Brazil and the Falkland Islands (Table 6.2) and suggesting a connection with seas advancing from the southernmost part of the Andean trough (Fig. 6.1). The upper parts of the Bokkeveld Series carry only plant remains and the succeeding *Witteberg Series* of shales and sandstones appears to be entirely non-marine, containing eurypterid, crustacean and plant remains. There is no obvious discordance between the top of the Cape 'System' and the base of the Karroo in the Cape belt (see Chapter 8). The succession of the foreland is less complete. The lower parts of the Table Mountain Sandstone wedge out northwards, while the higher beds become coarser and more feldspathic. The Bokkeveld and Witteberg are both absent, and the Karroo rests unconformably on the Table Mountain Sandstone or its basement.

Orogenic disturbances in the Cape belt affect not only the Cape 'System' but also the lower members of the Karroo (p. 227) and appear to date from about the end of the Upper Palaeozoic. The main fold-system follows the curved trend of the Cape belt. Near the northern foreland, the intensity of folding decreases sharply. Towards the interior of the belt, large folds overturned towards the foreland are defined by the competent Table Mountain Sandstone and thrusts are locally developed. In the extreme south-west the effects of deformation once more diminish and the Table Mountain Sandstone is seen almost horizontal in the Cape Peninsula and at the superb type-locality behind Cape Town. Although a crude cleavage is locally developed where folding is strongest, the Cape belt is singularly devoid of metamorphic effects and of granitic intrusions.

3 The Mauretanides

The Mauretanides of western North Africa extend almost due north for 1500 km from the vicinity of Freetown to the Moroccan border. Much of the belt is hidden beneath coastal sediments of Mesozoic and Tertiary age and much of its exposed part has, until recently, been assigned to the Precambrian basement. It now appears, however (Sougy, 1962), that folded and metamorphosed supracrustal rocks in the belt are equivalent in part to the flat-lying Infracambrian and Palaeozoic strata of the North African cratonic cover. The terminal phase of folding is regarded as Hercynian and apparent ages scattered between 500–200 m.y. suggest that this was preceded by early Palaeozoic episodes.

The metasedimentary cover in the marginal parts of the southern Mauretanides ranges from Infracambrian strata (including tillites) through Lower Palaeozoics (interrupted by disconformities above the Ordovician) to Frasnian. It is a marine sequence including dolomites and is associated with greenstones. The orogenic front is marked by a tract of mylonite zones to the east of which lies the equivalent foreland succession already described (p. 167). In the interior of the fold-belt more highly altered supracrustals of unknown stratigraphical position are represented. Towards the north, in Spanish Sahara and Morocco, an essentially similar succession is seen, often including Infracambrian tillites and thinning eastward onto the foreland. Granites which are regarded as Hercynian invade the folded cover in Morocco.

The relationships of the Mauretanide belt are far from clear. Its western margin lies beneath the coastal sediments or beyond the edge of the African continent; a glance at Fig. 6.6 shows that prior to the break-up of the continents the Mauretanides may have flanked the Appalachian belt of North America and could, indeed, have formed the eastern side of this belt. To the north, the Mauretanides are disrupted by fractures connected with the Atlas and Anti-Atlas system of Alpine mobile belts: but Sougy suggests that they may originally have continued northward into the Meseta of Spain.

IV The Permo-Carboniferous Glaciation and the Question of Climatic Zones

Towards the end of the Upper Palaeozoic era, important changes in the pattern of deposition took place in Gondwanaland. The huge cratonic region entered on a remarkable phase of deposition of continental facies during which new basins of deposition were defined not only in the marginal platform-areas but also within what had been the mid-Palaeozoic shield. The deposits of this phase form a number of celebrated systems — the Karroo of Africa, the Gondwana of India, the Santa Caterina of Brazil and the Beacon Group of Antarctica — which span the Upper Palaeozoic–Mesozoic boundary, ranging in age from late Carboniferous or Permian to late Triassic or Jurassic. The timing of the Gondwanaland phase of continental deposition does not fit the conventional stratigraphical time-divisions, and, since it seems best to treat the whole of this phase in a single chapter, we postpone our discussion until Chapter 8. The *Permo-Carboniferous glaciation* which opened the phase, however, was an important event in its own right and we shall end with a brief account of this episode.

Tillites and associated glacigene deposits have been recorded from localities within every one of the Gondwanaland continents, but not from any part of Laurasia. As the continents are arranged at the present day, the indications of Permo-Carboniferous glaciation are extraordinarily widely distributed, extending into tropical latitudes on both sides of the present equator and lying mainly between 30°N and 50°S of the equator. This remarkable distribution cannot easily be reconciled with any likely arrangement of climatic zones and would require the extension of glacial conditions over far greater areas than those affected by the Pleistocene glaciation. Restoration of the continents to positions

Plate II. Glaciated pavement beneath Permian tillite, Pemganga river near Chandra, India. The locality now lies close to latitude 20°N. (Photograph Pamela L. Robinson)

in a single supercontinent reduces the glaciated areas to dimensions comparable with those of the Pleistocene ice-sheets. The directions of ice-flow indicated by glacial striae, by till-fabrics, by the alignment of glaciated valleys and *roches moutonnées* and by glacial folding and thrusting show many anomalies on a map of the present-day continents; in India, for example, the direction of flow is predominantly northward, away from the equator, in South America it is

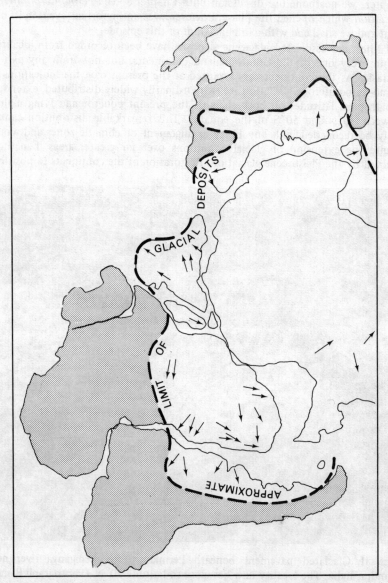

Fig. 6.5. Gondwanaland in Permian times, showing the approximate extent of traces of glaciation and the directions of ice-flow where recorded

westward, away from a source-region on the site of the Atlantic. Re-assembly of the continents removes these anomalies and reveals a picture of radial movement toward the margins of the glaciated area (Fig. 6.5).

The simplification of problems connected with the Permo-Carboniferous glaciation which follows when it is assumed that the continents affected were united in a single mass at the time of glaciation provided Wegener, du Toit and their successors with a persuasive argument in favour of the hypothesis of continental drift. This argument has been reinforced during recent years not only by the accumulation of far more detailed information concerning the distribution of ice-sheets and the directions of ice-flow but also by a line of evidence unknown to the original proponents. Measurements of the remanent magnetization of sedimentary rocks deposited at about the time of glaciation (Fig. 1.2) show that the regions affected were situated at high latitudes and occupied positions much closer to each other than they do at the present day.

Glacigene deposits often form the basal units of the Gondwanaland successions. Unlike those of the Infracambrian glaciation, they are predominantly terrestrial (Plate II) and may be seen resting unconformably on glaciated pavements or filling U-shaped valleys or other typical glacial erosion-forms. The deposits include tillites containing erratics of many sizes and kinds in an unsorted argillaceous or sandy matrix, as well as varved clays and sandstones regarded as fluvioglacial sediments and fine loess-like sandstones. These glacigene sediments sometimes reach extraordinary thicknesses. The Dwyka of the Great Karroo basin of South Africa, and the glacial sequences of southern Australia and Argentina are all commonly 800–900 m in thickness, while the Itarare Series of the Paraná basin reaches a maximum of 1600 m. Since the average thickness of Pleistocene glacial deposits in northern Europe is less than 100 m, these figures suggest that the Permo-Carboniferous glacial period was of very long duration. Palaeontological dating is hampered by the scarcity of marine zone fossils. In most regions, Lower Permian (Sakmarian) fossils have been reported in the upper parts of the glacial sequences. In the Paraná basin and in New South Wales, faunas regarded as Upper Carboniferous have been reported and, although the evidence is still inconclusive, it is probable that the period of glaciation as a whole straddled the Carboniferous–Permian junction. The Sakmarian stage of the Permian appears, from radiometric evidence, to have lasted some 10 m.y. and, consequently, it is not unreasonable to think that glaciation continued, or was repeated, over at least 5–10 m.y.

Alternations of tillites and fluvioglacials with non-glacial sandstones, shales or coal measures, suggesting the occurrence of a number of interglacial periods, are not uncommon. In a succession in Victoria, south-east Australia, for example, fifty-one tillite horizons have been counted. Elsewhere, for example in Brazil, thin seams and lenses of coal – often drifted accumulations – are interbedded with tillites, and in many localities more important coal measures follow the glacial groups. The coal measures which follow the Talchir Tillite of India, for example, include seams up to 30 m in thickness. These *Gondwanaland coal measures* are almost contemporaneous with the coal measures of Europe and North America dealt with in Chapters 3 and 4, but differ from them both in flora and in climatic setting. All authorities are agreed that they are closely connected with glacial deposits, whereas the coal measures of Laurasia are

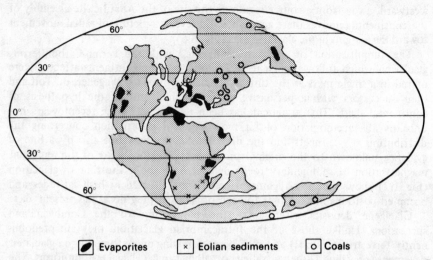

Fig. 6.6 Palaeoclimatic indicators for late Triassic times; the distribution of climate-sensitive sediments (evaporites, eolian sediments, coals) is plotted on a continental reconstruction according to a compilation by Pamela Robinson (1972)

regarded as accumulations of temperate or hot climates. The number of species in the southern coals is more limited and a few species, notably the Pteridosperms *Glossopteris* and *Gangamopteris*, are represented almost everywhere. Indications of growth rings in woody tissues suggest a seasonal climate consistent with the palaeomagnetic evidence of high latitudes and with the association with glacial strata. One is therefore confronted by the remarkable fact that towards the end of the Palaeozoic era, coal – a rare rock – accumulated on a vast scale in two very different environments over times which, geologically speaking, were not far separated. Some authors have seen significance in this association; the 'glacial control' hypothesis originally put forward by Wanless and Shephard in 1936 attributed the characteristic cyclic sedimentation of the coal measures to fluctuations of sea level caused by advance and retreat of ice-sheets.

The various lines of evidence mentioned – the distribution of glacial deposits and of coals, and the palaeomagnetic data – all have a bearing on the arrangement of Permo-Carboniferous climatic zones. They are, by good fortune, not the only lines of evidence which can be drawn on. Permian evaporites, the deposits of hot-arid zones, are developed in a number of regions in central Europe and southern North America. Certain fossil groups, whose distribution can be determined, had known or inferred climatic preferences. All these lines of evidence can be brought to bear on the problems of *palaeoclimatology*, a subject only occasionally touched on in earlier chapters. They depend to a great extent, for their validity, on the added precision of stratigraphical correlation which is supplied by the fossil record in Phanerozoic rocks. Some evidence of the climatic variations in late Triassic times is assembled for comparison in Fig. 6.6

7

The Alpine-Himalayan Belt and the Eurasian Craton

I Introduction

The Alpine-Himalayan belt which forms the zone of separation between continental masses derived from Laurasia and Gondwanaland has a history of mobility extending through the entire Phanerozoic eon. A glance at Fig. 7.1 shows that the belt began its evolution as a portion of the peripheral mobile system encircling the supercontinents. It was transformed to a largely intra-continental structure by the collision of northward-moving fragments of Gondwanaland – Africa, Arabia and Peninsular India – with the Eurasian continental mass.

The mobile belts. The *Alpine orogenic cycle* to which the mobile belts owe their present structure spanned the period from early Mesozoic times to the present day. The belts are, however, almost everywhere superimposed on tracts which had been mobile throughout Palaeozoic and often also late Precambrian times. These earlier complexes include portions consolidated at the close of the Precambrian (*Cadomian* or *Assyntian* cycles in Europe: *Baikalian* in central Asia), in mid-Palaeozoic (*Caledonian*) and in late Palaeozoic (*Hercynian*) cycles. Through most of the European belt and along the northern side of the system in Asia, rocks which had been folded and often metamorphosed and invaded by granites during one or more of these early cycles provide a basement to the Alpine cover-successions. In the Himalayas a less episodic history of sediment accumulation reaches back without fundamental breaks to early Palaeozoic times. These variations do not obscure the salient point: the evidence for crustal mobility in the Alpine tract continued over more than 600 m.y., that is, the greater part of the Grenville chelogenic cycle.

Fig. 7.1. Peripheral mobile belts of the supercontinents in early Mesozoic times

In this respect, a contrast may be drawn between the Alpine-Himalayan and Circum-Pacific mobile belts, which evolved at the peripheries of the super-continents, and the numerous Palaeozoic mobile belts which lay within the supercontinents (Chapters 2–5). Stabilisation of the interior belts at various times between 550 m.y. and 200 m.y. is seen by Sutton (1963, 1968) as an essential preliminary to the break-up of the supercontinents, which followed the development of vast interior cratonic land-areas. The peripheral belts at the leading edges of the migrating continental fragments retained their mobility and provided the necessary sites at which the crustal plates overridden by these fragments could be consumed. Although they have yet to be stabilised, the widespread development of terminal orogenic features such as regional moun-tain-building and molasse-formation, together with the transformation of the Alpine-Himalayan belt to a largely internal structure, suggests that its long period of mobility is nearing its end.

The northern craton. During the Alpine cycle proper, a single continental craton constituting the greater part of Europe and central and northern Asia flanked the northern part of the mobile belt. A number of 'swells' and fault-bounded massifs in the craton were separated by broad downwarps and more restricted troughs in which sediment accumulated during the cycle. Some of these *cratonic basins* contain Palaeozoic sediments and date back to very early stages in the history of the craton. Others were developed after the ending of the Hercynian cycle and contain Mesozoic and Tertiary sediments only. Among these younger structures, the *marginal basins* close to the Atlantic border are of special significance (p. 271).

The graben-systems of the northern craton. Rift-faulting and associated igneous activity took place in two major zones traversing the Eurasian craton during the Alpine cycle. The Baikal rift following a north-easterly course in central Asia is located within stabilised late Precambrian (Baikalian) and Palaeozoic complexes and was probably initiated very early in the Alpine cycle. The Rhine-graben which extends from the stabilised Hercynian belt of central Europe to the Baltic shield and North Sea is thought to be of Mesozoic and Tertiary age.

The southern cratons and oceans. The southern side of the Alpine-Himalayan belt is flanked by diverse geological units. Along a central tract the substantial cratons of peninsular India and Arabia border the mobile belt. To the east, the Indian Ocean flanks the belt where it passes laterally into the complex Indonesian island arc. To the west, the Mediterranean Sea intervenes between the mobile belt and the continental craton of Africa. The Mediterranean region, as will be seen, was distinguished by the occurrence of a number of small crustal units behaving more or less independently.

II The Mobile Belt in Europe and North Africa

1 Components of the belt

The numerous mountain tracts which flank the Mediterranean will be dealt with in two groups. The *Alpine system proper* (Fig. 7.2) forms a number of linked arcs looping about a discontinuous median tract. The *western and eastern Alps*, the *Carpathians* and the *Balkanides* border the European craton and their principal structures show a northward vergence towards this craton. The *Dinarides* and *Hellenides* lie to the south-west of the median tract and their principal structures show a south-westerly vergence (Fig. 7.2). The median tract, the *Pannonian block*, is floored by Hercynian or older rocks which are blanketed by a little disturbed cover in the Hungarian plain, but are exposed in the *Rhodope massif* of Thrace and the northern Aegean. Although this block is of continental type, it appears from seismic surveys that the crustal layer is only about 25 km, unusually thin for a unit of continental crust.

The *accessory belts* of the western Mediterranean form a number of short and sometimes strongly curved tracts which appear originally to have been portions of a more continuous system. They are separated by an assortment of small ocean basins and small continental blocks.

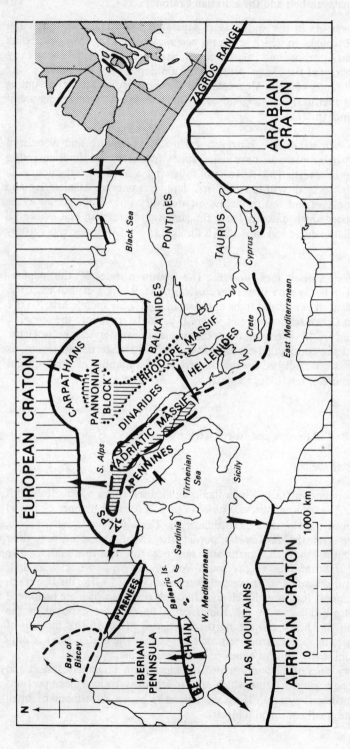

Fig. 7.2. Components of the Alpine-Himalayan belt in Europe: the arrows point from internal to external zones and indicate the orogenic polarity of each unit. The inset shows a possible reassembly of continental fragments displaced during the Tertiary era (based on Smith, 1971)

It seems certain that the Carpathians and the Alps represent the peripheral mobile belt of the European craton and it is at least probable that the Dinarides and Hellenides bore a similar relationship to the African craton. The Tethys ocean which formerly separated the converging portions of Laurasia and Gondwanaland has now almost disappeared, though a remnant may remain in the eastern Mediterranean.

The Alpine orogenic cycle

The Alpine cycle proper began in the Triassic period with the definition of geosynclinal furrows within or alongside the stabilised Hercynian tracts. After an early period of deposition, accompanied in some troughs by ophiolitic igneous activity, episodes of orogenic compression began in late Mesozoic times and continued through the early Tertiary, culminating in widespread mid-Tertiary mountain-building phases. Late-orogenic molasse-formation began in Miocene times and repeated uplift has continued, together with sporadic vulcanicity, to the present day. The earlier history of the Alpine tract has been touched on in previous chapters and will be mentioned only briefly: attention will be concentrated on the 200 m.y. span of the Alpine cycle proper (Table 7.2).

III The Alpine Orogenic System Proper

I General characters

Some elements of a common pattern can be recognised in the history of all the components of the Alpine mobile belt system dealt with under this heading. These common features, which are exemplified by the history of the Alpine chain itself (summarised in Table 7.2) are as follows:

(1) In each belt, a number of tectonic zones, parallel to the belt as a whole, are distinguished by the nature of the rock-assemblages and the tectonic style.

(2) A general distinction can be drawn between the *internal zones* character- ised by profound deformation involving the basement and usually by regional metamorphism: and the *external zones* in which metamorphism was lacking and deformation was generally of a more superficial type.

(3) One or more of the internal zones is occupied by Mesozoic assemblages including pelagic sediments and ophiolitic igneous suites. These zones are derived from basins of eugeosynclinal type.

(4) The pre-orogenic cover-successions (Triassic to early Cretaceous) both in the mobile belts and on the adjacent forelands consist largely of chemical sediments.

(5) The synorogenic successions of late Cretaceous and Palaeogene age are rich in detrital sediments and often dominated by turbidites. These successions constitute the *flysch.*

(6) Nappe-structures represented either by recumbent folds or by piles of thrust-sheets were developed in many parts of the internal zones, mainly during late Cretaceous and Palaeogene orogenic disturbances.

Table 7.1. THE MESOZOIC, TERTIARY AND QUARTERNARY STRATIGRAPHICAL COLUMNS

QUATERNARY		HOLOCENE OR RECENT		
		PLEISTOCENE		
TERTIARY or CENOZOIC	NEOGENE	PLIOCENE		
		MIOCENE	Upper	
			Middle	
			Lower	
	PALAEOGENE	OLIGOCENE	Upper	
			Middle	
			Lower	
		EOCENE	Upper	
			Middle	
			Lower	
		PALAEOCENE	Upper	
			Lower	
MESOZOIC	CRETACEOUS	UPPER	Danian	
			Maestrichtian	
			Campanian	
			Santonian	
			Coniacian	
			Turonian	
			Cenomanian	
		LOWER	Albian	
			Aptian	
			Neocomian	
	JURASSIC	UPPER	Purbeckian	
			Portlandian	
			Kimeridgian	
			Oxfordian	
			Callovian	
		MIDDLE	Bathonian	
			Bajocian	
		LOWER	Toarcian	
			Pliensbachian	
			Sinemurian	
			Hettangian	
	TRIASSIC	UPPER	Rhaetian	
			Norian	Keuper
			Karnian	
		MIDDLE	Ladinian	Muschelkalk
			Anisian	
		LOWER	Scythian	Bunter

Table 7.2. SEDIMENTATION AND MOVEMENT IN THE ALPINE GEOSYNCLINE

m.y.	System	Setting		Deposits
	[7]MIOCENE	elevation and erosion of high mountains	4	MOLASSE: impure sandstones, shales, coals, rare fresh-water limestones, thick conglomerates: piedmont deposits
30				
	EARLY OLIGOCENE ↑ LATE CRETACEOUS	compression and formation of new lands within mobile belt	3	FLYSCH: well-bedded micaceous sandstones and shales with slide-rocks containing fragments of all pre-Eocene rocks: unstable marine environment
100				
	EARLY CRETACEOUS ↑ JURASSIC	vertical and tensional movements allowing formation of subsiding trough bordered by more stable platforms	2	In eugeosynclinal troughs, BÜNDNERSCHIEFER or SCHISTES LUSTRÉS, including badly sorted calcareous, clayey and sandy sediments, deep-water starved successions, marine breccias along active fault-scarps: on platforms, well sorted limestones, sandstones and shales
195				
	TRIASSIC	widespread transgression, with sudden changes of depth inherited from Hercynian	1	Limestones, dolomites, sandstones and shales, widely deposited in geosynclinal region, facies varying with depth of sea
225				

(7) Regional metamorphism was largely confined to the internal zones and is usually of low-temperature, high-pressure type.

(8) Late-orogenic volcanic suites represented by Neogene and post-Tertiary assemblages of andesites, dacites and rhyolites are important in the eastern belts, less so in the west.

(9) Late-orogenic uplift is recorded by the restriction of marine Neogene and post-Tertiary sediments and the accumulation of thick non-marine detrital sediments which include the *Molasse*.

2 Pre-Mesozoic evolution

In most parts of the Alpine system proper, the Mesozoic cover rests with conspicuous unconformity on the eroded remains of a Hercynian orogenic belt. To the south-west of the Dinaride-Hellenide branch, however, the cover passes conformably down into a marine facies of the Permian, defining a region in which deposition continued without major interruptions through the later phases of the Hercynian cycle.

In the eastern part of the Carpathians and the Balkanides, a Precambrian basement to the north of the Alpine orogenic front has yielded radiometric dates of over 1700 m.y. In this region alone the Alpine front has trespassed into the pre-Hercynian craton of eastern and northern Europe. Elsewhere, the front lies well within the Hercynian belt and Alpine structures are theiefore overprinted on Hercynian patterns. The effects of the Hercynian cycle recorded in the basement accord very closely with those in the adjacent forelands. In the west Carpathians, for example, the units distinguished in the basement – an ancient highly-metamorphosed complex, a weakly-metamophosed assemblage affected by Assyntian disturbances, a suite of Palaeozoic greywackes and volcanics and a number of Hercynian granites – record a history which can be matched point by point with that of the Bohemian massif (p. 61).

Highly metamorphosed Precambrian complexes comparable with the Moldanubian of the Bohemian massif have been provisionally recognised in several other regions, though confirmation by radiometric dating is still required. Supracrustal assemblages assigned to the late Precambrian and showing effects of Assyntian (Cadomian) recrystallisation exist in several regions in the western Alps, west Carpathians and north-east Hellenides. Radiometric dates ranging up to 800 m.y. have been obtained from the basement in the Aiguilles Rouges. These dates provide the older end of a considerable age-range obtained from metamorphic and granitic rocks in the basement, which suggests that plutonic episodes recurred through much of the Palaeozoic. Material from the basement in the Austroalpine nappes, for example, has yielded apparent ages of up to 370 m.y. Fossiliferous *marine Palaeozoic sediments with volcanics*, ranging from Silurian to early Carboniferous, can be distinguished locally from the older units, though they are themselves folded and weakly metamorphosed. *Hercynian granitic intrusions*, occasionally associated in the western Alps with the characteristic Hercynian tin-tungsten mineralisation, are widely distributed. Finally, non-marine conglomerates, sandstones and coal-measures with associated acid, intermediate and basic volcanics provide a *post-orogenic assemblage* almost identical with that of the internal Hercynian zones on the foreland. The coal measures are principally Stephanian in age, the Permian includes the typical mottled *Verrucano* and both occupy intermontane basins, one of which is conspicuous in the Briançonnais zone of the western Alps.

If, as we have inferred, the Alpine system of mobile belts conceals the suture along which cratons derived from Gondwanaland were welded to the Eurasian continent, a contrast in the nature of the basement occurring on either side of this suture might be looked for. There is little evidence for the existence of any such contrast; the basement complexes which lie to the south of the Alpine-Carpathian tract appear to record a history of late Precambrian and Palaeozoic mobility terminated by late Palaeozoic granite-emplacement and Permian vulcanicity. Since the supercontinent of Gondwanaland was, like Laurasia, encircled by a peripheral mobile belt through most of the Phanerozoic eon, the general similarity of basement complexes across the width of the Alpine system need have little bearing on the problem of collision of continents.

3 Early stages of the Alpine cycle

The Alpine cycle proper may be said to have begun with the advance of Triassic seas over the eroded Hercynian complexes and the gradual definition of a number of geosynclinal troughs. The Triassic transgression defined two facies domains; those of the predominantly marine Alpine Trias and the predominantly continental Germanic Trias. The *Alpine Trias* was restricted to parts of the Alpine mobile belt, principally in the southern and eastern Alps where basal red-beds and quartzites are followed by limestones and dolomites reaching thicknesses of 2–3 km. The *Germanic Trias* laid down in the external zones of the Alps and Carpathians, in parts of the Adriatic region, and around the whole western border of the Mediterranean, was broadly similar to that of the northern foreland (pp. 95–6) consisting of non-marine detrital sediments and evaporites with intercalations of marine limestone. The widely distributed *Triassic evaporites* near the base of the cover-succession were destined to play a very important role in Alpine tectonics, since they provided surfaces at which detachment of cover from basement was easily achieved.

By the end of the Triassic period, marine conditions had been established through almost the whole of the Alpine belt as well as on many parts of the adjacent cratons. Deposition throughout these regions until mid-Cretaceous times was predominantly chemical and organic, the principal sediments being limestones, dolomites, marls, pelites and cherts. A differentiation of troughs and ridges in the mobile belts gave rise to a longitudinal zoning which is perpetuated in modified form in the tectonic zones of the orogenic belts. The relationship between supracrustal facies belts and tectonic zones in the western Alps is outlined in Table 7.2 (see also Fig. 7.3) and similar information concerning the Carpathians and Hellenides is summarised in Tables 7.3 and 7.4 and Fig. 7.4.

The key units in the early Mesozoic mobile belts were certain troughs on the site of the future internal tectonic zones which displayed *eugeosynclinal* characteristics. In the western Alps, the *Pennine zone* is of this type. The Triassic-Cretaceous *schistes lustrés* consist of rather deep-water calcareous pelites, pelagic limestones and cherts and are associated with basic and ultrabasic igneous rocks of the *ophiolite suite* (pillow-lavas, serpentines and other

Table 7.3. THE WEST CARPATHIANS: TECTONIC ZONES
(based on Mahel and others)

(North)

Foreland: basement complex resembling Bohemian massif, overlain by molasse

Flysch Carpathians: zone of late Cretaceous and early Tertiary flysch, folded and overthrust onto foreland during middle to late Tertiary episodes

Klippen zone: complex narrow zone along major dislocation (*Pieninian lineament*): tectonic inclusions of late Mesozoic rocks in marls and shales affected by mid-Cretaceous and mid-Tertiary disturbances

Inner Carpathians: geosynclinal tract with varied Mesozoic cover of limestones, dolomites, cherts, minor ophiolites: thrust-nappes and folds incorporating basement formed in Cretaceous episodes.

Pannonian or Hungarian massif: stable massif ('*zwischengebirge*'), Precambrian and Palaeozoic basement with cover of Mesozoic limestones and Tertiary molasse

(South)

Table 7.4. THE HELLENIDES: TECTONIC ZONES
(based on Aubouin, Mercier and others)

(South-west)	*Apulian zone:* stable basement with Mesozoic-early Tertiary cover of limestone microbreccias

EXTERNAL ZONES

Pre-Apulian (or Paxos) zone: marginal zone of Hellenide belt

Ionian zone: zone of miogeosynclinal character. Triassic evaporites, Trias-Eocene limestones including pelagic limestones followed by Oligocene and Miocene flysch: folded mid-Miocene without metamorphism

Gavrovo-Tripolitsa zone: geanticlinal zone with gently-folded cover of shallow-water Mesozoic-Eocene limestones overlain by early Oligocene flysch

Pindus zone: eugeosynclinal zone with cover of Mesozoic limestones, cherts, ophiolites followed by Cretaceous-Eocene flysch. Folds, thrust-nappes, low-grade metamorphism largely of Oligocene age

Sub-Pelagonian zone: margin of Pindus trough characterised by abundant ophiolites

INTERNAL ZONES

Pelagonian zone: internal zone, metamorphism and granites late Jurassic and early Tertiary

Vardar zone: internal zone, metamorphism and granites late Jurassic and early Tertiary

Almopias sub-zone: eugeosynclinal, deep-water Mesozoic cover, ophiolites, late Cretaceous flysch

Paikon sub-zone: geanticlinal

Peonian sub-zone

(North-east)	*Rhodope massif:* stable massif between Hellenide and Balkan belts

intrusives). The cover is not thick, indeed at some levels it forms a condensed sequence consisting mainly of cherts, and is almost devoid of coarse detritus except in the vicinity of the interior ridge which formed the Briançonnais zone. In the *Pindus zone* and *Almopias sub-zone* of the Hellenides an association of deep-water limestones, cherts and jaspers follows the basal Triassic detrital and evaporitic sediments. Pillow-lavas, serpentines and other basic and ultrabasic intrusive complexes are concentrated in two tracts at either side of the basement ridge (the *Pelagonian zone*) which separates the geosynclinal furrows. The voluminous ophiolites of the Dinaride-Hellenide belt and western Alps, with the rather restricted ophiolites of the Carpathians, are considered to be of Jurassic age, and to be related to an extensional phase preceding orogenic compression.

Carbonate sediments predominate in the Triassic to Cretaceous cover of the external zones and on the adjacent forelands above the level of the initial Triassic detrital and evaporitic formations (Table 7.5). The bulk of these sediments are of shallow-water types and are more richly fossiliferous than those of the

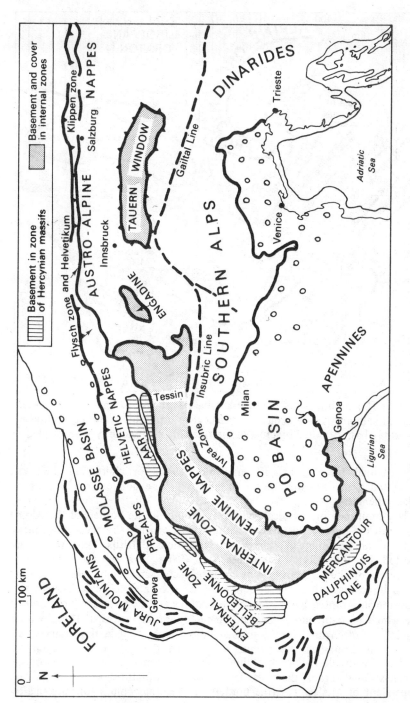

Fig. 7.3. The major tectonic units of the western, central and eastern Alps

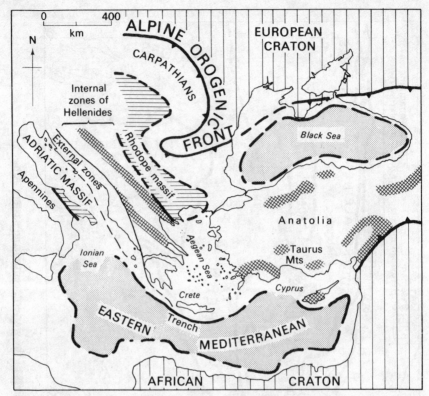

Fig. 7.4. The mobile belts and marine basins of the eastern Mediterranean: zones of abundant Mesozoic or Palaeogene ophiolites are cross-hatched; present-day basins floored by crust of intermediate or oceanic type are shaded

eugeosynclinal troughs. Pelagic limestones, shales or cherts are, however, represented locally towards the top of the cover, for example in the Ionian zone of the Hellenides.

The dominance of chemical sediments in and on the borders of the Alpine belts, with the modest thicknesses of the Triassic to early Cretaceous sequences, points to an environment starved of terrigenous detritus. Deep waters are indicated by the pelagic limestones and cherts, and a rugged submarine topography by the juxtaposition of facies belts characterised by these sediments with belts characterised by evidence of shallow-water, or even non-marine, deposition interrupted by erosion. Such features are shown in the *Briançonnais* zone which partially sub-divides the Pennine zone in the Alps. The old term 'geanticline', often used for the high-standing ridges such as the Briançonnais zone which diversified the mobile belt during the geosynclinal stage, was associated with the view that they were embryonic nappes. Most Alpine geologists now regard them as fault-bounded structures earlier than, and independent of, the nappes (the borders of the Briançonnais zone are slightly oblique to the Pennine nappes).

Table 7.5. COVER-SUCCESSION OF THE IONIAN (MIOGEOSYNCLINAL) ZONE IN ALBANIA

		metres
U. EOCENE-OLIGOCENE	Flysch	2000
CAMPANIAN-M. EOCENE	Micritic limestones with Trümmerkalke	300–700
BARREMIAN	Micritic limestones with calcarenite beds	20
	Limestones and marls, with Trümmerkalke	280
HAUTERIVIAN	Micritic limestones, pelagic faunas, with beds of calcareous turbidites (*Trümmerkalke*)	550
	Thin 'ammonitico rosso' followed by pelagic calcareous sediments	1000
LOWER-MIDDLE LIAS	Marly limestones, neritic facies Thick-bedded dolomitic limestones	400
	Limestones, bituminous shales	14
U. TRIASSIC	Dolomitic limestones ? Evaporites	100

The narrow deep basins of eugeosynclinal type such as those whose contents form the Pennine and Pindus zones invite comparison with certain modern structures. Some of their features recall the narrow extensional basins such as the Red Sea formed by separation of crustal blocks (compare Trümpy, 1960). Others recall the oceanic trenches developed at the junction of converging plates which are commonly backed by shallower marine basins providing a possible analogy with the miogeosynclinal basins represented in the external zones. Both alternatives imply that the eugeosynclinal troughs may have been floored by oceanic crust (see p. 201).

4 The flysch phase

Synorogenic marine formations in which terrigenous detritus is mixed with, or entirely overshadows, the chemical sediments of the early stages are represented in every component belt, though they seldom reach thicknesses of over 3 km. The great bulk of this flysch is Upper Cretaceous or Palaeogene. Its appearance, signalling the elevation of sourcelands in the interior zones of the mobile belts, coincided with the onset of orogenic compression.

The distinctive lithostratigraphical flysch-units are conspicuously diachronous, as is illustrated in Fig. 7.5 based on Wunderlich (1967). The earliest (late Cretaceous) deposits appear in the internal zones where the earliest orogenic disturbances were recorded; here, they usually rest conformably on the pelagic sediments of the eugeosynclinal tracts. In the external zones, the arrival of flysch was delayed until Palaeocene or Eocene times; towards the border-regions, flysch

the Alpine-Himalayan belt also the Uralian chain.

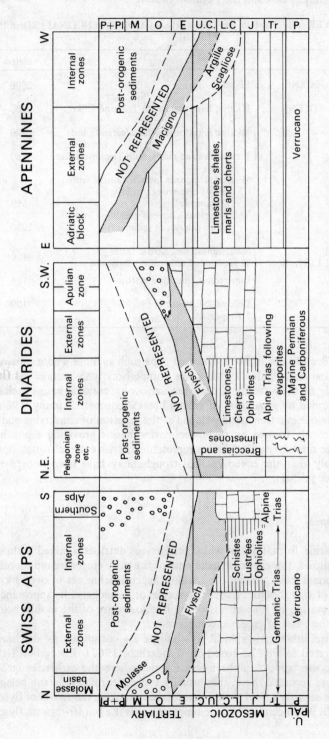

Fig. 7.5. The distribution in time and space of syn-orogenic sediments of flysch facies (shaded) in parts of the Alpine system of mobile belts (after Wunderlich, 1967)

is locally unconformable on the underlying successions and may extend up into the Oligocene.

The main sources of the flysch lay in the mobile belts themselves. The progression of flysch deposition from internal to external regions suggests that the bulk of the detritus was supplied from tectonically-elevated ridges in the internal zones. The general character of the sediments — the poor sorting and wide range of grain-size and the common (though by no means universal) occurrence of graded bedding, of sole-markings and of intercalated slide-breccias such as the *Wildflysch* — suggest that much of the flysch consists of turbidites redeposited from temporary resting places at the margins of unstable basins. The evidence of palaeocurrent markers, and that supplied by lateral and vertical lithological variations, indicate a complex pattern of deposition. Short-lived ridges within and on the northern side of the basins, in addition to internal sourcelands, have been shown to have contributed detritus to the Flysch of the Polish Carpathians (Ksiazkiewicz, 1963) and in both Carpathian and Hellenide basins redepositions was effected largely by turbidity currents flowing axially.

5 Tectonic evolution

Apart from some early (end-Jurassic) episodes recorded in the internal Hellenides (Mercier, 1966) the first major orogenic disturbances of the Alpine system proper took place in middle and late Cretaceous times. These disturbances were concentrated in the internal zones where extensive marine deposition was brought to an end in the Tertiary era. By Miocene times, the major structural patterns had been built up, regional elevation had led to the expulsion of the sea from the greater part of the mobile tract and late-orogenic (molasse-type) deposition had begun. Subsequent (Neogene, Pleistocene and Recent) vertical movements rejuvenated the mountain belts with consequent intensification of erosional and depositional activity. Nappe-formation and subsequent penetrative deformation were completed within a span of some 70 m.y. (Upper Cretaceous-Oligocene) and regional metamorphism reached its climax during or soon after the later part of this span.

6 The Alps

The demonstration that strata had been tectonically reduplicated in sheets piled one above the other in the Alps (Plate III) had a profound effect on geological thought. The scale of this reduplication became apparent in the 1880s at about the time when large-scale overthrusting on the Moine thrust-zone in Scotland was confirmed. Ideas concerning the importance of tangential compression in the formation of fold-belts owed much to these discoveries and within a few decades the classic *nappe-theory* of Alpine structure had been elaborated by Lugeon, Bertrand, Heim, Argand and their successors.

The essential basis of the nappe-interpretation, the existence of rock-piles in which the same succession is repeated several times in a vertical sequence of tectonically-accumulated sheets, is accepted by almost all modern workers. In

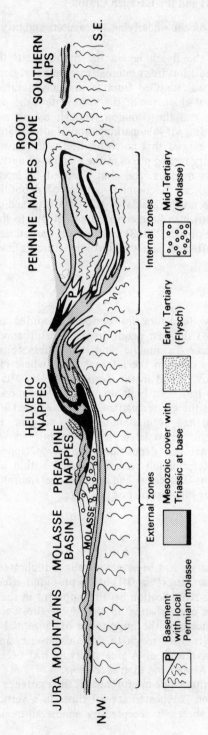

Fig. 7.6. Simplified cross-section of the western Alps (based on an unpublished compilation by J. G. Ramsay)

Plate III The Matterhorn: crystalline rocks in the core of a high Pennine nappe rest on rocks of its Mesozoic envelope (Copyright Aerofilms Ltd)

the rugged Alpine terrain the superposition of nappes is directly demonstrated by the occurrence of erosional remnants (*klippen*) of high nappes on mountain-tops (Plate III) and of *windows (fenster)* of low nappes in valleys and low ground. Not unnaturally, some aspects of the classic hypothesis have been modified or discarded; the view that most nappes originated as recumbent folds has given place to a concept of many nappes as simple 'right-way-up' sheets resting on dislocations; perhaps more important, the idea of a small number of nappes maintaining their identity for tens or hundreds of kilometres in the axial direction has been partly replaced by one involving larger numbers of nappe-units with lenticular cross-sections: and the emphasis laid on tangential shortening as the fundamental agent in nappe-formation has been questioned.

Early estimates of *horizontal shortening* ranged up to several hundred kilometres. Such figures, obtained by 'unfolding' recumbent structures and restoring the nappes to initial positions side by side, made little allowance for the effects of internal distortion, although evidence of distortion is provided, for example, by the stretching of belemnites in the *schistes lustrés*. A much greater source of uncertainty came from the fact that many nappes are no longer continuous entities. The frontal portions are commonly detached from their *roots* and a good deal of evidence suggests that certain rootless nappes accomplished part of their forward travel as isolated 'packets': the distance from root to tip in such nappes does not measure the shortening necessary for their formation. A minimum north–south shortening of 100 km is however, suggested by the distribution of the large, continuous and little distorted upper Austroalpine nappe of the Eastern Alps (compare Ramsay, 1969). Modern workers have tended to stress the importance of vertical crustal movements in the development of Alpine nappe-piles. The upheaval of one or more welts in the mobile belt is thought to have caused unstable units to slide under gravity, accumulating as cascades of recumbent folds, or detached sheets, near the foot of the tectonic slope.

The next few pages provide a brief commentary on the structure of the western and eastern Alps, beginning with the Pennine zone. It will be recalled (p. 181, and Fig. 7.6) that the *vergence* of almost all major Alpine structures is towards the north, and from this arrangement it follows that in each nappe-pile higher nappes usually root further south than lower nappes.

Table 7.6. NAPPE-SEQUENCES IN THE PENNINE ZONE OF THE SWISS ALPS

West (highest)		East		
Dent Blanche and associated nappes		Sesia zone Sella nappe Margna nappe	}	Lower East Alpine and transitional
Grand St Bernard Monte Rosa Gran Paradiso	} Mischabel nappe	Suretta nappe Tambe nappe Adula nappe	}	Upper Pennine
Monte Leone nappe Lebendun nappe Antigorio nappe	} Simplon	Ticino { Simano nappe Levantina (Lucomagna) nappe	}	Lower Pennine

The Pennine zone. The internal zone of the French and Swiss Alps is occupied by nappes in which cores of reworked basement are partially enveloped by a Mesozoic cover of eugeosynclinal character (p. 187). Towards the frontal (north-western) side of the zone, nappes consisting mainly of Cretaceous flysch are caught up in the pile. On a regional scale, the nappe-boundaries are flattish over most of the zone but turn downward near its southern border to define a narrow tract of steep, often northerly, dips regarded as the *root-zone*. On a smaller scale, complications are introduced by the effects of refolding, which was associated with the development of one or more sets of minor folds and related schistosities and lineations.

The lowest of the Pennine nappes exposed in the *Tessin culmination* of southern Switzerland are characterised by metamorphism of moderate grade and by fairly thorough regeneration of their basement cores. The higher nappes (Table 7.6) show only low grades of metamorphism and their basement cores are subject to retrogression and cataclasis. In most localities the junction of the basement with the basal (Triassic) units of the cover appears to represent a modified unconformity. In the Grand St Bernard nappe, however, the basement is in contact with *schistes lustrés* assigned to a more southerly source, its original cover (derived from the Briançonnais zone) being stacked independantly in the Prealpine nappes (see later). This arrangement is referred to as *substitution de couverture* by Ellenberger (1953).

The external zones. The Pennine zone is flanked on the north by the zone of Hercynian massifs in which the basement forms broad wedges and upfolds. A parautochthonous Mesozoic cover is pinched between these structures in tight folds or screens. The basement rocks are traversed by steep shear-zones and partial Alpine reworking is shown by overprinted metamorphic fabrics and by 'mixed' radiometric ages.

The zone in advance of the Hercynian massifs is almost wholly occupied by cover-rocks which appear to have been partly or completely detached from their basement. In the Dauphinois zone of the French Alps and the Helvetikum of the Eastern Alps, the Mesozoic limestones show broken disharmonic folds and minor thrust-slices, but are parautochthonous, having been displaced at a detachment surface in Triassic evaporites. In the Helvetic zone of the Swiss Alps on the other hand, Mesozoic limestones and flysch-units (sometimes with Permian strata adhering at the base) form wholly detached sheets stacked on flat dislocations. These Helvetic nappes, familiar to almost every geological visitor, are thought to represent the cover shed from the zone of Hercynian massifs which rose behind them. Consideration of the distribution of various Mesozoic facies has led Swiss geologists to place their source in what is now a narrow, highly-compressed cicatrice between the Aar and Gotthard massifs.

The Helvetic nappes are overlapped northwards by an extraordinary miscellany of nappe-fragments derived from sites further south — the *Ultrahelvetic nappes* from the Gotthard massif and the overlying *Prealpine nappes* from the tips of Pennine and Lower Austro-alpine nappes. Gravitational sliding may account for the emplacement of these rootless nappes. Finally, in advance of the whole Alpine structure and separated from it by the Neogene Molasse basin, lies the arcuate fold system of the Jura Mountains, formed by the disharmonic rucking of Mesozoic limestones above Triassic evaporite horizons.

In the Eastern Alps the external zones (consisting of the Mesozoic *Helvetikum* smothered by a thrust-unit of calcareous flysch, which forms the *flysch zone*) is separated from southern zones by a lineament which continues eastward through the Carpathians. This is the so-called *klippen zone* in which tectonic inclusions of Jurassic and early Cretaceous limestones are enveloped in a matrix of Upper Cretaceous-Eocene marls known as Buntmergel.

The Austroalpine nappes. In the Eastern Alps, thrust-nappes consisting of crystalline rocks and low-grade Palaeozoic supracrustals forming a basement to unmetamorphosed Mesozoic limestones and dolomites occupy a broad tract south of the klippen zone. In two critical regions — the *windows of Engadine and Hohe Tauern* — these Austroalpine nappes are seen to rest on an assemblage of reworked basement and metamorphosed cover resembling *schistes lustrés*; and in the west, the Pennine nappe-pile appears to plunge beneath the Austroalpine nappes. On this evidence, the Austroalpine nappes are assigned to a tectonic level above the Pennine nappe-pile and their roots are placed in a tract of steep gneisses to the south of the Pennine root-zone. These nappes are enormous: the Upper Austroalpine nappe measures 100 km from root to tip and extends for not less than 300 km in the axial direction. Some authors consider, furthermore, that Austroalpine nappes originally covered much of the Pennine zone (the erosional remnant of the Dent Blanche nappe (Table 7.6) and the highest Prealpine nappe both have East Alpine affinities). The vertical thickness of each nappe is no more than a kilometre or so; there is little or no Alpine metamorphism and internal distortion is slight, outside some spectacular mylonite-zones in the basement.

The supposed roots for the Austroalpine nappes on the south side of the Pennine root-zone consist of steeply dipping gneiss strips separated by metasedimentary screens. To the west of the Tessin, geophysical surveys have revealed the existence, at depths of only a few kilometres, of a rock-mass of high density and with high seismic velocities. This *Ivrea body*, possibly represented at the surface by basic and ultrabasic rocks of high metamorphic grade in the *Ivrea zone*, is underlain at depths of 20—30 km by material with density and seismic velocities appropriate to upper crustal rocks. The possibility exists that the Ivrea body is a flake of the lower crust or mantle driven up in the root of an Austroalpine nappe.

The Insubric line and the Southern Alps. The roots of the Pennine and Austroalpine nappes are bounded to the south by another major lineament marked by steep dislocations and belts of phyllonite or mylonite (the *Insubric* or *Tonale* line in the western Alps, the *Puster* or *Gail* line in the eastern Alps). This lineament forms an important divide: to the north lie rocks intensely deformed and metamorphosed during the Alpine cycle, to the south a Hercynian basement underlies a non-metamorphic and little disturbed Mesozoic cover. The 'knife-sharp jump .... between pre-Alpine and Alpine influences', as Gansser put it, shows that large displacements took place on the Insubric line. The nature of these displacements remains to be established: the undoubted northerly uplift may in some writers' view have been accompanied by large transcurrent displacements. Most of the small late-orogenic Alpine granites lie near the Insubric line. The Southern Alps present a relatively simple structural picture dominated by the effects of block-movements. In the Dolomites an essentially authochthonous cover made largely of Triassic dolomitic limestone of reef facies descends southward in step-like folds complicated by valley-bulges and other superficial structures. In the Bergomask Alps of Lombardy, the limestone cover is stacked in gliding-nappes facing towards the south.

Alpine metamorphism and crustal structure. Regional metamorphism in the western Alps is almost confined to the Pennine zone, extending only a short distance into the external zones north of the Tessin culmination. Both the mineral assemblages and the habitual fine to medium grain-size suggest that high temperatures were seldom reached. Only in the root zone and in the lower nappes are coarse textures and amphibolite facies assemblages developed, and it is only in these regions that pre-Alpine structures have been obliterated from the basement nappe-cores. Alpine migmatites are very restricted and Alpine granitic intrusions are few and small.

Total pressures, on the other hand, appear to have been unusually high, as is indicated by the occurrence of a high-pressure facies series. Eclogitic assemblages are developed in certain ophiolitic basic bodies, notably in the vicinity of Zermatt. A system of arcuate metamorphic zones centred about a higher-grade 'node' in the Tessin culmination is defined by the occurrence of stilpnomelane (developed in many parts of the external Hercynian massifs), lawsonite, and glaucophane and/or chloritoid. Towards the node of Tessin, zones are defined by garnet, staurolite or kyanite and finally sillimanite in pelitic rocks and by progressive increases in the anorthite percentage of plagioclase in equilibrium with calcite in calcareous rocks (Wenk, 1962).

The isograds marking the higher-grade zones outline a dome rising discordantly through the nappe-pile in the Tessin region (Fig. 7.7). From the fact that some characteristic minerals of the inner zones replace glaucophane and lawsonite, metamorphism in the Tessin node is ascribed to a comparatively late rise in temperature leading to a local steepening of the geothermal gradient.

The discordant relationships of the isograds with respect to tectonic boundaries and the fact that tectonite fabrics, where developed, generally relate to fold-sets overprinted on the nappes, show that metamorphism outlasted or followed nappe-emplacement. Apparent ages as high as 80 m.y. obtained from rocks from the Tauern window in the Eastern Alps suggest that metamorphism began before the end of the Mesozoic era. A minimum date for the ending of metamorphism is provided by Rb–Sr ages of up to 28 m.y. (roughly end-Oligocene) for muscovites from post-metamorphic pegmatites in the western Alps. Regional variations in the apparent Rb–Sr ages of micas are ascribed by Jäger *et al.* (1967) to variations in the history of uplift and consequent cooling. The youngest apparent ages (12 m.y. for biotites, about 20 m.y. for muscovites) were obtained from an area around the Simplon pass where unroofing was longest delayed; the highest figures for heat-flow in the Alps at the present day have been obtained from this same area.

The high-pressure facies series represented in the metamorphic domain of the Alps is characteristic of many of the regions affected by Alpine metamorphism. In the Alps themselves, the timing of metamorphism as well as the very modest thicknesses attained by the cover-successions seem to rule out the possibility that high pressures resulted from deep burial in a subsiding basin. Tectonic thickening due to the piling of Austroalpine on Pennine nappes to give an overburden of 10 km or more has been appealed to by some authorities but, again, the thicknesses represented by the existing units are not excessive. Burial beneath an overriding crustal plate and subsequent retrieval is rendered unlikely

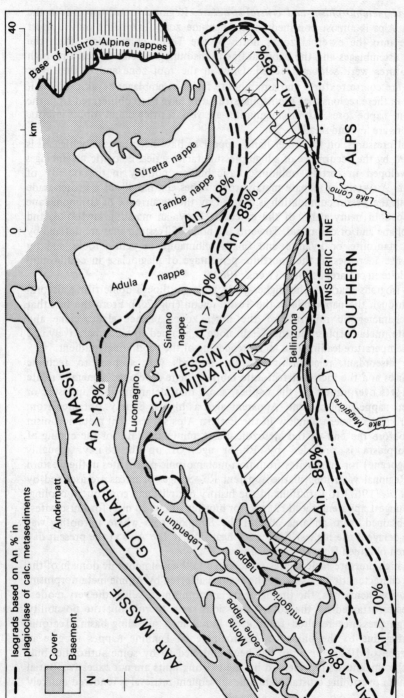

Fig. 7.7. Regional metamorphism in relation to structure in the western Alps; the isograds of higher-grade metamorphism defined by plagioclase composition transgress the Pennine nappes, whose structure is indicated by the arrangements of their basement cores (based on Jäger, Niggli and Wenk, 1962)

by the evidence that the internal Alpine zones stood high enough to provide erosion-debris during the flysch phase and to shed nappe-units downslope towards the northern foreland. The attainment of high P/T ratios by depositional or tectonic thickening of the overburden thus presents problems. Alternative explanations could be sought in tectonic 'overpressures' built up by the rapid approach of colliding plates or in the build-up of fluid pressures within the metamorphic complex.

One final point which may be recalled is the evidence for the existence of a Hercynian basement of continental type beneath the eugeosynclinal supracrustal assemblages of the Pennine zone. The crystalline cores of the Pennine nappes, from lowest to highest, are followed in apparently normal sequence by the Triassic formations and by *schistes lustrés*. Unless the phenomenon of *'substitution de couverture'* (p. 197) is invoked on an enormous scale, it seems indisputable that those parts of the Pennine trough from which the existing nappes were derived had a floor of continental crustal material. This conclusion introduces an element of uncertainty into the comparisons with oceanic trenches or embryonic oceans (p. 191).

7 *The Carpathians* (Table 7.7)

The internal zones of the Inner Carpathians are revealed in massifs such as the Tatry Mountains of Czechoslovakia but are enveloped by (and in the east largely smothered by) post-orogenic sediments and volcanics. The basement in these massifs consists of Hercynian complexes resembling those of the Bohemian massif. The cover-rocks show lateral facies variations which are attributed to the effects of vertical movements during deposition along faults parallel to the fold-belt. Such early faults are thought to have played a part in defining the tectonic units formed during the orogenic stage in late Mesozoic times.

The Inner Carpathians of Czechoslovakia are dominated by nappe-groups involving both basement and cover. Metamorphic effects are almost lacking in the Mesozoic cover, though glaucophane is locally developed in serpentine masses. From north to south the major nappe-groups are:

(a) *The Pieninikum* of the klippen zone (see below).

(b) *The Tatrikum:* low units arched over a possibly autochthonous basement dome in the High Tatra.

(c) *The Veporikum:* units characterised by fairly extensive retrogression in the basement.

(d) *The Gemerikum:* far-travelled units compared by some writers with the upper Austroalpine nappe.

The Inner Carpathians are separated from the Outer Carpathians, which suffered little disturbance till mid-Tertiary time, by the *Pieniny lineament* marked by a *klippen zone* probably continuous with that of the eastern Alps (p. 197). The klippen zone follows the Carpathian arc for over 600 km as a tract never more than 20 km in breadth in which repeated disturbances appear to have been registered. The Mesozoic rocks of the zone include limestones, clastic sediments

Table 7.7. STAGES IN THE EVOLUTION OF THE WEST CARPATHIANS
(based on Mahel, 1968)

8 *Late-tectonic stage:* uplift, faulting, erosion, basaltic-andesitic vulcanism (post-Miocene)

7 *Late molasse phase:* deposition partly in brackish or marine basins, rhyolitic-andesitic
    vulcanism

6 *Early molasse phase:* deposition in intermontane basins, rhyolitic-andesitic vulcanism

5 *Late flysch phase:* folding of outer zones, passage to non-marine deposition (mid-
    Miocene)

4 *Culminating flysch phase* (Palaeogene)

3 *Early flysch phase:* nappe-formation in Inner Carpathians (late Cretaceous)

2 *Deposition of early cover-succession:* mainly chemical sediments some ophiolites
    (Trias-mid-Cretaceous)

1 *Deposition of basal beds:* marine and lagoonal; Triassic quartzites, evaporites

of local derivation and pelagic cherts and marls topped by mid-Cretaceous
polymict conglomerates and tilloids of flysch facies. The limestones and other
competent formations occur as slices and tectonic inclusions within the softer
and more mobile materials; it is these slices to which the misleading term
'klippen' is applied.

The old-established Pieniny lineament forms the great tectonic divide of the
Carpathians. The *external zones* to the north are occupied, at the present level of
erosion, almost wholly by late Cretaceous and Palaeogene flysch deposited in
advance of the rising Inner Carpathians. Complex alignments of palaeocurrent
indicators show that additional detritus was contributed from sources within and
to the north of the flysch basins. These sediments are arranged in a number of
nappe-groups, generally containing no pre-Cretaceous components, which are
separated by flattish dislocations and often extensively refolded. In Poland, the
principal tectonic units, from below upwards, are the Marginal, Skole,
Sub-Silesian, Silesian, Pre-Magura and Magura nappes.

8 *The Dinarides and Hellenides* (Table 7.4, Fig. 7.4)

The *internal (north-eastern) zones* of the Dinaride–Hellenide system comprise at
least two tracts characterised by deep-water Mesozoic cover-formations with
enormous ophiolitic complexes in them, or at their margins, and one or more
'geanticlinal' zone occupied mainly by basement rocks. Compressional deformation and metamorphism began early (end-Jurassic?) and the principal nappe-
structures appear to be Cretaceous. The *Rhodope massif* separating the
Hellenides from the Balkanides at the eastern end of the Carpathian arc has a
thin, incomplete and little disturbed Mesozoic cover and may be regarded as a
relatively stable block. Portions of the cover appear to have been metamorphosed

and Alpine granites are emplaced in the massif; these somewhat incongruous features suggest that high crustal temperatures were attained beneath the massif.

The structures of the internal zones are interpreted in terms of nappe-complexes with south-westward vergence. The geosynclinal contents of the Pindus zone rest, with thrust contacts, on the cover of the Ionian zone and are themselves overridden by ophiolite nappes derived from the Sub-Pelagonian zone and by crystalline nappes derived from the Pelagonian zone. Around Mount Olympus, nearly 100 km within the internal Hellenides, limestones of types assigned to the Ionian zone are revealed in a structure interpreted as a window in the Pelagonian nappe.

Both basement and cover are affected in varying degrees by Alpine metamorphism in the internal Hellenides. Early metamorphic fabrics and minerals and granites dated at 150 m.y. are assigned by Mercier (1966) to a pre-Cretaceous episode. A Cretaceous or younger episode gave rise to phyllites, schists and marbles; glaucophane was formed in both ophiolitic and basement rocks in the mountains of Yugoslavia and in the Cyclades.

The external zones of the Dinaride-Hellenide belt lie on the side of the belt remote from the European craton and are bordered on the south-west by a narrow stable tract largely submerged beneath the waters of the Adriatic (see p. 205). The cover-rocks show broad folds disrupted by thrusts dipping north-eastward. Partial detachment at the level of Triassic evaporites is suggested by the fact that the basement is scarcely involved in the folds.

9 Late-orogenic and post-orogenic stages

The Cretaceous orogenic disturbances led (see Fig. 7.5) to the progressive outward displacement of marine basins in the Alpine fold-belts. By late Oligocene or early Miocene times, flysch deposition had been brought to a close by the elimination of the remaining seas and large highland tracts had been elevated. Only in the Aegean sector of the Hellenides did a predominantly marine regime persist through Neogene and more recent times. Elsewhere, block-uplift led to rejuvenation of the mountain-tracts, with the unroofing of lower tectonic units. The detritus removed during these events formed accumulations of *molasse facies* in marginal and intermontane basins some of which form lowlands, alluvial basins and marine gulfs today.

The sediments of these basins generally rest with profound unconformity on folded, metamorphosed or disturbed rocks, though the oldest Molasse is itself folded and dislocated. The sediments include non-marine polymict conglomerates, sandstone and shales, with coal-seams in cyclothems comparable with those of Carboniferous coal measures. Brackish-water and marine sediments are intercalated in some of the marginal basins. Marine formations of Middle Miocene age characterise the succession along the northern margin of the Alpine-Carpathian tract east of Munich and, of course, form the bulk of the post-orogenic deposits in the Adriatic and eastern Mediterranean. Miocene evaporites occur in Yugoslavia, Cyprus and the Balkans, attaining a considerable importance in the Tertiary oil-bearing regions of Transylvania.

The volume of sediment in the Neogene and post-Tertiary molasse basins north and south of the central Alps is estimated at approximately twice that of the Alps themselves. This estimate, implying that a few kilometres of overburden have been removed by erosion, is of interest in connection with the problem of the glaucophane-bearing metamorphic assemblages (p. 200). The recurrent uplift accompanying erosion may be seen mainly as an isostatic adjustment to thickening of the crust. The Moho is shown by seismic studies to descend from its average European level of 30–35 km to depths of 50 km beneath the Alps and Dinarides and to more than 60 km beneath the eastern Carpathians. Much of this thickening is accounted for by the lower (high-velocity) crustal layer.

Late-orogenic igneous rocks are patchily distributed. Dacites, andesites, rhyolites and basalts form large masses in the internal zones of the Carpathians, in the Vardar zone of the Hellenides and in the Rhodope massif. The majority are of Neogene age, though activity still continues in the Aegean. In the western portions of the Alpine tract, late-orogenic vulcanicity, like granite-formation, was notably meagre except in the Italian zone mentioned on p. 205.

IV The Accessory Belts and the Western Mediterranean

The distribution of the Tertiary mountain-belts flanking the western Mediterranean appears at first to defy rational synthesis (Fig. 7.2). A less disconnected picture has been obtained by reversing the postulated effects of disruption in Neogene times. Argand long ago suggested that Corsica, Sardinia and Italy had originally been positioned close to south-eastern Spain. More recently Carey (1955) envisaged an eastward displacement of these units in response to the rotational opening of the *Tyrrhenian Sea* and the *West Mediterranean*. These small marine basins are floored by thin crust of transitional character, heavily blanketed by young detritus. Although aeromagnetic surveys have not revealed typical oceanic patterns of linear magnetic anomalies (Vogt *et al.*, 1971), a suggestion of radial anomalies beneath the Tyrrhenian Sea and some details of the submarine topography would be consistent with an origin by sea-floor spreading. Palaeomagnetic pole determinations from late Palaeozoic and Triassic rocks in Corsica, Sardinia, northern Italy and Spain suggest that these continental units have suffered large rotations relative to the European craton. Finally, a late origin for the Tyrrhenian and west Mediterranean basins seems required to explain the emplacement of nappes in the Betic chain (p. 205) from a source now occupied by deep water, and the derivation of clastic sediments in the Apennines from a source south-west of Italy. A reconstruction based on the idea of an opening by anticlockwise rotation of the seas which separate Italy from Corsica and Sardinia, and these islands from Spain, not only straightens the sharp curve of the western Alps but also allows the Alpine belt to be continued through western Italy, Corsica, Sardinia and the Balearics into the Betic chain of south-eastern Spain and thence into Morocco and Algeria.

The *Pyrenees* and some subsidiary belts of Tertiary folding and faulting in south-east Spain find no place in the above mobile belt. These mountain-tracts do not incorporate geosynclinal Mesozoic successions: they are lacking in ophiolites and their fold-structures appear to result mainly from rucking or gravity-sliding of the cover associated with block-movements in the Hercynian basement. The axial zone of the Pyrenees appears to have been uplifted by several kilometres, while complementary depressions to north and south were filled with flysch- and molasse-type Tertiary sediments; and the Cretaceous-Palaeogene cover has collapsed outward in cascade-folds and slide-masses.

The *Betic chain* shows a closer resemblance to the Alpine belt. The internal (southern) zone consists of thrust-nappes made largely of basement material and exhibiting, at low structural levels, metamorphic assemblages of greenschist to amphibolite facies, replacing earlier glaucophane-bearing assemblages. The external or *Sub-Betic zone* is occupied mainly by Mesozoic limestones and pelagic sediments overlying Trias of Germanic facies. Disharmonic folds passing into thrust-nappes are developed above the evaporite horizons; the Rock of Gibraltar is a klippe of Lower Jurassic limestone resting on younger shales and flysch. Both zones of the Betic chain make a remarkable U-bend, slashed by the Straits of Gibraltar, into the Riff from which they extend eastward in reverse positions through north-west Africa: the crystalline nappes fringe the coastal region, and the external zone of Mesozoic nappes forms the northern parts of the Atlas Mountains of Morocco and Algeria. Towards the south of the Atlas plateau the effects of Alpine disturbances fade out through a zone in which a shallow-water and non-marine Mesozoic cover is rucked and disrupted above a block-faulted Hercynian basement.

The Apennines of Italy constitute an enigmatic and still-active sector of the Alpine mobile tract. The tectonic zones follow the length of the peninsula which, as already mentioned, has probably been rotated anticlockwise from an original site further west. Fragments of a zone with eugeosynclinal character-istics marked by ophiolite bodies and by low-grade metamorphism are preserved near the west coast from Rome northward, and in Elba and Corsica. The main Apennine mountain-tract represents an external zone in which limestones ranging up to Lower Miocene record a long history of shallow-water sedimentation. The 'heel' of the peninsula and the adjacent parts of the Adriatic form a small stable block, with a limestone cover, separating the Apennines from the Hellenides.

Orogenic uplift began in Palaeogene times in the western zone, leading to the formation of the late Oligocene—Miocene *Macigno*, a flysch-formation of graded turbidites which provided the evidence for Kuenen and Migliorini's classic exposition of the role of *resedimentation* in the formation of such rocks (1950). From this zone, disturbances extended eastward through Neogene times. The tectonic style is thought by Migliorini, Merla and their associates to have been controlled by vertical uplift of successive 'composite wedges' defined by faults converging downwards. Slumping of cover-units towards the north-east led to the development of a chaotic mantle of pelites (the *'argille scagliose'* or scaly clay) containing a jumble of limestone, basement and ophiolite inclusions.

The sense of migration of orogenic disturbances, and the positioning of internal and external zones, are the opposite of those recorded in the Dinarides and do not correspond to those of the Alps. The Apennines remain as a late and aberrant portion of the mobile belt. They are, moreover, associated with a longitudinal tract of exceptional late-orogenic vulcanicity in which basalts occur together with trachytes and highly alkaline lavas whose composition is only partially explained by appeals to limestone contamination. The historic and still-active volcanoes of Vesuvius, Stromboli and Etna, with other centres in this tract, exhibit a wide geochemical range from leucite-rich lavas to the sodic pantellerites.

V The Eastern Mediterranean and Middle East

1 The mobile belt

The belt of Alpine mobility passes eastward into Asia via the *Caucasian Mountains* of the U.S.S.R., the *Anatolian peninsula, Cyprus*, the *Elburz and Zagros ranges* of Iran and the desert mountains of *Afghanistan* (Fig. 7.8). As in the regions already considered, the components resolve themselves into northern and southern units with contrasting tectonic symmetry. The Caucasus have much in common with the Carpathians and continue the tract bordering the Eurasian craton. Their structures show a northward vergence and they are fronted towards the north by foredeep basins. In the Taurus Mountains and adjacent parts of Anatolia, nappe structures suggestive of southward transport have been described. The Zagros Mountains of Iran show a southward vergence, expressed by the overthrusting of rocks of an internal zone over those of a more southerly external zone along the *Zagros lineament*; while in Oman, Mesozoic ophiolites and cherts representing ocean-floor material derived from the north form exotic sheets resting on the cratonic Mesozoic cover in Arabia. In the complex region between these outward-facing components, some blocks in which Mesozoic and Tertiary rocks remain little disturbed are seen in central Anatolia and possibly also in Iran.

In large basement outcrops of the Great Caucasus, and the Menderes, Kischir and Bitlis massifs of Anatolia, late Precambrian metamorphic complexes are quite widely preserved, together with Palaeozoic supracrustals, often showing low grades of metamorphism, and with Hercynian granites — as in the European sectors, mobility from earliest Phanerozoic times seems to be indicated. The Mesozoic cover-successions are once again dominated by limestones and other chemical sediments (except in certain Caucasian basins mentioned below) and the existence of deep-water troughs of eugeosynclinal type is suggested by thin starved sequences of cherts associated with basic lavas, serpentines and other ophiolitic intrusives. These assemblages are exposed in the Taurus Mountains of southern Anatolia and in the internal (north-eastern) zone of the Zagros Mountains. The timing of ophiolite-emplacement provides a contrast with the European sectors in that ophiolite suites appear to have been emplaced in Cretaceous and Eocene as well as in Jurassic times.

Other contrasts are indicated by the timing of orogenic disturbance and metamorphism. Mid-Mesozoic (principally late Jurassic and early Cretaceous) episodes were responsible for establishing the main structures of the Caucasus and were associated with the emplacement of granodiorite plutons and with related acid-intermediate vulcanicity (Table 7.8). These episodes, which constitute the *Kimmerian orogeny*, were reflected in the development of deep external basins in which up to 15 km of marine and non-marine synorogenic detrital sediments accumulated in Jurassic times. Low-grade metamorphism preceding

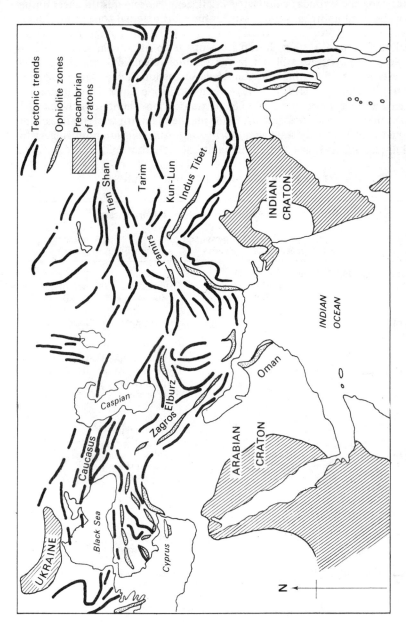

Fig. 7.8. The Phanerozoic mobile belts of the Middle East and central Asia (based on a compilation of Gansser, 1966)

the deposition of Upper Cretaceous but extending into Triassic and locally Jurassic rocks is developed in parts of Anatolia.

Later orogenic episodes culminated in Miocene-Pliocene disturbances during which widespread uplift took place. At this time, the external zone of the Zagros belt was thrown into regular folds and was overridden by the rocks of the internal zone along thrusts with displacements of up to 100 km. The effects of these Neogene pulses are still felt in the seismic Iranian and Anatolian areas. Fold-crests in the oil-bearing limestone-shale succession of the external zone in the Zagros Mountains (Table 8.3) have risen in historic times. More important displacements take place along systems of transcurrent faults roughly parallel with the fold-trend. The *Zagros fault*, following the old north-westerly lineament mentioned above, and the east-west *North Anatolian fault* are both subject to dextral displacements today. With other faults, notably those of the Jordan valley

Table 7.8.. MAGMATISM IN THE CAUCASUS
(based on Dzotsenidze, 1968)

	Great Caucasus		Minor Caucasus	
	geosyncline	peripheral zones	geosyncline	marginal zones
NEOGENE TO RECENT	andesite-rhyolite assemblage		andesite-rhyolite assemblage	local dolerites and basalts
PALAEOGENE			andesite-basalt assemblage (Oligocene) geosynclinal porphyrites (Eocene) dacites, tuffs (Palaeocene)	local potassic basalts, leucite basalts, trachytes (U. Eocene)
CRETACEOUS	tuffs and porphyrites		tuffs porphyrites, serpentines of ophiolitic type	
		olivine-basalts trachytes	granodiorites and acid volcanics	
MIDDLE JURASSIC	granodiorites preceded by ophiolites, keratophyres		granodiorites preceded by ophiolites, porphyrites	
LOWER JURASSIC	keratophyres			
TRIASSIC	minor diabases			

(p. 280) they have a role as accommodation structures allowing the northward displacement of Arabia in response to the oblique opening of the Red Sea (p. 241). *Igneous activity* during the Alpine cycle was characterised by late-orogenic andesitic-rhyolitic vulcanicity on a scale not equalled further west. Table 7.8 summarises the magmatism in various zones of the Caucasus; further south and east, huge outcrops of Neogene and post-Neogene volcanics extend from eastern Turkey into Iran, including such famous centres as that of Mount Ararat.

2 Marine basins

The *south-east Mediterranean* differs from most of the marine basins along the Alpine-Himalayan tract in that it is not entangled in the mobile zone – it occupies a huge embayment in the northern front of the African–Arabian craton and appears to form an integral part of the southern crustal plate (Fig. 7.4). The crust appears to be of transitional rather than of strictly oceanic thickness changing from about 20 km near the edge of the Nile delta to about 30 km further north. Gravity anomalies indicate that much of the area is out of isostatic equilibrium and a number of submarine ridges and furrows of roughly east–west elongation are tentatively attributed to distortion of the crust. The crustal unit which underlies the south-eastern Mediterranean is bordered on the north by the region of the Aegean Sea and the shallow waters around Rhodes and Cyprus, which have a very different character. The orogenic structures of the Hellenide mobile belt are carried through this region by chains of islands, notably the Cyclades, Crete, Rhodes and Cyprus, which unquestionably belong to the mobile belt. It has been suggested that the Hellenide belt overrides the floor of the south-east Mediterranean along a curved line passing through the *Pliny trench* south of Crete and along the southern side of Cyprus, the line towards which the crust thickens. On this interpretation, the south-eastern Mediterranean can be regarded as a *remnant of the Tethys* sheltered in the re-entrant in the African craton. The submerged condition of the Aegean sector of the Hellenides may be due to the fact that the collision of continental masses has not yet taken place along this sector.

The *Black Sea* and the southern part of the *Caspian Sea* are small deep basins entangled in the Balkanide-Caucasian mobile belt. Both have been found from seismic studies to be underlain by crust with transitional or oceanic characteristics (Fig. 7.9); beneath a mass of young sediments up to 10 km in thickness lies a 10 km layer of material with seismic velocities appropriate to lower crustal (that is, non-granitic) material. Beneath the Caucasian mountains, the Moho descends to at least 60 km and the crust includes a 'granitic' layer. The development of the basins in late Neogene times must therefore have involved great changes in crustal structure. The mechanism favoured by several Soviet authors is one of vertical movement associated with *oceanisation* of the crust, a process which Beloussov (1969, p. 710) defines as 'destruction of continental crust by splitting the crust into blocks which sink and are dissolved in the mantle'; an alternative origin involving lateral displacements of the continental crust might be suggested by analogy with the western Mediterranean basins (p. 204).

VI The European Craton

The European craton remained as a low-lying, often partially submerged, continental raft through Mesozoic and Tertiary times. The Baltic shield, to which the Caledonides of Scandinavia and Britain are welded, functioned as a positive region, while the fragmented Hercynian tract to the south foundered beneath a blanket of platform-sediments from which the positive Hercynian massifs blocked out at the end of the Palaeozoic era projected as positive elements.

Three principal tectonic developments dominated the evolution of the craton. The first was the differentiation of a number of *basins of deposition* between the positive massifs. Some of these basins were already in existence at the start of the Alpine cycle – the *Donetz basin* (Donbass basin) on the flank of the Hercynides in Russia, for example, already contained up to 10 km of Upper Palaeozoic including important coal-measures. Others developed above the New Red Sandstone (for example, the *North Sea basin*, see p. 271) and still others were Tertiary structures. The *Rhine-graben* is a distinctive fault-bounded trough of Tertiary age. The second major tectonic event was the development of narrow basins of deposition at the future *Atlantic margin* and the subsequent opening of the Atlantic, with associated marginal igneous activity (Chapter 9). Finally, the *Alpine orogenic phases*, especially the mountain-building mid-Tertiary episodes, brought about changes in palaeogeography, with the elevation of mountain-tracts to the south, and led to broad warping and localised rucking of the cover as far north as southern Britain. Sporadic igneous activity, mainly with an alkaline cast, followed in Neogene times.

1 The cover-successions

The Mesozoic and Tertiary cover from southern Britain to the Urals is generally only a kilometre or so in thickness and is almost devoid of volcanic components. Six or seven kilometres of sediment are contained in the deeper basins, such as that beneath the North Sea and Holland. The bulk of this material consists of fine detrital sediments and limestones of shallow marine or non-marine facies; arenaceous beds are largely confined to the margins of positive areas. The history of deposition can be assigned to five stages:

(1) Deposition of *continental formations* (with evaporitic or marine inter-calations) on the land-mass created by the Hercynian orogeny: Permian-Triassic (pp. 95–6).

(2) Deposition of *varied neritic sediments* interfingering with non-marine formations: Jurassic-L.Cretaceous. The deposits of this stage, clays, shelly, coralline or oolitic limestones, ironstones, greensands, vary laterally in thickness and lithology, and appear to have accumulated on irregular and mildly unstable

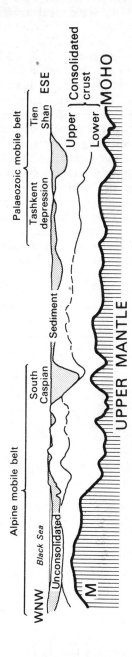

Fig. 7.9. Crustal structure inferred from seismological observations in a section of Eurasia from the Black Sea to Tien Shan (based on Kosminskaya *et al.*, 1969)

foundations. Shoreline facies and non-marine formations fringe some of the Hercynian or pre-Hercynian massifs on and near which the successions are incomplete. The accurate chronostratigraphic subdivisions of the Jurassic made possible by the employment of ammonites as zone-fossils have enabled the history of deposition to be established in unusual detail.

(3) Deposition of uniform limestones – *the Chalk* – following a widespread marine transgression: Upper Cretaceous. This *Cenomanian transgression* flooded almost the whole of Europe south of the Baltic shield and a large part of North Africa, and can be regarded as one of the great transgressions of the stratigraphical record. Although the life-span of the extended Upper Cretaceous sea overlapped with the initial phases of nappe-formation in the Alpine belts and with a phase of vigorous sea-floor spreading in the Atlantic, its principal deposits indicate a remarkable uniformity of conditions. *The Chalk*, which reaches thicknesses up to several hundred metres, is a soft, fine-grained white limestone with a very small content of detritus and frequent layers of nodular flint.

(4) Deposition of *neritic and non-marine sediments* in more restricted basins: Paleogene-Neogene. The continued Alpine disturbances were associated with a slow emergence of the craton. By Miocene times the sea was more or less restricted to tracts bordering the Alpine belt, to basins close to the Atlantic margin (notably the deep *Aquitaine basin*, p. 271) and to the environs of the

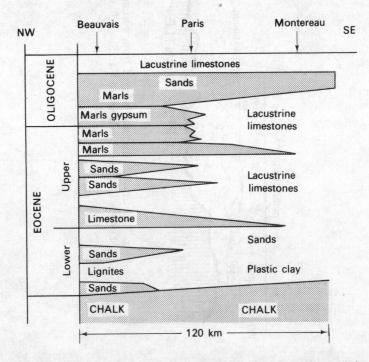

Fig. 7.10. The relationships of marine (shaded) and non-marine Palaeogene sediments near the southern border of the Paris basin (after Gignoux, 1950)

North Sea and English Channel. Marine deposits in these regions interfingered with non-marine deposits as is illustrated by a well-known section of Gignoux (Fig. 7.10).

(5) Deposition of *glacigene, fluvioglacial and periglacial sediments:* Pleistocene-Recent. The advance of polar ice-sheets into Europe and North America took place considerably later than the expansion of the Antarctic ice-sheet (p. 247) and the continents still endure the aftermath of glaciation. Over the whole of Fennoscandia, most of Britain and much of the Alpine mountain-tract, the land-forms are those resulting from glacial erosion and deposition, and from the changes of sea-level brought about both by the abstraction of water and by the subsidence and recovery resulting from loading of the continent by ice. Boulder clay forms irregular patches, drumlin-swarms, valley-fills and moraines, perhaps the most striking of which is the *Ra moraine* of Scandinavia, marking the limit of one glacial advance. Eskers, fluvioglacial sands and varved lake-deposits were spread over the lowlands and windblown fractions (*loess*) settled out over parts of the craton.

The *Great Ice Age* of the northern hemisphere was interrupted by several interglacial episodes which are reflected, especially, by variations in flora. Dating and correlation of these episodes was first attempted by stratigraphical studies including the counting of annual varves; more recently radiocarbon dating combined with pollen analysis has provided a new means of approach.

2 The Rhine-graben

In central Europe a domed 'swell' is defined by a cluster of Hercynian massifs – the Vosges, Black Forest, Ardennes and Rhenish Schiefergebirge – off which the lower formations of the Mesozoic cover dip outward. This dome came into existence in late Mesozoic times and stood above the transgressing Cenomanian sea. By early Tertiary times it had been cloven by a north-north-east rift-valley splaying out at either end in complex fractures towards the Jura and the plains of North Germany. This Rhine-graben acted as a sediment-trap, receiving some 3000 m of river alluvium with marine (Oligocene) intercalations, and remains as a dramatic topographical feature today (Fig. 7.11). The swell and graben are pierced by a famous suite of alkali-basalts, phonolites, trachytes, tuffs and agglomerates. Among the most remarkable of these rocks are the *tuffisites* of the Swabian pipes east of the Black Forest, which represent pulverised country-rock with alkali-basalt emplaced by a gas-streaming mechanism set in motion by volatiles released from the magma. A similar igneous suite (also emplaced in a Hercynian massif which had risen above the Cenomanian sea) occurs in the *Auvergne* district in the eastern part of the Massif Central of France; the youngest (Holocene) centres include the steep-sided little cones and domes of the *Puys*, as yet almost untouched by erosion.

The Rhine-graben illustrates on a limited scale an association between upwarping, rift-faulting and alkaline igneous activity, which is repeated on a far larger scale in the African and Baikal rifts. The fundamental importance of this association of phenomena was demonstrated by Hans Cloos in the title of his great paper on the Rhine-graben: *'Hebung – Spaltung – Vulkanismus'*. Cloos

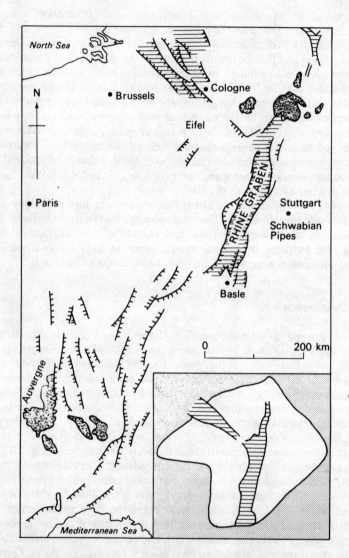

Fig. 7.11. The Rhine graben with associated faults and Neogene volcanics in the European craton. *Inset:* the graben in relation to the land-area which existed at the beginning of Upper Cretaceous times (based on H. Cloos)

was able to reproduce the structure of the graben experimentally by up-arching a 'crust' made of layered clay, a process which agreed with the observation that warping had preceded rifting.

VII Central Asia and the Himalayas

1 The mountain-block

From the Middle East, the Alpine mobile tract swings north-eastward to merge into the earth's greatest highland block. A vast complex of mountains and plateaux occupies central Asia, Mongolia, Tibet and the Himalayas. Over at least 2 500 000 km^2 the land-surface is on average 4–5 km above sea level, and scattered geophysical observations suggest that the crust is almost twice its normal thickness, the Moho descending to 70–75 km beneath the High Himalayas and 65 km beneath the Pamirs.

Belts formed directly as a result of Alpine orogenic activity make only the southern and western parts of this highland block (Fig. 7.8). The mountain regions to the north and east are made up of consolidated late Precambrian and Palaeozoic fold-belts and older tectonic units whose abnormal elevation is the result of vertical block-movements in late Tertiary and Quaternary times. It is evident that regional elevation and thickening of the crust during the late stages of the Alpine cycle affected not only the Alpine mobile tract but also a large part of the adjoining northern craton; the southern craton of peninsular India lies at a much lower level.

The far-reaching effects of mountain-building in this Asian sector of the Alpine-Himalayan belt are attributed to the collision of two continental crustal plates. The Indian craton, advancing rapidly northward at the leading edge of a spreading Indian Ocean plate, is thought to have ploughed into and under the main Eurasian continent which was buoyed up isostatically by the resulting thickening of its granitic layer (p. 218).

The *Himalayas* incorporate a portion of an uplifted marginal mobile belt, together with huge thrust-masses sliced from the frontal parts of the Indian craton. They are separated from the elevated Eurasian craton by a major lineament – the *Indus suture* – which marks the junction of materials derived from southern and northern plates. Fold-belts generated during the Alpine cycle on the north and west side of the Indus suture forming the *Pamirs* and the ranges of the *Karakorum* and *Hindu Kush* appear to link up through Afghanistan with the orogenic belts of the Middle East derived from the mobile border of the Eurasian continent. The trends of all these ranges curve sharply around the projecting northern salient of the Indian craton to produce the western *Himalayan syntexis*. With a similar curvature at the eastern end of the Himalayas (Fig. 7.8), this structure suggests that the entire border-zone has been distorted by, and moulded on, the invading Indian craton.

2 The pre-Mesozoic belts and massifs

The early history of the huge mountain regions of Tibet, Mongolia and central Asia had much in common with that of the Hercynide zone at the southern border of the European craton. An anastomosing system of fold-belts made almost wholly of Precambrian and Palaeozoic rocks, loops its way between the Siberian craton and the Alpine tract, about ovoid massifs which probably represent older stable units (Fig. 7.8). Late Precambrian sequences folded and often metamorphosed in Baikalian episodes (Part I) are conspicuous in many regions and Proterozoic granites form enormous areas east of Lake Baikal. The succeeding Palaeozoic sequences, with thick volcanic intercalations in some areas, show effects of Caledonian and/or Hercynian orogenic processes, and are associated with Palaeozoic granites. The Mesozoic and early Tertiary deposits in and alongside the fold-belts are relatively undisturbed, though they may show open folds and are strongly faulted. They are largely of non-marine facies and have much in common with equivalent rocks on the Siberian platform.

The distinctive characters of the region are those related to Neogene and post-Neogene block-movements which resulted in differential uplift, with the development of enormous fault-scarps and warps between adjacent units. The main parts of the Palaeozoic fold-belts moved upward and shed debris into the *Tarim* and *Tsaidam basins* on the site of old stable units and into narrow intermontane troughs. The Tarim basin contains Pliocene—Pleistocene conglomerate wedges up to 6 km in thickness. Its present surface lies 2—3 km below the adjacent mountain-blocks, which are thrust over it in places. The *Tibetan plateau*, also founded on an old stable block, now forms a high tectonic unit, though it was partially submerged until early Palaeogene times. That elevation of this unit preceded the final upheaval of the Himalayas is shown by the astonishing behaviour of the huge rivers which drain the plateau. The Indus, Sutlej and Brahmaputra (Tsangpo) all flow southward through the Himalayan range in gorge-like valleys 4 or 5 km in depth. Continuous downcutting enabled the *antecedent* rivers draining the Tibetan plateau to maintain themselves during the rise of the mountain barrier.

Another remarkable late Tertiary structure is the arcuate *Baikal rift-system* which follows the border of the Angara-Baikal mobile belt against the Siberian craton for almost 2000 km. This system of fault-troughs and downwarps follows the crest of a broad arch along which the land-surface is tilted outward on either side. A linear negative bouguer anomaly (a feature which distinguishes the Baikal system from parts of the African rift-system) is partly accounted for by the presence of 2 or 3 km of Neogene and post-Tertiary sediments in the fault-troughs. Alkaline basalts and trachytes resembling some of the African rift-volcanics occur along the rift-zone. Rift-faulting appears to have begun in Miocene times, following on the development of a linear upwarp. Subsidence of the central strip amounted to several kilometres — Lake Baikal is more than 1600 m deep and its floor lies over 3 km below the shoulders of the tilted blocks on either flank.

Both structural and geophysical evidence suggest that the Baikal rift-system is principally an extensional feature, though horizontal displacements are also recorded along several faults. The overlap in time between the development of the rift-faults and the regional elevation following on the collision of India and Asia suggests that the Baikal structures may have played a part in adjustment to this collision, though the connection remains obscure.

3 The Himalayas

The Himalayan belt incorporates rock-units derived from the basement and fill of one or more marine basins which appear to have formed part of the Tethyan Ocean (p. 11). These units, however, are confined to the most northerly tectonic zones (Table 7.9); the greater part of the belt is made up of rock-units closely resembling the basement and cover of the Indian craton, which appear to have been sliced off and piled up during the northward advance of the craton. Throughout the belt the vergence of folds and thrusts is towards the south, recording the effects of *underthrusting* by the craton (Fig. 7.12).

The geological cycle recorded in the northern Himalayan zones had a much longer time-span than the Alpine cycle proper. No major tectonic breaks intervene between the lowest (late Precambrian or Cambrian) members of the cover-sequence and the late Mesozoic strata laid down immediately prior to the first episodes of Himalayan orogeny. The Himalayan belt as a whole is superposed on Precambrian tectonic provinces and suffered little Palaeozoic orogenic disturbance. Syn-orogenic *flysch* within the belt, ranging from mid-Cretaceous to Eocene, indicates that Alpine disturbances began in late Mesozoic times. The late-orogenic stages were significantly later than those of the Alps, the main period of *molasse-formation* being Pliocene-Pleistocene.

Table 7.9. THE TECTONIC ZONES OF THE HIMALAYAS
(based on Gansser, 1964)

(South)	*The Indian craton:* crystalline basement Precambrian and early Palaeozoic: cratonic cover formations include Vindhyan, Gondwana System, Deccan Traps
SOUTHERN FORELAND	*The Indo-Gangetic trough:* basement depressed beneath a thick Tertiary and post-Tertiary cover of detritus derived from the Himalayas

The Sub-Himalayas: zone of folded and thrust Palaeogene and Neogene detrital sediments

Main boundary thrust

The Lower Himalayas: thrust-nappes and fold-complexes composed of crystalline basement and late Precambrian-Palaeozoic cover-units resembling those of the Indian craton, Alpine metamorphism low-grade or absent

Main central thrust

The Higher Himalayas: complex nappe overlying Lower Himalayan zone on the main central thrust: crystalline basement overlain by Phanerozoic cover partly of Tethyan facies: Alpine metamorphism and migmatization in basement, late-orogenic granites

The Tibetan Himalayas: basement overlain by Phanerozoic cover of Tethyan facies, folded or thrust

The Indus suture-zone and Indus flysch: mélange of Cretaceous-Palaeogene flysch and ophiolites with exotic blocks of eugeosynclinal cover-formations

The Tibetan plateau: elevated stable massif with little disturbed Phanerozoic cover

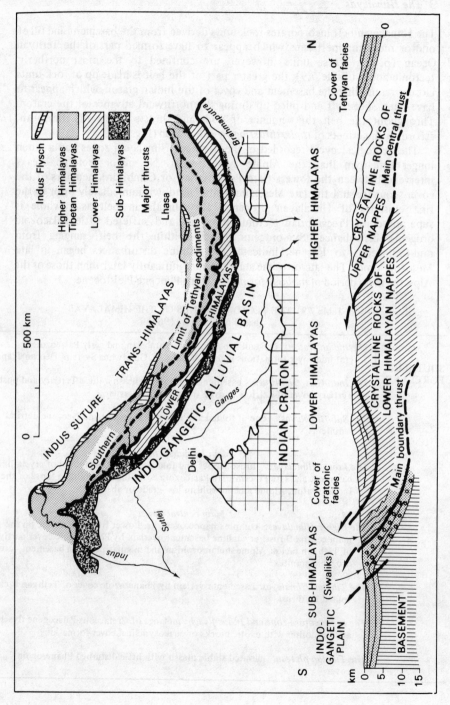

Fig. 7.12. The principal tectonic units of the Himalayas (map and section based on Gansser, 1966)

The Sub-Himalayan zone at the south of the mountain-belt is occupied by syn-orogenic and late-orogenic detrital sediments folded, faulted and overridden along the main boundary thrust. These Tertiary and Pleistocene sediments (Table 8.6) were derived from the rising mountains and culminate in the molasse-like *Siwaliks*.

The Lower Himalayas are made largely of thrust-nappes derived from the Indian craton. Units from the cover-sequence form the frontal nappes which are commonly overlain by slabs of basement gneiss. Although the scarcity of fossils and the prevalence of tectonic junctions obscure the stratigraphy, there is evidence that the cover ranges from late Precambrian to Mesozoic. Formations identical with the non-marine Gondwana System of the craton, together with shallow-water detrital sediments predominate, and a tillite (for example, the Blaini boulder-bed of Simla) equivalent to the Talchir tillite is widespread.

The Higher Himalayas and Tibetan Himalayas incorporate a cover-sequence of a very different type, consisting of shelf-sea detrital and pelagic sediments which carry rich faunas showing a strong affinity with faunas of the *Tethys* elsewhere. This contrast leads Gansser to infer that displacements on the *main central thrust* may have been as much as 100 km. The principal tectonic unit resting on this thrust is a vast slab of basement gneisses and metasediments surmounted in natural succession by the lower units of the cover. Higher cover-groups frequently form disharmonic structures or detached nappes.

The Higher and Tibetan Himalayan structures are invaded by late-orogenic tourmaline-granites and pegmatites of the Alpine cycle. The extent of Alpine metamorphism in the zones is problematical, since the cover-rocks are generally little altered. Alpine migmatisation is recognised in the basement, however, and metamorphism up to kyanite grade in certain metasedimentary units; K–Ar dating of micas gives a scatter from 17 m.y. downward.

The zones considered above provide no evidence for the former existence of a eugeosynclinal trough on the site of the orogenic belt. Such evidence comes from the narrow *Indus suture-zone* lying directly against the northern border of the belt and from a few thrust-sheets derived from it. The principal components of this zone are Cretaceous-Eocene flysch-like sediments (the *Indus Flysch*) associated with ophiolitic basic lavas, pyroclastics and jaspers. In this assemblage is contained exotic blocks, some of tremendous size, which include ophiolitic lavas, pyroclastics and serpentines associated with limestones, cherts and marls of Permian, Triassic and Jurassic age. The sediments of these enclaves include a deep-water pelagic facies not represented elsewhere, and carry distinctive ammonite faunas.

The mélange revealed in the Indus suture-zone is thought by Gansser to include remnants of a eugeosynclinal assemblage, the bulk of which has been eliminated by shortening across the Indus suture-zone. This zone today is backed almost directly by the Tibetan plateau, carrying a little disturbed Phanerozoic cover. If the Eurasian craton was fringed by geosynclinal troughs in Mesozoic times, their contents too must have been eliminated. The sheaf of mountain-belts opening south-westward in the Hindu Kush from the *Karakorum* and *Pamir* ranges, however, lie on the Eurasian side of the suture and appear to represent the mobile border of the northern craton. The Himalayan tract with all its unusual features is localised at the front of the battering-ram represented by the Indian craton where extreme shortening might be anticipated.

VIII The Siberian Craton

The Ural Mountains, which cross the Eurasian craton like a raised scar, underwent sporadic rejuvenation during the Mesozoic and Tertiary eras and consequently emerge at the present day as a mountain-tract occupied almost entirely by Palaeozoic and older rocks. The foundered south-eastern portion of the Hercynian Uralides (p. 93) and the much older crystalline complexes beneath the Siberian platform itself are blanketed by a little disturbed cratonic cover, generally a few kilometres in thickness but increasing to as much as 6 km in the West Siberian Lowlands which flank the Urals. The Precambrian basement emerges in a number of broad swells to form the *Aldan shield* and the *Anabar* and *Yenisey massifs*. There is an extensive superficial cover of Pleistocene and Holocene material — largely boulder-clay, fluvioglacial and permafrost deposits in the west and north; largely loess in the deep interior of the continent.

Much of the basement underlying the Siberian platform was stabilised in mid-Proterozoic times. The overlying cover has a time-range of over 1500 m.y. and consists mainly of shallow-marine deposits of shelf facies, evaporites and non-marine deposits. The Proterozoic successions which border the Aldan shield in the east (Part I, pp. 94–5) are principally pelites, dolomites, limestones and sandstones. The cambrian, which follows conformably on late Precambrian in many localities, is seen to be largely calcareous above a basal detrital and evaporitic division, but includes thick detrital wedges along the southern border of the platform, which was flanked by the Angara-Baikal mobile belt. The higher Ordovician rocks are largely red-beds and from this level upward, non-marine deposits predominate over much of the platform, interfingering with limestones and other shelf-sea sediments towards the north and west. Coals occur at several levels, notably in the late Palaeozoic *Tunguska suite* and in the Jurassic of the *Irkutsk* and *Kanskbasins*. Cretaceous strate are restricted and there is no record of a major Cenomanian transgression.

The principal geological event in the Siberian craton during the Mesozoic era was an igneous episode which had no parallel in the European craton — the eruption of plateau-basalts and emplacement of basic dykes and sills. These *Siberian Traps* (Permian and Triassic) cover 1500 000 km^2 and reach thicknesses of up to a kilometre towards the western part of the platform. They are broadly coeval with the first plateau-basalts of the Gondwanaland craton (Chapter 8), but unlike the latter have no spatial relationship with ocean-forming fracture-systems.

Kimberlite pipes and dykes form several clusters in and near the basaltic province in Siberia. Apart from some early (Devonian) examples, these bodies are mainly Triassic in age and post-date the bulk of the Siberian Traps. They closely resemble the African kimberlites (p. 238) and carry eclogitic and peridotitic inclusions. Diamond-bearing pipes are confined to the cratonic swells, occurring especially on the flank of the Anabar massif. The Aldan shield, another swell, is perforated by carbonatites and alkaline plugs possibly related to the Baikal rift-system.

IX The Far East

The Siberian platform is flanked on the north-east by a broad tract in which Mesozoic and Tertiary cover-sequences (thicker than and containing a higher proportion of marine sediments than those of the platform) are affected in varying degrees by late Mesozoic or Tertiary tectonic disturbances. Immediately adjacent to the platform the very thick *Verkhoyansk* sequence consisting mainly of marine detrital sediments (Permian–Jurassic) exhibits large and not very complex folds. On the eastern side of its main outcrop, the small *Kolmya* stable massif is ringed by folded strata which include thick Jurassic porphyrites and tuffs. Late Mesozoic granites associated with pneumatolytic and hydrothermal deposits of gold, tin and tungsten intrude these strata; the granites and the principal structures are assigned to the pre-Cretaceous *Kimmerian* orogeny (p. 206); palaeomagnetic data suggest that the Kolmya massif was not united with the Siberian craton until the time of this orogeny. To the south-east of the Siberian platform, a rather similar zone of Kimmerian folds, granites and ore-deposits flanks the Angara-Baikal Palaeozoic mobile belt.

Towards the continental margin, effects of profound disturbances lasting on into the Tertiary era and assigned to the *Alpine orogeny* become increasingly apparent. Most of the marginal parts of mainland Asia from the root of the Kamchatka peninsula in the north to Cambodia in the south is occupied by a mobile belt consolidated during the Alpine cycle; in this belt two or three small massifs which escaped severe Alpine disturbances can be identified: such massifs occupy parts of southern Manchuria and south-eastern China.

The eastern mobile belt is floored in many places by folded and metamorphosed early Palaeozoic and late Proterozoic rocks, which suggest a long history of Phanerozoic mobility. Low-grade Alpine metamorphism is developed in some tracts while Tertiary granites and Tertiary–Quaternary dacites, andesites and basalts are widely distributed. Lines of serpentines define one or two lineaments roughly parallel with the coast. The mobile belt, with its early Phanerozoic basement, is not seismically active but is fringed on the east by the active island arcs of Kamchatka, the Kuriles, Japan and the Philippines which mark the active oceanic margin today (Chapter 10).

The south-eastern corner of mainland Asia is occupied by the ranges which run down from the eastern Himalayan syntexis (Fig. 7.8) through Burma and Malaya to the Indonesian arc. The relationships of these ranges with the Himalayas are doubtful for although, like the latter, they result mainly from Cretaceous, Tertiary and Quaternary orogenic events their structures do not appear to turn westward to join those of the Himalayas. Complexes of folded and sometimes metamorphosed rocks including Palaeozoic and Precambrian components form a narrow tract extending southward from the eastern side of the Tibetan plateau. To the west of this tract in Burma lie outcrops of Mesozoic strata with Tethyan affinities bordered on the east by a zone of Cretaceous flysch with serpentine bodies (possibly analogous to the Indus suture) and smothered by huge Tertiary detrital formations derived from the emerging mountains. Further east, the Malayan peninsula is distinguished by granite-intrusions from which important *tin placers*, may have been derived.

8

Gondwanaland: Disruption of a Supercontinent

I The Craton of Gondwanaland

By the end of the Palaeozoic era, orogenic activity in the interior parts of Gondwanaland had died away, leaving the supercontinent as a vast cratonic mass bordered by still active mobile belts (Fig. 8.1). To the west and south, marginal belts extending from the western margin of South America through west Antarctica to eastern Australasia formed part of what was to become the *Circum-Pacific orogenic system* (Chapters 10, 11). To the north and east, the belts of the *Alpine–Himalayan* orogenic system skirted Africa, Arabia and India (Chapter 7). The main purpose of this chapter is to examine the history of the cratonic regions of Gondwanaland from the onset of the late Palaeozoic glaciation to the present day. The marginal mobile belts will be referred to only where their evolution is reflected by events within the cratonic regions.

The post-Palaeozoic history of the cratonic regions of Gondwanaland falls into three stages:

(a) *Before disruption:* The greater part of the cratonic area stood above sea-level and the characteristic sedimentary formations were continental deposits of interior basins.

(b) *The initial phases of disruption* (late Jurassic-early Cretaceous): the future continental fragments were blocked out by faulting, by flexuring of the basement or by the development of narrow marine troughs. Basic and alkaline igneous activity was widespread (Plate IV).

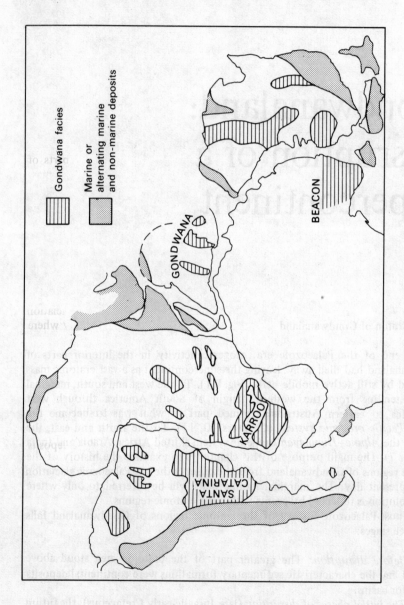

Fig. 8.1. The distribution of Permian and early Mesozoic non-marine formations (the Gondwana facies) in the supercontinent of Gondwanaland

Plate IV Wase Rock, a volcanic plug south of Jos, Nigeria: igneous activity related to continental disruption. (Copyright Aerofilms Ltd)

(c) *The phase of dispersal:* as the separating fragments took on their own identities, deposition and igneous activity became related to the structure and evolution of each continent.

II The Craton Before Disruption

1 The Gondwana facies

The accumulation of formations of continental facies which began during the late Palaeozoic glaciation (p. 174) continued in many interior basins without serious interruptions until late Triassic, Jurassic or even early Cretaceous times, spanning a period of the order of 150 m.y. The extent and thickness of these formations suggest that the land-mass on which they accumulated was little smaller than the supercontinent itself. During a part of the period of accumulation, sedimentation of continental facies was also taking place on an unusual scale in Laurasia (p. 95). The emergence of vast land-areas in both supercontinents, as Sutton has pointed out (1968), was an event unique in the Grenville chelogenic cycle, following the cessation of orogenic mobility within the supercontinents and preceding their disruption.

The late Palaeozoic and Mesozoic continental successions in Gondwanaland are best developed in a number of large cratonic basins, though remnants of thinner sequences are spread more widely. Correlation within and between the basins is aided by the occurrence of a few marine bands as well as of marker horizons of other kinds. A number of common features may be said to define a *Gondwana facies.*

Glacigene sediments are represented at or near the base in every continent. These sediments are all late Palaeozoic in age and were dealt with in Chapter 6: taken together, they record a long period in which much of Gondwanaland lay in high latitudes and was subjected to glaciation on a scale comparable with that of the Pleistocene glaciation (Fig. 6.5). They are associated with, or closely followed by, Permian *coal measures* of sub-arctic facies characterised by a *Glossopteris* flora, which is the principal source of coal in the southern hemisphere (p. 178). The incoming of abundant reptilian remains, with a common change of sediment-colour from blue or grey to yellow, green or brown, provides evidence of progressive warming-up, culminating in late Triassic times in the widespread appearance of eolian sandstones. Palaeomagnetic evidence suggests that the climatic changes indicated by the climate-sensitive sediments were due to a northward drift of Gondwanaland as a whole from antarctic latitudes towards the tropical regions. The Triassic desert sandstones of Gondwanaland represent deposits of the southern arid belt, those of the corresponding northern arid belt being represented in the New Red Sandstone of Laurasia.

The bulk of the Gondwana successions consists of detrital sediments, predominantly shales, mudstones or fine sandstones, Coarser deposits occur locally, especially near the margins of the basins, while thin coals, bituminous shales and limestones occur at some levels. Alluvial fan and piedmont deposits have been distinguished in addition to the eolian sandstones mentioned; lacustrine deposits are marked by fresh-water limestones or by mudstones carrying fish-remains; much of the sediment in the larger basins seems to have accumulated on broad plainlands of low relief where conditions were remarkably uniform. *Marine bands* have been recorded at more than one level, even in regions which lay well within the supercontinent. The Lower Permian *Eurydesma band*, for example, represented both in south-west Africa and in the Paraná basin of Brazil, appears to mark an extensive but short-lived transgression, possibly associated with a phase of glacial retreat.

Towards the margins of the craton, continental and marine sediments interdigitate, for example along the north-western border of peninsular India, in western Australia and in the Tasman belt of eastern Australia (p. 171). It should be emphasised that the shorelines defined by the distribution of marine strata of late Palaeozoic and early Mesozoic age bear little relationship to those of the present continental fragments. Deposits of terrestrial facies extend to the east coast of South America, to the east and west coasts of southern Africa and (with exceptions to be mentioned later) to the east coast of India, suggesting that all these regions were parts of a larger land-area.

In many regions, the uppermost Gondwana sediments alternate with or give place to *plateau basalts* and other volcanics. Those volcanic groups may be Triassic, Jurassic, Cretaceous or Eocene in age according to locality, and are associated with the second stage of our threefold history, the disruption of Gondwanaland.

Africa south of the Sahara. Here the *Karroo System* provides the most famous of the Gondwana successions. The *Great Karroo basin* of southern Africa, measuring at least 1000 x 500 km contains a maximum thickness of about 8 km. Its southern boundary is made by the Cape fold-belt in which folding, uplift and erosion took place during the later stages of Karroo deposition. Its eastern border lies beyond the present limits of the continent but its western border is marked by a tract of elevated basement rocks in South-West Africa which appears to have constituted a highland-region during the period of deposition. To the north-east of the main basin, Karroo sediments extend intermittently to the Transvaal as a thinner cover lacking the middle groups of the full succession (Table 8.1). In East Africa and in Madagascar, Karroo outcrops form narrow tracts bounded by early faults or warps related to the rift-valley system; variations of thickness and facies suggest that the sediments accumulated in elongated troughs or graben. In the western part of the continent, Karroo sediments underlie younger terrestrial sediments in the *Kalahari basin* and in the *Congo basin* both of which were to have a long history of subsidence. Far to the north, continental deposits of Nubian Sandstone facies (p. 168) include early Mesozoic formations.

Glacigene sediments form the bulk of the *Dwyka* division (see p. 177). The *Eurydesma* marine band of the Upper Dwyka in south-western Africa carries a Lower Permian fauna which dates the later stages of glaciation. A marker-horizon of astonishing persistence is provided by the *White Band* characterised by reptiles of the short-lived genus *Mesosaurus*. This 50 m band of white-weathering bituminous shale extends through much of the Great Karroo basin at the top of the Dwyka and also appears in South-West Africa and in the Paraná basin of

Table 8.1. THE KARROO SYSTEM
(based on du Toit, 1939)

	Cape		metres	Central Transvaal	metres
Stormberg Series	Drakensberg basalts		1500	Bushveld basalts	330
	Cave Sandstone		270	Bushveld Sandstone	100
	Red Beds		500	Bushveld Marls	130
	Molteno Beds		670		
Beaufort Series	Upper	(red, brown or green mudstones and sandstones)	670	(absent)	
	Middle		330		
	Lower		2000		
Ecca Series	Upper	(sandstones and shales)		Upper (shales)	100
	Middle		2000	Middle (coal measures)	70
	Lower			Lower (shales)	70
Dwyka Series	Upper shales		220		
	Tillites		800	Tillites	1
	Lower shales		250		

South America. It is followed by the predominantly shaly and dull-coloured *Ecca*. In the Transvaal, Rhodesia and East Africa, the equivalent groups are coal measures (of importance, especially, in the *Wankie basin* of Rhodesia) and at the margin of the Cape belt they include alluvial sandstone wedges. The overlying *Beaufort* division, which is complete only in the Great Karroo basin, consists of fine sandstones and mudstones, often rather brightly coloured and containing abundant reptile remains. The many large herbivores and the bright tints of the sediments suggest that deposition took place on vegetated plainlands under temperate or warm conditions. The *Stormberg* Group includes the late Triassic eolian sandstones already referred to and may indicate a further stage in the drift to low latitudes. The *Cave Sandstone* which underlies Stormberg volcanics in the Drakensberg range is a fine-grained massive rock regarded as a loess.

South America. Gondwana formations are preserved in and around the Amazon basin, in the Paraná basin of Brazil and adjacent territories, and in the Falkland Islands, as well as in widely scattered smaller outcrops. The *Santa Catarina System of the Paraná basin*, provides an astonishing parallel with the Karroo successions of southern Africa. Its time-span is from late Carboniferous to late Triassic and its maximum known thickness about four kilometres. The main divisions, beginning with glacial groups such as the Itarare group (p. 177), progressing through coal measures to shales, mudstones and sandstones and terminating with eolian sandstones interleaved with volcanics, correspond with those of the Karroo. Still more remarkable is the recurrence of the *Eurydesma* marine band, and of an equivalent of the White Band with species of *Mesosaurus*. Since the Paraná basin is regarded by Martin (1961) as an entity separated from the Great Karroo basin by the crystalline ridge of south-western Africa, these similarities emphasise still more the general stability and uniformity of conditions over the craton of Gondwanaland.

Antarctica. The *Beacon System* forms an intermittent cover over the rocks of the Ross geosyncline in the Trans-Antarctic Mountains. Locally-developed tillites such as the Buckeye tillite of the Horlick Mountains are closely followed by coal measures dated as Upper Permian and containing elements of the *Glossopteris* flora, and by sandstones and shales including desert sandstones. The youngest of these sediments carry Lower Jurassic fossils.

Australia. Here the sequences which follow the tillite groups are less consistent in facies than those already discussed and record incursions of shallow seas. Within the *Tasman belt* of eastern Australia, where orogenic disturbances continued throughout the period of deposition (p. 171), thick and variable successions (including important groups of acid volcanics) accumulated in the narrow *Bowen basin* and *Yarrol trough* of Queensland where marine conditions alternated with, and finally gave place to, non-marine conditions. Permian coal measures are important in some of these basins.

Near the western and north-western borders of the continent the *Carnarvon, Canning and Bonaparte Gulf basins* which had come into existence in mid-Palaeozoic times received further deposits at intervals from the Carboniferous to the Cretaceous. The Permian period was marked here, as elsewhere, by glacial deposition which often spread beyond the previous limits of the basins. Many of the tillites appear to have been marine, and at higher levels marine Permian of shallow-water facies alternates with non-marine beds; the rich Permian marine faunas of Western Australia have affinities with those of the eastern Tethys (Teichert 1958). Triassic rocks are hardly recorded in the western basins, but Jurassic and Cretaceous rocks of both marine and non-marine facies are well represented. The Carnarvon basin, for example, contains up to 4 km of largely non-marine Jurassic claystone and siltstone and some hundreds of metres of largely marine Cretaceous ending, surprisingly, with a chalk-like limestone unit. The interfingering of marine and continental deposits suggests that western Australia was bordered throughout the Mesozoic by epicontinental seas possibly in direct communication with the eastern Tethys.

In the interior of the Australian craton, early deposits of the Gondwana period are thin and incomplete. Above the customary tillite groups, brackish-water Permian detrital sediments are known in the Murray basin and in the region of Lake Eyre, but Triassic deposits are almost absent. The *Great Artesian basin*, draining towards the east and opening into some of the remaining troughs in the south-eastern part of the Tasman belt, was established in Jurassic times and has received fluviatile and other non-marine sediments at intervals until the present day. Early Jurassic deposits were limited to the south-eastern parts of the basin, but later Jurassic and Cretaceous sediments spread westward and northward over low-lying plainlands. A brief incursion of shallow seas in Lower Cretaceous times led to the deposition of extensive calcareous claystones (the *Roma Formation*), but the bulk of the sequence consists of non-marine shales with sandstone units and occasional sub-bituminous coals. The total thickness of the Mesozoic sequence is usually less than 200 m; Jurassic sandstones provide the main aquifers of the artesian basin.

India. The *Gondwana 'System' of India* is preserved in a number of narrow outcrops situated mainly in the north-eastern part of the peninsula. Although many of these outcrops are now fault-bounded, they appear to be remnants of an originally more extensive cover, rather than deposits laid down in fault-troughs. At the north-western margin of the peninsula, deposits of marine facies laid down in or at the border of the Tethys are seen in the Salt Range and in Kashmir. Shallow-water marine sediments with continental intercalations are contained in some of the thrust-nappes of the Lower Himalayas; in much of the Lower Himalayas, however, the nappes contain rocks of Gondwana facies, showing that orogenic disturbances encroached on the continental parts of the Indian craton (pp. 217–19).

The Gondwana succession of India was traditionally sub-divided on a floral basis. The *Lower Gondwana* (Permian) characterised by a *Glossopteris* flora, begins with widely-developed tillites (the Talchirs) and fluvioglacial deposits which pass up into 1–2 km of coal-measures, barren sandstones and shales. The bulk of India's coal comes from Lower Gondwana measures in the north-eastern part of the peninsula. The *Upper Gondwana*, sometimes unconformable on the older division, ranges from early Triassic to mid-Cretaceous, though Jurassic strata are usually missing. It is made up largely of fluvatile sandstones and shales, some of them red in colour and some containing a varied reptilian fauna.

III Later Mesozoic Events: Fragmentation

In mid-Mesozoic times, the long period of quiescence in the craton came to an end. Systems of narrow marine troughs, crustal warps and fractures were developed which defined the margins of the future continents, while other factures, warps and rifts came into existence within the future continents. The uprise of basic magma led to the eruption of plateau-basalts on a continent-wide scale and more localised alkaline igneous activity took place along some of the newly formed lineaments. These interconnected events foreshadowed the disruption of Gondwanaland. For the sake of convenience, they will be dealt with under separate headings.

1 Marginal features

Marine sediments of Jurassic, Cretaceous or Tertiary age fringe the continents derived from Gondwanaland in many places and wedge out or interfinger with continents. deposits towards the interior. These marine successions, which are of considerable economic interest as proved or potential sources of oil, occupy long narrow troughs from which a few offshoots extend into the interiors of the continents. Both in form and in arrangement, the *marginal marine basins* differ conspicuously from the earlier continental basins and can be regarded as the first geological structures which defined the borders of the future continental fragments.

The marginal basins, however, follow structural features of much earlier date; they are located preferentially within the mobile belts developed during the late Precambrian and early Palaeozoic cycle and only locally extend into older tectonic provinces. This control is especially clearly displayed in Africa where the formation of the coastal basins seems, in Kennedy's phrase (1964), to have been predetermined by structural conditions established during the cycle culminating at about 500 m.y.

Some characteristic features of the marginal basins can be illustrated by reference to those of Africa south of the Sahara which have been explored as potential sources of oil (Fig. 8.2). Along the *Atlantic coast of Africa* a number of basins appear to have been separated by incomplete barriers in which subsidence was slight and igneous activity sometimes took place. In the north, the basin of *Senegal* overlies the Mauretanide belt which had remained mobile until late Palaeozoic times. Thin late Jurassic deposits are followed by thick calcareous sandstones and shales of Cretaceous age and by thinner, mainly calcareous, Tertiary deposits. The maximum thicknesses of about 5 km have been recorded offshore; the sequence thins rapidly eastward where marine and coastal-plain sediments interdigitate. Salt domes rising through the Cretaceous indicate the occurrence of evaporites near the base of the succession, a feature which is characteristic of many basins.

The *Benue trough* of Nigeria and the adjacent territories extends inland from the Niger delta as a branching trough some 200 km broad flanked by crystalline massifs. It is filled by Cretaceous and younger sediments reaching total thicknesses of well over 6 km. The bulk of the deposits are sandstones and

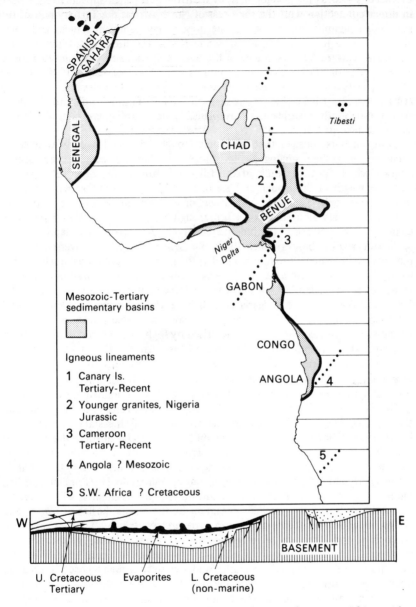

Fig. 8.2. The marginal Mesozoic and Tertiary basins of western Africa with a cross-section of the Gabon basin; the dotted lines indicate schematically the sites of Mesozoic-Tertiary igneous centres

shales, but coal-seams occur at more than one horizon. Shallow-water marine Cretaceous records the development of a narrow gulf which appears to have been in direct connection with the shelf seas of North Africa; the Tertiary part of the sequence is mainly non-marine. At the present day, sediment is being laid down along the trough by the Niger and Benue Rivers and the trough-filling is being extended seaward by the advance of the *Niger delta* whose submarine part forms a great bulge on the sea-floor and contains some 8 km of sediment. The nature of the gravity anomalies suggests that the distal part of the delta rests on oceanic crust (Hospers, 1965); this part is seen on reconstructions to disturb the fit of Africa and South America and may plausibly be regarded as a late addition to the African continent.

Some authors consider that the Benue trough originated in early Cretaceous times as a graben analogous to those of East Africa. The Cretaceous strata — which form the bulk of the fill — are thrown into open folds parallel with the length of the trough and are invaded by dykes and bosses of basic and intermediate rocks associated with a belt of lead-zinc mineralisation. A positive Bouguer anomaly along the axis of the trough has been tentatively attributed to thinning of the crust or to the presence of larger igneous masses at depth.

The crystalline blocks which flank the Benue trough are traversed by two remarkable lineaments broadly parallel with the long axis of the trough. On the west, the *'Younger Granites'* of Nigeria form part of a north—south tract of igneous centres (plate IV) which can be traced intermittently northward for 1000 km. The centres of Nigeria (which have given Jurassic radiometric ages) form ring-complexes of alkali-granites associated with acid volcanics. They carry an unusual trace-element suite characterised by high fluorine, niobium, uranium, thorium, yttrium and lanthanides and are associated with cassiterite-bearing greisens. The occurrence of a late Phanerozoic tin province in this region is of especial interest in view of the Precambrian tin mineralisation in the same area; it is possible that the later activity mobilised tin already contained in the crust.

On the eastern side of the Benue trough, a line of Tertiary alkali-granite and gabbro ring-complexes locally carrying tin occurs in the state of Cameroun. Larger areas of lavas lying on the same line are seen as far north as the border of the Chad basin. The south-western extension of the lineament crosses the continental margin east of the Niger delta and is continued into the Atlantic basin by a chain of volcanic islands including Fernando Po, Principe and Sao Thome; Cameroon Mountain, at the landward end of this chain, is still active.

To the south of the volcanic lineament lie the coastal basins of *Gabon* and *Congo* in which sediments of Karroo type are overlain by Lower Cretaceous evaporites and then by Cretaceous and Tertiary shales, sandstones and limestones. The Gabon basin is partially divided by a horst into an eastern trough whose fill is mainly non-marine and a western trough whose fill is mainly marine; a swarm of salt-domes invades the western trough (Fig. 8.3). A progressive westward shift in the zone of maximum deposition in the marine tract is indicated by the variations in thickness of successive formations. The sedimentary mass as a whole thins seaward, so that its basement is now most deeply depressed beneath the present coastal plain. In the *Cuanza* basin of Angola, early Cretaceous evaporites are followed by a complex of shallow-water marls, limestones and deltaic sandstones ranging in age from mid-Cretaceous to

Upper Miocene. Here too there are indications of repeated shifts in the site of maximum deposition, and of a general thinning of the sequence towards the Atlantic ocean. An oblique feature marked by a line of alkaline igneous centres lies close to the south-east side of the basin.

Along the east coast of Africa and in Arabia, Mesozoic and Tertiary rocks overlie rocks of the Mozambique belt and occupy a tract which is very broad in Arabia, Somalia and Ethiopia, but narrower and less continuous further south.

Jurassic strata (including highly fossiliferous limestones) form extensive, though relatively thin, outcrops in the north, where they appear to have accumulated in shelf-seas connecting with the Tethys. They are represented as far south as the southern border of Tanzania and in western Madagascar. In Mozambique, South Africa and eastern Madagascar, the oldest marine strata are Cretaceous and the bulk of the basin-fill is Tertiary; Triassic evaporites occur locally in southern Tanzania. These variations suggest that the continental margin was blocked-out relatively early in the north — perhaps by the extension of a marine gulf opening from the Tethys — and only later defined in the south; we may recall that no pre-Cretaceous marine strata appear in the more southerly basins of the west coast.

Evaporites are locally developed in East Africa near the base of the sequence and are followed by shallow-water sandstones and limestones interfingering with non-marine shales and marls. Seismic and gravity surveys indicate that the floors of the basins are faulted and that horst blocks, roughly parallel with the coast, existed during deposition; one such block is represented by the islands of Zanzibar and Pemba.

The fault-systems associated with the marginal basins appear to link up with early fractures defining the rift-valley system in the region of Laka Nyasa. Further south, the *Lebombo monocline* connects with rift structures in the Limpopo area and continues southward to become the marginal structure in Swaziland and Natal (Fig. 8.3). Along this vast structure the entire Karroo succession is tilted eastward at angles of up to $20°$ and is brought down in the coastal tract beneath shoreline sediments of Cretaceous, Tertiary and Quaternary age. Along the monoclinal hinge, the Stormberg volcanics (elsewhere almost exclusively basaltic), contain a prominent group of acid lavas and ignimbrites, together with intrusions of granophyre. The extent of the structure and the localisation of acid igneous activity along it suggest that the monocline overlies a major crustal fracture: movement on this feature is dated at late Stormberg (that is, early Jurassic).

Marginal features of the other continental fragments of Gondwanaland can be referred to very briefly. Along the Atlantic coast of South America, several basins closely resembling those of western Africa fringe the Brazilian shield: they rest mainly above the relatively young Brazilide belt. Evaporites of mid-Cretaceous age mark the first incoming of marine conditions in these basins and are followed by Cretaceous carbonate sediments. The east Antarctic craton has no Mesozoic marine deposits. In Australia, marine basins had been in existence from late Palaeozoic times along the northern and western margins of the craton and there was little change in their distribution during the Mesozoic. The Fitzroy basin is of interest for the occurrence of a post-Permian suite of plugs and pipes of highly potassic leucite-rich rocks. In peninsular India, marine sediments were laid down in late Jurassic or Cretaceous times at the margin of the craton in the Cutch, at the head of the Bay of Bengal and in south-east India and Ceylon. They interfinger towards the interior with continental deposits of Gondwana facies.

2 Fracture-systems

The cratons of Africa, Australia and South America are traversed by anastomosing faults which collectively outline a number of polygonal blocks showing evidence of broad warping. The fault-patterns which define these blocks are, as we have already seen, often linked with fractures defining the marginal features of the continents (for example, Fig. 8.3) and, like the latter, appear to have been established in middle to late Mesozoic times. Many of the faults continued to move in late Tertiary or Quaternary times and are marked by conspicuous topographical features. Others suffered little further movement in post-Mesozoic times and consequently have more subdued topographical expressions.

Although the cratonic fracture-system as a whole appears to have originated in Mesozoic times, some lineaments follow more ancient structures. In eastern Africa, for example, many of the major rift-faults are parallel to the dominant trends in the crystalline basement and the rift-valley system as a whole keeps within Proterozoic or Palaeozoic mobile belts and avoids the more ancient massifs of Rhodesia and Tanzania. Such relationships emphasise the extraordinary continuity of the history of the oldest cratonic massifs in Africa. As will be noted in the next section, igneous activity of types associated with the Mesozoic and Tertiary fracturing dates back as far as 1700 m.y., while the still older lineament produced by the Great Dyke and related intrusions (Part I) would fit readily into the pattern of Mesozoic and Tertiary fractures.

3 Igneous activity

Magmatic activity during the period which led up to the fragmentation of Gondwanaland was of two principal kinds, the first predominantly basic and giving rise to *plateau-basalts* and *basic dyke swarms*; the second predominantly *alkaline* and giving rise to an assortment of plugs, ring-complexes and allied intrusions, usually arranged in clusters or along lineaments.

The plateau-lavas and associated minor intrusions (Table 8.2). The continental successions of Gondwana facies are commonly penetrated by dolerite dykes and sills and are overlain, conformably or unconformably, by flood-basalts. These basic rocks are associated locally with acid intrusions or with minor amounts of acid or alkaline lavas or pyroclastics. Although the plateau-basalts tend to be at their thickest near the continental margins (the Deccan Traps, for instance, thicken towards the west coast of India) they nevertheless extend for hundreds of kilometres across the interior parts of southern Africa, South America and peninsular India; the dolerite dykes and sills which accompany them are even more widely distributed.

Table 8.2. MESOZOIC AND EARLY TERTIARY PLATEAU-BASALTS OF GONDWANALAND

	SOUTH AMERICA	AFRICA	ANTARCTICA	AUSTRALIA	INDIA
FORMS OF ACTIVITY	lavas, dykes and sills, mainly tholeiitic	Stormberg lavas (tholeiitic) Karroo dolerites (dykes and sills)	Ferrar volcanics (tholeiitic) Ferrar dolerites (mainly sills) Dufek Mountains layered intrusion	Dolerite sills and dykes	Deccan Traps (mainly tholeiitic) Rajmahal lavas, dolerite dykes and sills
MAIN OUTCROPS	plateau-lavas in Paraná and Amazon basins; dykes and sills extend beyond these basins	plateau-lavas in Great Karroo basin (Drakensberg lavas), remnants extend north as far as the Zambesi, dykes and sills throughout southern Africa	mainly in Trans-Antarctic Mountains	mainly in Tasmania	northern part of peninsular India
SCALE OF ACTIVITY	65000 km³ of lavas in Paraná basins, areal extent 1000000 km²	over most of southern Africa; lavas reach >1 km thickness		aggregate thicknesses of sills 450 m	Deccan traps cover 500000 km² maximum thicknesses 2 km
AGE	Jurassic to Lower Cretaceous. Radiometric datings suggests main eruptions at 130–120 m.y.	late Triassic to Jurassic. Radiometric dating gives ages of 190–154 m.y.	Jurassic. Radiometric dating gives ages of 190–154 m.y.	Jurassic? Radiometric dating gives ages of 167–143 m.y.	Rajmahal lavas Jurassic. Deccan Traps, latest Cretaceous–early Tertiary

Dates from McDougal, *Nature*, 1961; *J. Geophys. Res*, 1963, 68(5)

By far the most important rock-types are basalts and dolerites, usually tholeiitic but including olivine-basalts and a variety of minor types. In Antarctica, the Dufek Mountains layered complex of pyroxenite, gabbro, anorthosite and syrenite with strong iron-enrichment is sited alongside the Filchner Ice-shelf, close to the west border of the craton. The occurrence of acid volcanics and granophyres along the Lebombo monocline of southern Africa has been mentioned already (p.233). The ratios of strontium isotopes, $^{87}Sr/^{86}Sr$, of these acid rocks fall in the range 0.704–0.709 which is generally shown by material derived from sources in the mantle. In India, trachytic lavas, alkaline plugs and ring-complexes occur near the border of the Cutch marine basin. All these igneous centres appear to be younger than the bulk of the basic lavas.

The very wide distribution of basic lavas, dykes and sills over the cratonic areas of Gondwanaland suggests a regional abnormality in the mantle. Stratigraphical and isotopic dating suggests (Table 8.2) that conditions favourable to magma-production were not attained simulataneously in all regions: the lavas of south-east and central Africa, Tasmania and Antarctica are probably mainly Jurassic, those of south-west Africa and Brazil mainly late Jurassic or early Cretaceous and those of India mainly Cretaceous or early Tertiary. Volcanic activity overlapped in time with the evolution of the marginal basins of deposition, with the initiation of fractures and lineaments in the crust and with the early stages of continental disruption.

Alkaline intrusions, carbonatites, kimberlites. Igneous bodies of the second group mentioned on p. 234 were emplaced at intervals over a considerable period in several cratons. The style of igneous activity which they represent characterised the cratons over a very long period (see Part I), but attained some kind of peak in Jurassic–Cretaceous times. A second (late Tertiary– Recent) peak is recorded in the rift valley system of Africa (pp. 242–5). Alkali-complexes and carbonatites tend to be concentrated near major crustal fractures. In Africa, they are clustered along the southern branches of the rift-valley system and along other lineaments such as those which extend north-eastward from the coast in Angola and South West Africa (Fig. 8.3). In eastern South America, similar clusters are associated with features oblique to the Atlantic coast. In India, there are groups of plugs and ring-complexes near the west coast and locally at the northern border of the Deccan Traps. Intrusions of this type have been recorded from Australia (p. 233) but none, so far, from Antarctica.

The *alkaline complexes* are usually no more than a few kilometres across. Many are ring-structures and consist of concentric bodies becoming more under-saturated and more melanocratic towards the centre. The majority, including almost all those of Mesozoic or pre-Mesozoic age, are sodic and are made up of such rocks as syenite, nepheline-syenite, ijolite, urtite and pyroxenite. *Carbonatites* form central plugs or small independent bodies. Metasomatic alkaline rocks (*fenites*), ranging from saturated syenites to feldspathoidal or melanocratic types, are arranged concentrically around some carbonatite plugs and are considered to have been formed by desilication of the country-rocks, with the addition of alkalis, alumina, iron and phosphorus,

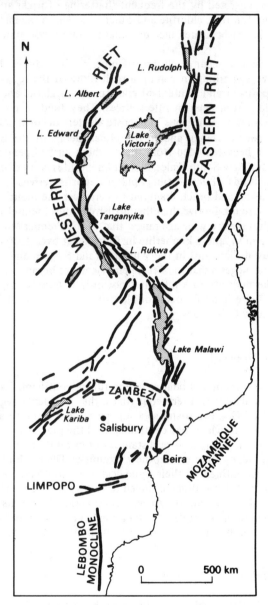

Fig. 8.3. The rift-valley system of eastern Africa

Many complexes are associated with remnants of lava-piles or with explosion-breccias and were evidently related to volcanic centres. The importance of volcanic gases is suggested by the frequent shattering of rocks and by accessory minerals carrying fluorine, chlorine and other volatiles. A characteristic suite of accessories in the carbonatites includes apatite, rare-earth minerals and pyrochlore.

Kimberlites usually occur as clusters of pipes or short dykes. Their habit and distribution have something in common with those of the rocks just discussed and their composition and content of enclaves of ultrabasic and eclogitic rock suggest derivation from deep mantle sources. They tend to concentrate away from major lineaments. Numerous kimberlite clusters occur in southern Africa, western Africa and Tanzania, and in the Minas Gerais of the Brazilian shield. Single pipes have been recorded from the northern part of peninsular India and (doubtfully) from New South Wales. Not all kimberlites carry diamonds and still fewer carry diamonds of gem quality; placer deposits derived from kimberlites (notably the alluvial and beach diamond placers near the mouth of the Orange River) may be more productive than the intrusive bodies themselves.

The kimberlites show a wide age-range: that of the Premier Mine near Pretoria is dated at 1700 m.y., and a kimberlite in Tanzania at over 2000 m.y., while the presence of alluvial diamonds in the Witwatersrand System and in the Roraima Series of Guyana suggests that kimberlite emplacement began in Archaean times. Nevertheless, the majority of kimberlites not only in Gondwanaland but also in Laurasia (p. 220) are late Palaeozoic or Mesozoic in age; those of Gondwanaland appear to be mainly Cretaceous.

IV The Stage of Dispersal

By the late Mesozoic the boundaries of the future continents had been almost completely blocked out. In Cretaceous times, the continents began to separate and the history of Gondwanaland as a supercontinent came to an end. During the phase of dispersal, which has continued until the present day, the fragments of Gondwanaland moved apart over distances of several thousand kilometres and some – notably India – entered into new groupings. Old-established patterns of crustal activity still influenced their geological evolution; many of the marginal and interior basins of deposition continued to receive sediment in Tertiary times, while the rift-valley system of Africa was extended and its characteristic vulcanism was renewed. Although the continents have established their separate identities, therefore, their later histories have many features in common.

1 Africa and Arabia

The later history of the African continent has been dominated by two principal themes: the moulding of a vast cratonic land-mass and the evolution of the rift-valley system which traverses it. Orogenic activity associated with the development of the Alpine-Himalayan fold-belts affected only the extreme north-west of Africa. Epicontinental seas invaded the adjacent foreland region from time to time and overlapped the edge of the continent in some of the old-established marginal basins; in the main, however, the history of Africa after the disruption of Gondwanaland is the history of a land-mass.

At the present day, Africa stands at an average height of about 800 m above sea level. The main watershed lies well to the east where a broad north–south tract of exposed basement-rocks is traversed by the rift-valley system. From this watershed the land level descends fairly steadily towards the east. To the west, on the other hand, numerous interior sags are separated by 'swells' marked by exposures of basement-rocks. Several of the great *interior basins* such as the *Kalahari* and the *Congo basins* are old-established downwarps in which Quaternary and Tertiary accumulations are underlain by late Mesozoic or Karroo successions; in parts of Libya, Egypt and Sudan, too, continental formations similar to the Cretaceous Nubian Sandstone appear to have accumulated intermittently since Cambrian times. Although the total thicknesses recorded are seldom great, the superposition of successive formations of continental facies suggests repeated downwarping in basins, some of which had been defined well before the break-up of the supercontinent. The absence of both Karroo and post-Karroo sediments from parts of the intervening *swells*, as well as from much of the continental watershed, also suggests a continuity in the processes of epeirogenic warping.

The sediments of the interior basins are naturally mainly detrital and vary in facies according to the local environment and climate; fluviatile or lacustrine marls and sandstones, eolian sandstones and loess-like sediments are the principal deposits. Mantles of laterite, or of silcrete, or calcrete, formed by the deposition of siliceous or calcareous cement from evaporating ground-waters extend far beyond the basins proper. At the continental margins, terrestrial sediments commonly interdigitate with littoral and neritic deposits. Late Cretaceous and Tertiary marine sequences are recorded in the coastal plain and continental shelf, east of southern Africa where earlier marine deposits are lacking. Near the northern border of the continent, shallow-water Upper Cretaceous or Eocene sediments extended over many areas which had previously received sediments of Nubian Sandstone facies. In Libya, Egypt and the Arabian peninsula, these marine sequences were mainly limestones and marls, their most distinctive members being the Eocene *nummulitic limestones* used in the construction of the pyramids. Near the mobile regions of the Atlas Mountains to the north-west and Zagros and Elburz Mountains to the north-east, calcareous sediments containing evaporite horizons alternate with and pass up into detrital wedges derived from newly-elevated orogenic lands. In the great *oilfields of the Middle East* the principal reservoir-rocks are Tertiary limestones on the southern flank of the Zagros range (Table 8.3).

A more localised marine trough was developed in Oligocene–Miocene times on the site of the Red Sea, heralding the separation of Arabia from Africa. Miocene and Pliocene deposits in this trough include clays, marls, evaporites, algal limestones and dolomites which had begun to accumulate in and alongside a graben opening into the Tethys; changes in the affinities of the faunas suggest the opening of a connection with the Indian Ocean in Pliocene times and at some stage a strip of oceanic crust came into existence beneath the floor of the trough (p. 281).

Table 8.3. THE MESOZOIC AND TERTIARY SUCCESSION OF SOUTH-WEST IRAN

		metres
	Bakhtiari Conglomerates	
MIOCENE	*Fars:* sandstone, silt, marl, anhydrite	>3000
	Asmari limestones: massive limestone followed by bedded limestone	>300
EOCENE AND LATE CRETACEOUS	*Pre-Asmari marls:* marls with thin marly limestones	600–700
MIDDLE CRETACEOUS	*Middle Cretaceous limestone:* massive or thin-bedded	500–1000
LOWER CRETACEOUS	*Albian marls:* marls, shales, thin limestones	>300
LOWER CRETACEOUS, JURASSIC, TRIASSIC	*Pre-Albian limestones:* mainly limestones with Jurassic evaporites	>1000

The later history of the African land-mass is recorded not only by the sedimentary formations described here but also, perhaps more interestingly, by the erosional land-forms. Interpretation of the *erosion-surfaces of Africa* is due largely to the work of F. Dixey (for example 1956) and L. C. King (for example 1962) on which the following summary is largely based. Over much of eastern and southern Africa, a number of gigantic topographical 'steps' are defined by flattish plains connected by relatively short and steep slopes. The higher steps represent old erosion-surfaces elevated by regional uplift, while successively lower steps represent surfaces of younger age. According to King, the dominant process in the development of the erosion-surfaces has been the concentration of erosive activity on the slopes, which has caused them to recede inland, extending the younger plains and eating into the older and higher plains. Although, from the nature of the mechanism, it is clear that each surface was developed over a considerable time-span, a younger age-limit can be obtained by reference to the age of sediments resting on it. The sequence of erosion-surfaces recognised by King in southern Africa is given in Table 8.4 and the possible extent of each surface is shown in Fig. 8.4.

Table 8.4. EROSION-SURFACES OF SOUTHERN AFRICA
(L. C. King)

(highest) restricted to high divides	'Gondwana' surface: underlies early Cretaceous (late Jurassic in East Africa) 'Post-Gondwana' surface: underlies Cenomanian or Senonian
the main plateau-surface	'African' surface (= Sub-Miocene surface); usually stands about 1300 m, underlies Miocene
coastal regions and major river-valleys	Younger surfaces of late Tertiary and Quaternary, two or three distinguished (e.g. Coastal Plain surface, mid-Miocene and later, Congo surfaces, late Pliocene and later).

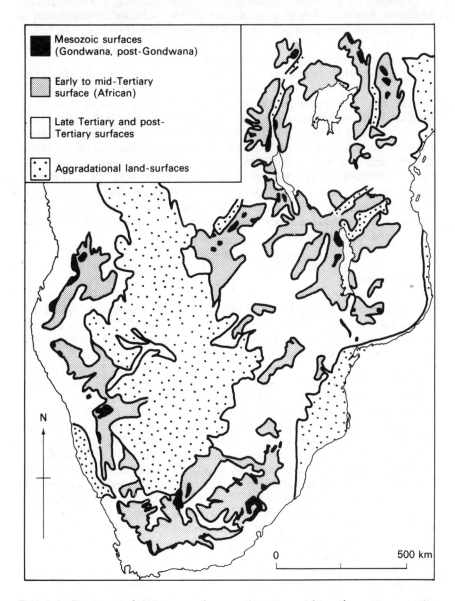

Fig. 8.4. Erosion-surfaces in southern and eastern Africa (according to the interpretation of L. C. King)

Erosion-surfaces dating back to Mesozoic times are deeply dissected and can be recognised only in high plateaux such as the Drakensberg. The early Tertiary African surface (the Sub-Miocene surface of East African literature) can be recognised over large areas as a remarkably smooth and often lateritised feature at heights of 1300—2000 metres. The main drainage pattern is cut in this surface and younger erosion-levels eat into it near the coast.

The characteristic faulting and warping of the African craton is reflected by variations in the level of the older erosion-surfaces which are downwarped beneath the contintental formations of interior basins such as that of the Kalahari; King considers that they also slope down to underlie the marine fill in some of the late Mesozoic marginal basins. The African and older surfaces are upwarped to heights of more than 2000 m along the shoulders of the rift valleys in East Africa (Fig. 8.4). A late Cretaceous surface in the vicinity of Nairobi rises some 500 m over 80 km to the lip of the Eastern Rift. On the elevated *Ruwenzori massif* within the Western Rift, the African surface is arched up to a height of over 4 km. Between the Western and Eastern Rifts, rivers originally flowing east were ponded back by the differential movements to form Lake Victoria and Lake Kyoga.

2 The rift-valley

The African rift-system is part of a 5000 km tract of fracturing which extends northward from the Limpopo Valley to East Africa, Ethiopia, the Red Sea and the Jordan Valley (Fig. 8.3). This long tract is a component of the craton-wide basin-and-swell structure and links up at several points with the marginal crustal features formed prior to the disruption of Gondwanaland; the Lebombo monocline, for example, can be regarded as a marginal feature in the south and as a unit of the rift-system at its northern end.

The southern portions of the tract of rift-faulting, in and around the Limpopo and Zambesi Valleys, came into existence in late Mesozoic times immediately prior to disruption of the supercontinent. The main phases of movement and igneous activity in these portions were completed before the end of the Mesozoic era. As we have already seen (p. 234), still older lineaments exist in southern Africa which have some affinities with the rift-fracture pattern. The northern portions of the system in East Africa, Ethiopia and the Near East appear to have originated only in mid-Tertiary times (Table 8.5). Fault movements took place especially in Miocene and Pleistocene times; igneous episodes were repeated from the Miocene to the present day and high heat flow, seismic and volcanic activity still characterise the region.

The structural details of the system are varied. In addition to the 'ideal' form of a fault-strip about 40 km wide bounded on both sides by faults, there are one-sided structures such as the Ethiopian rift in which a normal fault is paired with a monocline, troughs bounded by echelons of faults and regions of irregular block-faulting. Vertical displacements are considerable; the floors of Lakes Nyasa and Tanganyika, for example, are depressed to 2—3 km below the level of the valley rims and spectacular scarps are developed at their sides. Modest thicknesses of Tertiary or late Mesozoic sediments occupy the rift-valleys in

Table 8.5. SEQUENCE OF VULCANISM IN THE EASTERN RIFT VALLEY
(based on B. C. King)

QUATERNARY	{ Trachytes, phonolites, rhyolites Basalts, basanites	} (mainly central volcanoes and small cones)
M. MIOCENE TO PLIOCENE	{ Trachytes, trachytic, pyroclastics (mainly central volcanoes Basalts Trachytes, trachytic pyroclastics (mainly central volcanoes) Plateau phonolites Basalts, basanites, nephelinites	

some regions. Tertiary and post-Tertiary igneous rocks are concentrated in the northern sector, from Tanzania to Ethiopia, whereas Mesozoic igneous centres are scattered down the southern part of the system.

As has been noted above, the rift-structures tend to avoid the Archaean basement provinces in which the tectonic 'grain' is not far from east–west and also tend to keep within the Proterozoic and early Palaeozoic mobile belts which enclose these provinces. In East Africa, two principal branches, the *Western and Eastern Rifts*, skirt the ancient Tanzania massif and in southern Africa, branches of the system similarly skirt the Rhodesian and Transvaal massifs.

Many of the fractures which bound the rift-valleys proper have been proved directly, or from borehole evidence, to be normal faults, as are most of the associated fractures. This evidence (which is in line with that obtained from the Rhine and Baikal rifts) was questioned for a time before the Second World War when geophysical surveys by Bullard and Wayland showed that some valley-floors were characterised by negative gravity anomalies; the apparent need to envisage a mechanism whereby a light crustal strip was prevented from achieving isostatic equilibrium led to an idea of steep reverse faults masked at the surface by the effects of slumping. Although later work has confirmed the frequent lack of isostatic balance, this seems attributable to abnormal density-distributions in the crust and mantle rather than to the forcing-down under compression of a narrow crustal strip. The predominance of normal faulting and the connection of rift-valleys with embryonic oceans in the Red Sea and Gulf of Aden combine to suggest that the rifts are fundamentally extension-structures. Evidence of major horizontal displacements is found in regions where complex block-movements have taken place: a sinistral displacement of some 100 km along faults following the Jordan Valley, for example, appears to have resulted from the oblique opening of the Red Sea which involved a relative northward shift of the Arabian block. In the context of the system as a whole, such adjustments are of small importance.

The African rift-valleys, as we have noted, follow the main crustal swell of the craton, resembling the keystone of an arch. The formation of an arch involving the full thickness of the crust could, in fact, achieve roughly the extension needed for the development of the graben as was suggested by Cloos' models based on the Rhine graben (pp. 213—15). The formation and maintenance of such a crustal feature requires compensating abnormalities of density or composition in the mantle.

Igneous activity in the rift-system began in the southern sector in Mesozoic times (p. 236). In the rift-valleys of East Africa and Ethiopa, which were not initiated until mid-Tertiary times, Tertiary and post-Tertiary igneous rocks are widely, if patchily, distributed. The huge fault-troughs containing Lakes Nyasa and Tanganyika are almost devoid of igneous rocks, apart from the small Rukwa lava-field on the culmination which separates the lakes. The Western Rift is peppered with small igneous centres and explosion-vents. The Eastern Rift and its borders are flooded by lavas in Ethiopia and northern Kenya and the rift is flanked further south by the huge shield-volcanoes of Kilimanjaro, Mweru and Mount Kenya. The initial phases of Tertiary volcanism preceded the main Tertiary episode of rift-faulting and may be attributed to preliminary changes taking place in the mantle.

The principal early products of magmatism in Kenya and Ethiopia (Fig. 8.5) were *plateau-lavas*, well over half of which are basalts. Olivine-basalts and alkali-basalts, the dominant types, are associated with phonolites, trachybasalts and trachytes. Some of the *central volcanoes* and associated intrusive complexes are made up of a similar basalt-phonolite-trachyte assemblage, but more strongly alkaline components and carbonatites are not uncommon. Phonolites, nephelinites, melilite-basalts and other feldspathoidal varieties are represented, together with basalts, in the shield-volcanoes of Kilimanjaro, Mweru and Mount Kenya. In the Western Rift, alkaline volcanics of similar types are associated with strongly potassic varieties rich in leucite. Plutonic rocks, forming the roots of central volcanoes and other ring-complexes or small plugs, and occurring rather widely as blocks in lavas and pyroclastics, are predominantly alkaline; they include pyroxenites, peridotites, alkali-gabbros and syenites of many descriptions and carbonatites.

The assemblages of alkaline affinities here mentioned, which were emplaced from the Miocene to the present day, may be contrasted with basaltic assemblages without conspicuously alkaline affinities which appeared in the rift valley terrain over the same time-span. Olivine-basalts are the principal components of late Tertiary and post-Tertiary lava-plateaux in Ethiopia. Tholeiitic types are almost confined to certain tracts in the Afar Depression, where the Ethiopian rift meets the Red Sea and Gulf of Aden and to the embryonic oceans themselves (p. 281).

The rift-valley volcanic association is characterised by widespread indications of the activities of volcanic gases. Pyroclastic rocks, including ignimbrites, are abundant in some central volcanoes, minor explosion-vents are very numerous and alkaline lakes such as Lake Natron in the Eastern Rift contain much soluble material of juvenile origin. The nature of some of the alkaline lavas and especially the occurrence of carbonatites in subsurface complexes and occasionally as surface effusions have led some authors such as Holmes (1965, p. 1067—78) to assign a critical role in rift-valley vulcanicity to the rise of gas-fluxed carbonatite from sources near the base of the crust. Such accumulations of material of low density would provide a possible means by which the characteristic unwarping of the crustal tract containing the rift-valleys was attained, while the reaction of carbonatite with crustal material, leading both to

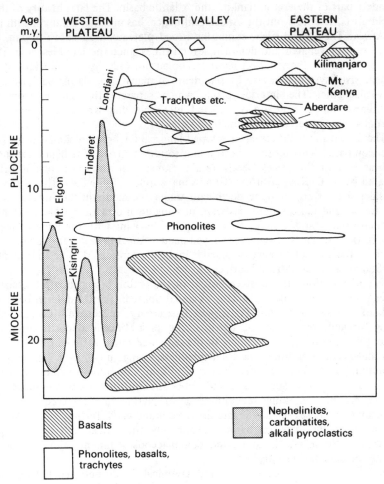

Fig. 8.5. The distribution in time and space of volcanics associated with the Eastern Rift in Kenya (based on Baker and others)

fenitisation and to the mobilisation of alkaline magmas, is held to account for the generation of alkaline volcanics and central complexes. Other authors, impressed by the volumetric importance of basaltic rocks in the rift-province, prefer to derive the alkaline assemblages from basaltic or alkaline peridotitic source-material or from the parent-material of kimberlites.

3 South America

The craton of South America has evolved in the shadow of the Andean orogenic belt and has received much erosion-debris from it. The two principal areas of exposed crystalline rocks — *the Brazilian and Guyanan shields* — lie in the broadest part of the craton flanking the Atlantic basin. The later history of these shield areas, remote from the Andean influence, has much in common with that of Africa. They stand high, exhibiting broad plateaux well above 500 m, and have been rejuvenated and denuded several times since the late Mesozoic. They are separated by the *Amazon basin*, occupied not only by the alluvium of the present river system but also by sediments of many ages, dating back as far as the Palaeozoic. The interior plains which separate the shields from the Cordilleran ranges are largely covered by coalescing wedges of non-marine detrital sediment derived from the mountains.

The Brazilian shield reveals a stepped landscape whose erosional forms have been compared with those of central and southern Africa. The high watersheds, such as that of the Minas Gerais, retain remnants of a smooth erosion-surface equated by L. C. King with the 'Gondwana' surface (Table 8.4). This surface is considered by King to slope northward under sediments in the Amazon basin and to descend more rapidly eastward in a marginal downwarp. More extensive erosion-surfaces at lower levels are compared by King with the late-Cretaceous and Tertiary surfaces of Africa. The well-planed *Sub-Americana surface* in which much of the present scenery of Brazil is incised is considered to be equivalent to the Sub-Miocene or 'African' surface.

Deposition along the western border of the craton began in late Cretaceous times soon after the first episode of regional uplift in the Cordilleran belt. The bulk of the sediments are Tertiary or Quaternary, successive orogenic pulses being marked by wedges of detritus; the post-Miocene deposits often rest unconformably on older members of the sequence which suffered gentle folding in a mid-Tertiary episode. The total thicknesses range up to 6 or 7 km near the mountain front and decrease eastward. From Colombia to the Argentine, lenticular units of conglomerates and red-beds form the bulk of the succession. A few groups containing brackish-water or marine fossils record short-lived incursions of the sea, mainly in late Cretaceous and early Tertiary times. Around the head of the Amazon basin exceptionally thick piedmont deposits pass eastward into shales, siltstones and carbonaceous sediments along the lower reaches of the river system.

In Bolivia, which occupies the topographical divide between the drainage-basins of the Amazon and the Paraguay, piedmont deposits are thin and discontinuous, as they are on the Pampean massif further south. They thicken, however, on the plainlands of Argentina where they unite into a continuous tract extending eastward to the Atlantic coast.

In the extreme north, the Caribbean mobile belt is flanked by a mass of largely marine Cretaceous, Tertiary and Recent sediments. These are the deposits of the *East Venezuela basin*, important during the last few decades in the search for oil. The maximum thicknesses of about 12 km are found near the northern side of the basin where Cretaceous and early Tertiary sediments were laid down in marine gulfs opening towards Trinidad. The higher Tertiary and Quaternary sediments, including the deposits of the Orinoco River which overlap southward onto the Guyana shield, are largely non-marine. Most of the oilfields of Eastern Venezuela and Trinidad lie near the junction of the East Venezuela basin with the more strongly disturbed rocks of the Caribbean mobile belt to the north. Those of western Venezuela derive oil from Cretaceous limestones underlying a

Tertiary sequence of marine and non-marine detritals in the isolated *Maracaibo basin* within the mobile belt itself.

4 Antarctica

The geological record of the history of the Antarctic craton after the disruption of Gondwanaland is fragmentary. No supracrustal deposits younger than the Jurassic Ferrar lavas and older than the first glacigene sediments have been found on the craton, though such deposits are present in the mobile belt of West Antarctica. L. C. King (1965), considers that the subglacial geomorphology of the craton is consistent with a history of erosion and warping similar to that recorded over much of southern Africa. Igneous activity of basic or alkaline type is recorded in the Ross Sea area where the basalt-trachyte-phonolite-kenyite *McMurdo Volcanics* range up to the present day.

The principal events in the later history of Antarctica were of course the onset and progress of *glaciation*. The present ice-sheet, some 27 km^3 in volume, extends in most places out to or beyond the margin of the land and rises inland to well over four kilometres above sea level. It has been estimated that total melting might raise world sea level by about 60 metres. The occurrence of moraine, or of erratics on mountains now clear of ice, indicates that there has been a considerable retreat from the period of maximum glaciation, while the structure of tillites in the region of McMurdo Sound suggests that there have been several advances and retreats. New information as to the total length of the period of glaciation has come recently from the sampling of sediments on the floor of the Southern Ocean. Cores recovered from sites in the South Pacific and Scotia Sea contain ice-rafted pebbles and detrital grains derived from continental sources and showing features attributed to glacial action. The magnetic reversals recorded in the cores show that their oldest material at depths of about 25 m dates back at least 5 m.y. and the distribution of glacial detritus suggests that a glacial maximum was reached during the Gauss palaeomagnetic episode (2.55–3.35 m.y.). These findings would suggest that the Antarctic ice-sheet had reached its greatest extent well before the Pleistocene glacial advance in the northern hemisphere.

5 Australia

By late Mesozoic times, orogenic activity in the Tasman belt had almost died away, though mobility continued in the archipelagoes to the north and east. The later history of the Australian continent was marked by warping and dislocation and by vigorous, if localised, volcanic activity. The final elevation of the Eastern Highlands on the site of the Tasman belt did not take place until the late Tertiary.

Deposition in and around the continent took place in two main environments: in relatively narrow marginal basins fringing the oceans and in broad downwarps partly or wholly enclosed by continental rocks. Some *marginal basins* date back to the late Palaeozoic. We may cite the *Carnarvon basin* of Western Australia which is delimited by the Darling lineament of persistent downwarping. The fill of this basin includes both marine and non-marine sediments ranging from Ordovician to Quaternary. It is divided by minor unconformities, related in some instances to transgressions and regressions at the continental margin, and includes shallow-water carbonate sediments, sandstones and shales with little or no volcanic material. The maximum known thicknesses are: Lower Palaeozoic, >3000 m; Devonian and Carboniferous, > 2000 m; Permian, 5000 m; Jurassic, > 4000 m; Cretaceous > 900 m; Tertiary, 500 m.

In addition to such long-standing basins, new marginal areas of deposition were established in Cretaceous or Tertiary times. The *Eucla basin* bordering the Great Australian Bight received a thin but extensive veneer of early Tertiary limestones. Outliers of early Cretaceous and early Tertiary marine beds far from the coast suggest that short-lived seas spread well beyond this basin. The smaller but deeper basins bordering the Bass Strait contain detrital Cretaceous and Tertiary sediments up to several kilometres in thickness. The east and north coasts are fringed by areas of carbonate deposition, most notably the platform, sometimes more than 200 km broad, which supports the *Great Barrier Reef* of Queensland. Several hundred metres of limestones and sandstones ranging back to early Miocene have been proved on this platform.

The *interior basins* are roughly equidimensional and are defined at many points by narrow rims which are not only topographically high but which also reveal the Precambrian or early Palaeozoic basement. Their arrangement recalls the 'basin-and-swell' structure of the African craton and appears to have been determined largely by recurrent uplift of the rims; the Mount Lofty and Flinders Ranges on the western rim of the Murray basin are among the seismically active regions of the craton at the present day. The main sites of deposition from late Cretaceous times onwards, as in earlier Mesozoic times, were in low-lying areas west of the Tasman belt; the *Carpentaria basin* opens northward to the sea, the *Lake Eyre basin* is a region of internal drainage on the site of the Great Artesian basin and the *Darling–Warrego* and *Murray basins* lie on the course of the Darling River and its tributaries (Fig. 8.6). Beyond the limits of these basins, a veneer of Quaternary deposits covers much of the arid central part of the craton, often taking the form of widely spaced and very long ridge-dunes.

The Tertiary and Quaternary deposits of the interior basins are not as a rule more than a few hundred metres in total thickness. They include gravels, sandstones, siltstones, occasional lignites and remarkable silicified limestones often capped in outcrop by a crust of chalcedony. Weathering deposits in areas of erosion or slow deposition include laterites and silcrete or 'billy'. Some idea of the sedimentary facies is given by a succession from the Birdsville area of the Great Artesian basin (Fig. 8.6). In the far west of the continent the ancient Yilgarn and Pilbara blocks make a broad plateau almost devoid of Phanerozoic cover-sediments. This 'ageless and undateable old-land' as it is called by E. S. Hills appears to have been planed off in, or before, the early Tertiary and to have been subsequently elevated by a few hundred metres.

In sharp contrast, the *Eastern Highlands* – the other major region from which a late Phanerozoic cover is absent – display a young topography fretted by developing river-systems. Post-orogenic uplift began spasmodically early in the Tertiary, reaching a first climax in late Miocene, when associated elevation of the craton expelled the sea from a number of marginal basins. Later phases of uplift

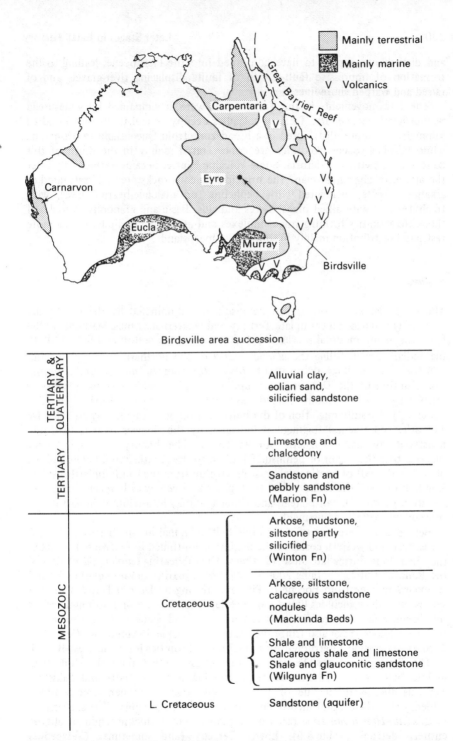

Fig. 8.6 Areas of Tertiary sedimentation in Australia, with a representative succession from the Eyre basin

and disturbance appear to have continued into the Pleistocene, leading to the formation of impressive fault-scarps, of faults displacing river-gravels and of raised and warped shore-lines.

The late movements in the Tasman belt were accompanied by widespread *post-orogenic vulcanicity*. Tertiary volcanics occupy several tracts of a hundred kilometres or more, distributed in a broad zone from Queensland to Tasmania, while Plio-Pleistocene volcanics are concentrated mainly in the north of this zone and in southern Victoria. Some volcanic centres became extinct only after the arrival of aboriginal man. The principal igneous rocks are undersaturated or alkaline basalts, associated with nepheline- or analcime-bearing lavas, with trachytes and with allied plugs, dykes and sills. Basic and granophyric rocks of tholeiitic affinities forming ring-complexes and associated lava-piles have a more restricted distribution mainly in south-east Queensland.

6 India

The later history of the Indian peninsula was dominated by the rise of the massive Himalayan ranges in late Tertiary and Quaternary times. Material eroded from the rising mountains was swept down onto the craton and beyond it to the Indian Ocean, filling the *alluvial basins* of the northern Indian plains and constructing the huge deltas of *the Indus, the Ganges and the Brahmaputra*. The eruption of the *Deccan Traps*, unlike the comparable igneous episodes in other fragments of Gondwanaland, continued well into the Tertiary (p. 235, Table 8.2). A bodily migration of the Indian craton on a scale hardly equalled by any other continental fragment is indicated by the changes of palaeolatitude registered by palaeomagnetic studies (p. 7). The history of displacement suggests that the union of peninsular India with the Asiatic continent as a result of which the raft of continental crust moving up from the south underthrust the mobile border of the Asiatic plate took place at a geologically recent time. The elevation of the Tibetan plateau and Himalayas may be attributed in part to the consequent isostatic adjustment.

Before the emergence of the mobile belt, marginal marine basins at or near the eastern and western coasts of the peninsula continued to receive sediments as they had done during late Mesozoic times. The oil-bearing *Cambay basin*, east of the Rann of Cutch, contains 2000 or 3000 m of marine and non-marine detrital sediments ranging from Eocene to Pliocene, resting on Deccan Traps. A thinner succession, which includes limestones, overlaps onto the craton north and east of this basin. Shallow-water limestones, sandstones and shales of early Cretaceous to Lower Miocene age also fringe the south-east coast and extend into Ceylon. A Tertiary succession interrupted by several unconformities is seen in West Bengal.

The Tertiary and post-Tertiary sediment-masses which flank the Himalayan mobile belt occupy an arcuate tract crossing northern India and Pakistan. Towards the northern side of the arc, syn-orogenic sediments are strongly folded, often thrust, and incorporated in the Himalayan ranges. The successions of this *Sub-Himalayan zone* reach more than 10 km in thickness and are almost entirely detrital (Table 8.6). Lower Tertiary (and sometimes Cretaceous) members are partly marine, whereas Upper Tertiary and Quaternary formations

Table 8.6. THE TERTIARY SUCCESSION OF THE SUB-HIMALAYAN ZONE IN THE PUNJAB (Fig. 7.12)

PLIOCENE	Upper Siwalik: conglomerates, grits, sandstones, brown clays
	Middle Siwalik: coarse grey micaceous sandstones, dull-coloured clays
UPPER MIOCENE	Lower Siwalik: red shales and clays, brown sandstones
LOWER–MIDDLE MIOCENE	Muree (Dharmsala) Series: red and grey-green clays and shales, grey and purple sandstones, 4000 m brackish and fresh-water
←——————————— *(unconformity)* ———————————→	
EOCENE	Subathu Series: red and green shales and calc-shales, thin limestones, rare sandstones, 1800 m, marine

Mather and Evans, 1964, Oil in India, *Int. Geol. Congr.*, 22nd Session.

are non-marine. The incoming of the Upper Siwalik conglomerates reflects the vigorous stages of uplift and erosion to the north. On the plains south of the mountain-front, the corresponding successions of the *Indo-Gangetic basin* consist mainly of Upper Tertiary and Quaternary fluviatile sediments whose latest units constitute the alluvium of the Ganges and other modern rivers. The youngest formations overlap southward to rest directly on the basement.

Perhaps even more remarkable for bulk are the deposits which underlie the lower reaches of the rivers draining the Himalayas and which form enormous deltas. Both the *Indus basin* on the west of the craton and the *Assam basin* on the east are underlain by late Mesozoic and Tertiary sequences locally reaching more than 10 km in thickness. These sequences thin rapidly into shelf-facies towards the peninsular craton. That of the Indus basin is gently folded, that of Assam is interrupted by several unconformities. Virtually all of the basin-fill consists of detrital material, with minor coals and limestones; the lower members are partly marine, but the later Miocene, Pliocene and Quaternary are almost entirely non-marine, laid down on advancing delta-plains. Recent surveys show that the sub-aerial deltas are fronted by abyssal cones channelled by many submarine canyons and passing into blankets of sediment which extend for at least 1000 km southward from the mouths of the rivers (see Chapter 9 p. 277).

9

The New Ocean Basins

I Introduction

A feature common to the history of all the oceans has been the extension of the oceanic crust as a result of the repeated introduction of new material along the mid-oceanic ridges. In the young oceans widening of the ocean basins themselves since Mesozoic times has been made possible by the lateral drift of the adjacent continental fragments: oceanic and continental crustal masses have moved together as *composite crustal plates*. There is a fundamental contrast between the tectonic setting of these widening oceans and that of shrinking oceans, such as the Pacific, which are dealt with in Chapter 10. The former are coupled to the bordering continents; the latter are detached from the bordering continents along the world-wide system of mobile belts at which the surplus crustal material is returned to depth (Fig. 9.1).

II The Atlantic Ocean

1 The age of the Atlantic

The structure of the bordering continents and the characters of widely distributed formations such as those which make up the successions of Gondwana facies in South America and Africa provide no indication of the existence of a major ocean on the site of the Atlantic in late Palaeozoic times. The earliest structures related to the ocean are narrow marine basins and the associated fractures and igneous centres which were formed in Triassic, Jurassic or early Cretaceous times (pp. 230–3). The earliest basins of this system – or, at any rate, those parts of them which are accessible – appear to be floored by basements of continental type and their formation probably preceded the emplacement of the first strips of oceanic crustal material. The oldest sediments, which rest on a crust of oceanic type, are Upper Jurassic limestones found in the Cape Verde Islands. Cretaceous sediments are quite widely distributed not far

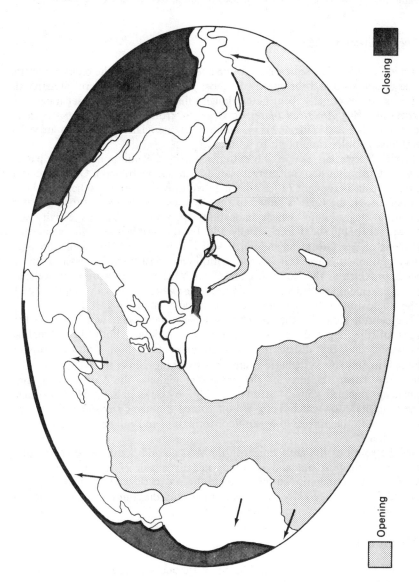

Opening

Closing

Fig. 9.1. The new and old ocean basins

from the base of the continental slope and this fact, with the dating of linear magnetic anomalies in the sea-floor, suggests that the older parts of the Atlantic basin are Cretaceous (Fig. 9.2). We have already seen that the siting of the oceanic margins was related to structures of much earlier date (p. 234).

2 The mid-Atlantic ridge

The mid-oceanic ridge represents, at one and the same time, the oldest structure of the ocean basin and the site of the most recent additions of crustal material. It forms an essentially continuous longitudinal feature which separates the westward-moving *American plates* on the west from the *Eurasian and African plates* on the east (Fig. 9.2). A number of associated structures may be mentioned in order from north to south.

In the Norwegian Sea of the extreme north, the mid-oceanic ridge passes between Greenland and Spitsbergen. An aseismic ridge between Greenland and Arctic Canada is thought to have been active in early Tertiary times. At about Latitude 65°N, the ridge is linked to the adjacent continents by broad aseismic submarine ridges along which indications of vigorous igneous activity are provided both by the Faroe Islands and by the plateau-lavas and associated intrusions of East Greenland and north-west Britain. Iceland sits astride the ridge at its junction with these features and, with them, constitutes the classic Tertiary igneous province of the North Atlantic (Plate V). South-westward from Iceland, the mid-oceanic ridge forms the *Reykjanes Ridge* which played an important part in the elucidation of the structure of the oceanic crust (see Fig. 9.2).

The central sector of the ridge follows a sigmoidal curve which reflects the shapes of the opposed coasts of Africa and America. Both the submarine topography and the distribution of earthquake foci show that curvature is achieved mainly by systematic displacements along numerous transverse fractures. The larger fractures of the equatorial region — the Vema, Chain and Romanche fracture zones — offset the ridge by several hundred kilometres. These fractures, for which Tuzo Wilson coined the name *transform faults*, are bound up with the process of ocean-floor formation and will be considered in the next section.

To the south of the great bulges in the African and South American coasts, the ridge straightens out and is approached by a pair of aseismic submarine ridges: the Rio Grande rise on the west and the Walvis ridge on the east. At about Latitude 55°S, it apparently turns eastward around the southern tip of Africa to link up with the ridge-system of the Indian Ocean.

3 The crust of the Atlantic

Linear magnetic anomalies of the characteristic oceanic type run parallel to the mid-Atlantic ridge and can be seen in some areas to be symmetrically arranged with respect to it. The strikingly regular and symmetrical pattern obtained from the vicinity of the Reykjanes Ridge by Heirtzler, Le Pichon and Baron in 1966 (Fig. 9.3) played a major part in convincing geophysicists of the reality of sea-floor spreading. The outward succession of anomalies, marking successively

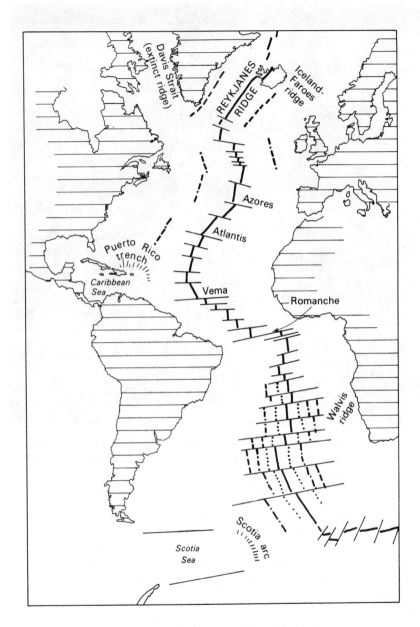

Fig. 9.2. Structural elements of the Atlantic Ocean

Plate V. A fissure eruption close to the Atlantic spreading-centre in Iceland; small scoria- and spatter-cracks mark the eruption of 1783 at Laki. (Reproduced by permission of the Director, Lanmaelingar Islands.)

older rock-strips, can be matched with that encountered on the flanks of the East Pacific Rise and can be dated by reference to the known sequence of magnetic reversals (p. 283).

From the breadth of the anomalies, it has been inferred that about 100 km of crust has been added to the Norwegian Sea on either side of the ridge during the last 10 m.y. The implied rate of movement for each plate — about a centimetre per year — is slow compared with movements recorded elsewhere. The apparent rate of spreading in the Southern Atlantic was somewhat faster, in conformity with the much wider gape of the ocean.

Independent evidence of sea-floor spreading is provided by lateral variations in age of the oldest sedimentary and igneous rocks obtained from the sea-floor. The distribution of sediments dated by their fossil contents (Funnell and Smith,

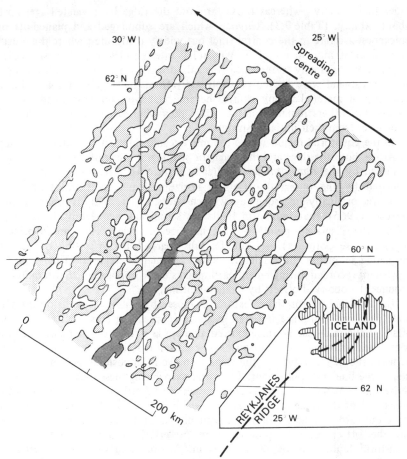

Fig. 9.3. Magnetic anomalies over the Reykjanes Ridge, based on Heirtzler and others

1968) suggests that early Tertiary and Cretaceous deposits are confined to the sea-floor adjacent to the continental margins, whereas late Tertiary and Holocene material is distributed over almost the full width of the ocean; the mid-Atlantic ridge and its immediate environs are almost bare. The total thicknesses of sediment are small. The greater part of the terrigenous detritus is trapped along the continental margins (see later), and the more widespread *pelagic sediments* appear to have accumulated very slowly, except beneath the productive polar waters where oozes of organic origin are more bulky

The dating of igneous rocks on the sea-floor has been attempted in a few areas. Dredge samples obtained by the Geological Survey of Canada on traverses of the mid-Atlantic ridge near Latitude 45°N yielded ages ranging from 13 000 years in the central rift valley to 8 m.y. some 65 km to the west. Oceanic islands which lie close to the ridge – Bouvet Island, Tristan da Cuhna, Ascension, the Western Azores, central Iceland and Jan Mayen – have yielded no isotopic dates older than 20 m.y., whereas those far from the ridge have yielded ages up to about 60 m.y. (Table 9.3). *Guyots*, which are submerged and planed-off old volcanoes, are rare in the central tract but widely distributed where the crust is more than 30 m.y. in age. J. T. Wilson has inferred (1963) that most volcanic islands were originally formed near the mid-oceanic ridge above the hot zone of active crust and were moved passively away from the ridge by the addition of new crustal material. Many centres became inactive, and submerged to form guyots, but others remained active while they were carried many hundreds of kilometres from their initial positions.

In the equatorial region where the mid-oceanic ridge curves round the bulge of Africa, both the ridge and the magnetic striping of the sea-floor are repeatedly displaced by the *transform faults* already mentioned. The sense of movement indicated by the observed offsets is horizontal, and transform faults were therefore originally regarded as wrench-faults. J. T. Wilson pointed out, however (1965), that transform faults are seismically active only in the tract linking the displaced points on the mid-oceanic ridge. They are continued beyond these points as aseismic features marked by offsetting of oceanic magnetic anomalies but seldom enter continental regions. The sense of movement recorded during recent earthquakes is the opposite of that required to produce the observed offsets by wrench-faulting.

The explanation for these features proposed by Wilson depends on the fact that transform faults are initiated in zones of spreading crust. Where an initial offset of the mid-oceanic ridge existed, perhaps as a result of irregularities in the continental margins, the addition of new crustal material must cause the rocks between the offset points on the ridge to move in opposite directions on either side of the line of displacement. The surface separating these moving blocks is an active fault. As new material is added, each crustal strip in turn travels past the offset end of the ridge into a region in which blocks on both sides of the fault move outward together. Relative movement and seismic activity therefore cease, but the offsets already produced are preserved in the magnetic pattern. Transform faults may be regarded as integral parts of the oceanic structure formed at an early stage and often concentrated in zones where the supercontinents ruptured irregularly. The trace of the faults is necessarily parallel to the direction of movement of the oceanic plates; the faults themselves therefore lie on small circles about the 'pole of spreading' and their orientation may provide the clearest evidence as to the direction of movement of the plates (Fig. 9.2). In the Atlantic between 30°N and 10°S, transform faults trace out small circles about a pole of spreading near the southern tip of Greenland; the fastest rates of opening should therefore be expected in a zone about 30°S, an expectation which agrees reasonably well with rates calculated from the width of the magnetic stripes by Le Pichon (Table 9.1).

4 Atlantic igneous activity

We may draw a first distinction between forms of igneous activity which were essentially marginal and those which were truly oceanic. *Marginal vulcanism* dated largely from the early stages of ocean-formation. In mid-Mesozoic times,

Table 9.1. RATES OF SPREADING IN THE ATLANTIC OCEAN

Latitude	Spreading rate (cm/year)
60 N	0.95
28 N	1.25
22 N	1.40
25 S	2.25
28 S	1.95
30 S	2.00
38 S	2.00
41 S	1.60
50 S	1.53

ring-complexes and other central intrusions, with relatively small accompaniments of lava, were emplaced within the continental crust along lineaments associated with some of the marine basins developed on the site of the Atlantic. Reference has already been made to centres of this type, often alkaline in character, situated near the coasts of south-western Africa and Brazil (Chapter 8). Rather similar ring-complexes of Tertiary date are situated along a lineament running obliquely northward from the eastern margin of North America (the *Monteregian magma series*). Predominantly basic assemblages occupy the continental margins in several well-defined regions — on either side of the Davis Strait, where they are of late Mesozoic and early Tertiary age; and near the ends of the aseismic Iceland-Faroes ridge in East Greenland and north-west Britain (pp. 272—4).

The floor of the ocean basin proper and the oceanic islands that rise from it are made principally of basic igneous materials. Although it is tempting to treat the islands as samples of oceanic material it must be borne in mind that islands are anomalous structures in ocean basins: that many are situated on dislocations and that convenience rather than logic has given them an undue importance.

Sampling the ocean floor — by dredging and coring — has revealed igneous rocks falling into several categories:

(a) *Basaltic lavas* and *dolerites* appear to be very widely distributed in the higher parts of the crust. They have common chemical characteristics and are lumped together at the present stage of knowledge under the term *oceanic or abyssal tholeiites*. In their high silica, lime and alumina and low potash content they are distinct from continental basalts and approach the composition of achondritic calcium-rich meteorites. A few variants have normative quartz; a few have normative nepheline and can be classified as alkali-basalts; spilites seem to be rare. Their apparent chemical consistency, not only in the Atlantic but also in the other oceans, suggests that the oceanic tholeiites are derived from a widely-available source; many authorities regard them as derivatives of the fluid fraction produced by partial melting of peridotitic material in the upper mantle.

(b) *Gabbros* which resemble the oceanic tholeiites in composition have been recorded, especially where submarine fault-scarps expose deep sections through the basaltic pile. Igneous layering suggests that some belong to intrusive bodies differentiated *in situ*. Enrichment in iron and titanium has been recorded in differentiates of oceanic gabbros and a few dioritic samples have been recovered.

(c) *Peridotites*, usually heavily serpentinised, include many variants – olivine, orthopyroxene, clinopyroxene and spinel are combined in various proportions in samples from single dredge-hauls suggesting that they may be derived from layered bodies. The textures often indicate partial mylonitisation. Most of these ultramafic rocks have been found where deep sections are exposed on fault-scarps. They appear at the surface, however, in *Saint Paul Rocks*, a group of islets close to the mid-Atlantic ridge in the dislocated equatorial region. The prevalence of mylonitic textures in these islets and in dredged samples has led to the suggestion that masses of ultrabasic material have been emplaced within the basaltic crust in a solid condition; such masses might be derived from residual mantle material in the zone of partial fusion in which the oceanic tholeiites were generated.

(d) *Metamorphic derivatives* of the igneous rocks mentioned above have been recorded from several areas of high submarine relief where faulting reveals rocks originally buried to a depth of 2 km or more. The extent of oceanic metamorphism is not yet clear; there are indications that high geothermal gradients exist beneath the mid-Atlantic ridge, but since most specimens obtained so far have come from the vicinity of faults, the effects of local dislocation metamorphism cannot be discounted. Most of the metamorphic products are of zeolite or greenschist facies, but a few characterised by hornblende and plagioclase appear to be of amphibolite facies. Preliminary studies suggest that the lower parts of the oceanic crust may be composed largely of metabasalts, metagabbros and serpentines metamorphosed before they left the hot mid-oceanic ridge.

The *oceanic volcanic islands* form large and small groups scattered widely over the ocean floor. By far the largest single unit is Iceland, which straddles the mid-Atlantic ridge and is here considered at some length. Details of other island groups are summarised in Table 9.3 and will be referred to again later.

Iceland. The mid-oceanic (Reykjanes) ridge comes ashore in south-western Iceland and is continued as a bifurcating zone of sub-aerial volcanic and seismic activity for a distance of over 400 km before plunging again below sea-level off the north coast (Fig. 9.4). The land-mass coincides with a negative bouguer anomaly which suggests the existence of upper mantle material of a less than average density at depth. Similar anomalies have been recorded from other oceanic islands.

The geological structure is broadly symmetrical, Pleistocene and Holocene volcanic rocks occupying a central tract enclosing the bifurcating active zone and Tertiary lavas cropping out to east and west. On the eastern side of the island the lavas dip westward at angles of up to 8°. The lowest exposed groups have been dated isotopically at 12.5 m.y. (mid-Miocene) and no major stratigraphical break is apparent between the Tertiary lavas and those still being erupted in the central zone. Successive groups of lavas thin rapidly up-dip and probably terminated not far east of their present outcrops. On the western side of Iceland, the structure is complicated by warping of the lava-pile, but here too the oldest rocks are remote from the active zone; they have yielded isotopic ages of about 16 m.y. The entire

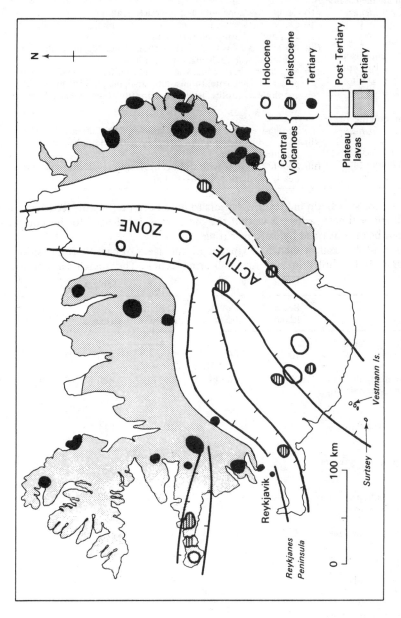

Fig. 9.4. Sketch-map of Iceland showing the distribution of plateau-lavas and central volcanoes (based on maps of Walker and Sigurdsson)

Table 9.2. ZEOLITE ZONES DEVELOPED IN BASALTS OF EASTERN ICELAND
(based on Walker, 1960)

(a) In olivine-basalts
highest

↓ increasing metamorphic grade ↓

(i) Zeolite-free zone with empty vesicles, 200 m
(ii) *Chabazite-thomsonite zone* with levyne, phillipsite (calcite, apophyllite, gismondite) *c.* 44 m
(iii) *Analcime zone* with chabazite, thomsonite, levyne, phillipsite (mesolite, stilbite, heulandite) *c.* 160 m
(iv) *Mesolite-scolecite zone*, widespread zeolitisation, many species present — chabazite, thomsonite, analcime, stilbite, heulandite, apophyllite, girolite, levyne, laumontite at low levels, 800 m

(b) In tholeiites
highest

increasing metamorphic grade

(i) Zeolite-free zone with empty vesicles, 600–700 m
(ii) *Chalcedony-quartz-mordenite zone* with chlorophaeite and celadonite, 500 m
(iii) *Zeolite-rich zone* with stilbite, heulandite, scolecite, epistilbite, quartz, chalcedony, 300–400 m

lava-pile, some 400 km in breadth, appears to have been built up in Neogene and post-Tertiary times; earlier views that the older lavas dated back to the Eocene have not been confirmed by isotopic dating.

Some 88 per cent of the whole igneous assemblage is of basaltic composition, including both tholeiitic and olivine-bearing types. The remainder consists

Table 9.3. OCEANIC

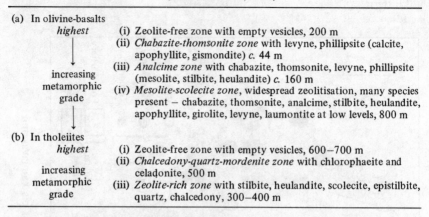

	Bouvet Island	Gough Island	Tristan da Cuhna	St Helena	Ascension Island
Number of islands	1	1	3	1	1
Area (km²)	70	68	115	120	96
Approximate distance from mid-oceanic ridge (km)	0	560 E	480 E	800 E	110 E
Age of oldest dated rocks, m.y.			18	14.5	
Centres active in recent times			X		
Approximate percentage basaltic/all igneous rocks		45	90	90	90
Alkali-basalts present	X	X	X	X	
Trachytes present		X	X	X	X
Alkaline central complexes			X	X	
Granite or granophyre present					as ejected blocks

*from faunal evidence in associated sediments

predominantly of rhyolites, acid pyroclastics, felsites, granophyres and granites, with little intermediate material. The proportion of acid to basic rock is not far from that which might theoretically be obtained from the fractionation of basic magma; the isotopic ratio $^{87}Sr/^{86}Sr$ in certain acid rocks of 0.700–704 δ, is similar to the ratios in basic rocks derived directly from sources in the mantle (cf. p. 236). These geochemical considerations suggest that the Icelandic massif is wholly of oceanic origin, an inference supported by the fact that the standard reconstruction of the supercontinent of Laurasia prior to the formation of the Atlantic leaves virtually no room for Iceland (Fig. 1.1).

The lavas include plateau-basalts produced by fissure-eruptions, and assemblages related to central volcanoes which form lenticular masses intercalated in the lava-pile; they include most of the acid volcanics and are pierced by major and minor intrusions of both basic and acid composition. Secondary changes amounting to low-grade regional metamorphism are recorded in eastern Iceland by the development of zeolite assemblages in vesicles and throughout the rock-fabric. A vertical metamorphic zoning (Table 9.2) appears to be controlled by the thermal gradient.

Basic dyke swarms roughly parallel with the mid-oceanic ridge are important components of the structure. These swarms, which include the feeders of the plateau-lavas, increase in density with depth; at sea-level in the Reydarfjördur area of eastern Iceland, for example, Walker recorded a thousand dykes with an

ISLANDS OF THE ATLANTIC

St. Paul Rocks	Cape Verde Islands	Canary Islands	Maderia	Azores	Faroes	Iceland	Jan Mayen
(1)	>10	>10	4	>10	>10 1400	>10 100 000	1 350
0	1500 E	1500 E	1000 E	130–0 W 0–400 E	600 E	0–200 E and W	0
	Jurassic	20	pre-Upper Miocene*		60	16.0	Late Tertiary?
		X		X	X	X	X
0 (ultrabasic)	High	80	High	High	100	90	High
	X	X	X	X	X		X
	X	X	X	X			X
	X	X	(X)				
	(X)					X	

aggregate thickness of 3 km in a section of 53 km. Walker has calculated that emplacement of the dykes needed for eruption of a lava-pile with the breadth of that forming Iceland should involve a lateral crustal extension of 38 km. Extension due to this cause, combined with that resulting from the eruption of successive groups of lavas with lenticular cross-section, would seem competent to bring about a broadening of the Icelandic massif at about the rate suggested from the magnetic striping on either side of the Reykjanes Ridge.

Iceland is by far the largest land-mass made entirely of oceanic igneous material. Its position at the meeting-point of the mid-oceanic ridge with submarine ridges which have been marked by exceptional igneous activity, suggests that it is situated on a part of the ridge with a long history of unusual activity. Some other large island groups are close to transform faults; the Azores, for example, are spread along an ESE line of seismic activity which extends toward the Straits of Gibraltar. The distance of any island from the mid-oceanic ridge, measured parallel to the direction of spreading, should, as J. T. Wilson has pointed out, be proportional to its age. The available isotopic data (Table 9.3), though consistent with this inference, are of limited significance because they refer to the islands themselves, and not to their much more bulky submarine foundations. Baker and others (1967), for example, point out that the sub-aerial part of St Helena, which had a minimum active life of 7.5 m.y. forms only about 5 per cent of the entire volcanic pile. Geochemical variations related to distance from the ridge have been postulated by Wilson who suggested that active centres remote from the ridge have a more alkaline cast than those close to the ridge, perhaps reflecting an increase in depth of the source-region. Not all workers, however, accept the reality of an outward increase in alkalinity.

5 The western margin

The regions bordering the western Atlantic fall into four principal sectors:

(a) The northern region: a stable continental margin characterised by thick accumulations of sediments.

(b) The Gulf of Mexico and Caribbean arc: a mobile region incorporating deep sedimentary basins, a volcanic island arc and associated oceanic trenches.

(c) The southern region: a stable continental margin locally carrying thick accumulations of sediment.

(d) The Scotia arc: a fragmented arc closely related to the Andean mobile belt.

(a) The northern region. In the far north, marine and non-marine sediments, mainly sandstones, shales and bituminous shales, underlie Tertiary basalts on both shores of Baffin Bay and probably extend onto the continental shelf. They are principally late Cretaceous (West Greenland) or Palaeocene (Baffin Island) and suggest a late Mesozoic date for the onset of ocean-formation.

Alongside the mainland of North America, tracts of Mesozoic and Tertiary sediments occupy much of the continental shelf and slope and form the coastal plain which fringes the Appalachians; almost the whole of the continental shelf may have been above sea-level at the maximum extent of the Pleistocene ice-sheets. In these areas, well-known seismic, magnetic and gravity surveys show that two parallel prisms of sedimentary material are separated by a narrow tract in which basement rocks lie closer to the surface (Fig. 9.5). A linear magnetic anomaly, the *Atlantic shelf anomaly*, which appears to follow the continental

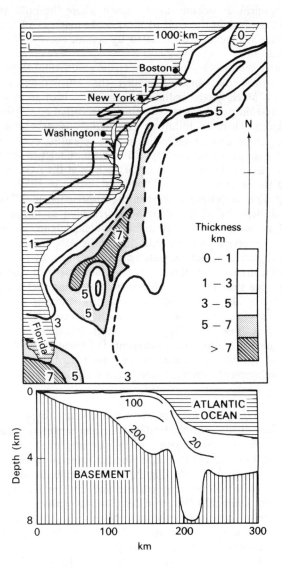

Fig. 9.5. Map and section, based mainly on seismological studies, of the Mesozoic-Tertiary sedimentary prisms deposited at the Atlantic margin of North America; the figures on the section give, in m.y., the inferred ages of certain reference-surfaces (based on publications of Ewing, Drake and others)

margin not far from the shelf-break, may indicate the existence of a belt of igneous rocks.

The inner sedimentary pile which occupies the coastal platform and much of the continental shelf includes Jurassic rocks and reaches thicknesses of well over 5 km. It consists of stable 'platform' sediments, mainly detrital in the north but mainly chemical or organic in the south where the bulk of the land-derived detritus is trapped in the Gulf of Mexico. As much as 11 km of late Mesozoic and younger sediments are indicated by seismic surveys in the region between Florida and Cuba.

The outer sedimentary pile occupies the continental slope and extends onto the continental rise, beneath which the crust thins sharply to oceanic proportions. Much of the sediment in this zone is of greywacke type and may represent resedimented material originally deposited on the shelf. At the present day, turbidity currents following the numerous *submarine canyons* in the continental slope deposit sediment in fans spreading out over the continental rise. In the south, carbonate deposition predominates in the shallow waters of the Bahama Banks. The two sedimentary prisms at the Atlantic margin have been compared (for example Drake, Ewing and Sutton in 1959) with the fillings of miogeosyncline and eugeosyncline in the Appalachian mobile belt, the relatively well-sorted deposits of the shelf being regarded as miogeosynclinal and the 'overspill' turbidites of the slope as eugeosynclinal. Igneous contributions, though not known for certain, may be indicated by the Atlantic shelf anomaly.

The recognition of the characteristic arrangement of sedimentary deposits at the Atlantic margins has played an important role in the development of ideas concerning the significance of Phanerozoic orogenic cycles. Since these deposits accumulated at the trailing edge of a continent during the Mesozoic–Tertiary period of sea-floor spreading, many authors have inferred that geosynclinal prisms such as those of the Appalachians were laid down in a similar setting after earlier phases of continental disruption. Deformation and metamorphism, not recorded at the present Atlantic margin, are regarded as phenomena associated with the subsequent closing of the ocean and consumption of crust at Benioff zones.

(b) *The Gulf of Mexico and the Caribbean arc.* Between North and South America there is no cratonic land-mass on the western border of the Atlantic. Here, a complex of small marine basins partially divided by the peninsulas of Yucatan and Florida, and the Caribbean arc intervene between the main ocean and the western mobile belt (Fig. 9.6). The Caribbean island arc is a volcanic island arc convex toward the Atlantic and flanked on this side by oceanic trenches and sediment-filled troughs. Seismic activity continues in a zone dipping westward beneath the islands, and both islands and trenches show large departures from isostatic equilibrium. The *Gulf of Mexico* and the *Caribbean Sea* appear to be floored by crust of varied, but partly oceanic, thickness.

The coastal plains and continental shelf bordering the Gulf of Mexico are underlain by thick Mesozoic to Recent sediments which are of exceptional importance on account of their proved and potential oil-reserves. Evaporites, probably formed in Jurassic times in early basins related to the definition of the Atlantic margin (compare West Africa, p. 232), occur at a low level in the

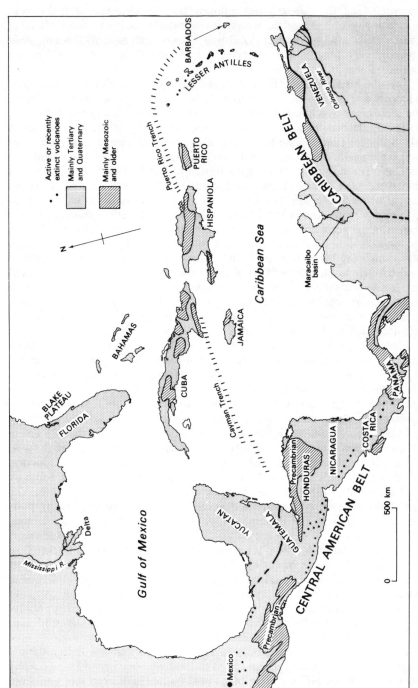

Fig. 9.6. Locality map of Central America and the Caribbean

succession, and the *salt-tectonics* resulting from diapiric rise of these rocks provide many of the oilfield structures. Salt-domes occur in hundreds in an arcuate belt curving through Louisiana, Texas and eastern Mexico, while submarine salt tongues are squeezed out on the floor of the Gulf at the continental margin.

Near the eastern side of the Gulf, in and near Florida and Yucatan, carbonate sediments and evaporites predominate, indicating a rather stable environment. The successions which occupy the western part of the Gulf are often thicker and are largely detrital, as a consequence of their proximity to the western mobile belt. Sediment from this western part is today swept by turbidity currents down the exceptionally steep continental slope into the deep waters of the centre of the Gulf. The northern border is dominated by the Mississippi delta, receiving detrital sediment from a vast drainage basin in central North America. The delta consists of many overlapping lobes and its history is recorded in sediment-piles of formidable thickness; a borehole near the coast of Louisiana penetrated 3 km of Pleistocene sediment and another in the same region over 7 km of Miocene and post-Miocene sediments. In each successive group of Tertiary delta-formations a southward transition can be observed from the non-marine facies of the delta-plain to the marine facies of the submerged delta. Some eight short-lived transgressions complicated the general southward advance of the delta and cause marine and non-marine strata to interfinger. A rather abrupt thickening of each formation near the transition to marine facies marks the position of the delta-front and produces a bulge known as a 'flexure'.

Along the western border of the Gulf of Mexico, Jurassic and Cretaceous carbonate sediments were laid down in a 'miogeosynclinal' environment related to the Cordilleran mobile belt. A late Cretaceous—early Eocene wedge of flysch-like detrital sediments extends eastward over them to interfinger with limestones on the Yucatan peninsula: its proximal parts are folded by Eocene movements and post-tectonic marine sediments extend over it into the Gulf.

The Caribbean Sea is floored by a crust not much more than 10 km in thickness and made largely of high-velocity material. It is bordered on the west by the narrowest part of the land-bridge connecting North and South America, (which was probably not completed until Pliocene times), and on the south by the Miocene—Pliocene Caribbean orogenic belt of Colombia and Venezuela. The thick Cretaceous—early Tertiary sequences of this belt are mantled by post-tectonic accumulations of Pliocene and younger age which occupy the Orinoco valley and delta.

The Caribbean or Antillean arc stands on a welt of crust much thicker than that beneath the ocean basins on either side; beneath Puerto Rico, the crust appears from seismic refraction and gravity surveys to be some 30 km in thickness, thinning both northward and southward to about 10 km. Positive bouguer anomalies mark the islands of the arc, while a belt of strong negative anomalies follows the *Puerto Rico trench* and extends eastward and southward on the convex side of the arc for over 2000 km (Fig. 9.6).

The Caribbean arc is one of only two island arcs associated with the Atlantic Ocean. Recent earthquake foci have been concentrated in a Benioff zone dipping westward at about 60° and deep-focus earthquakes have been recorded from depths of 200 km (Sykes and Ewing, 1965). Transcurrent east—west dislocations tend to detach the central sector of the arc from its moorings on the continent; one of these dislocations crosses the arc between Cuba and Hispaniola to follow the line of the Cayman trough.

The islands of the Greater Antilles are founded on Cretaceous volcanic edifices formed at, or soon after, the opening of the Atlantic. The main period of vulcanicity in these northern islands was late Cretaceous and Eocene. The Lesser Antilles consist of an outer line of low islands in which degraded early volcanoes are partly covered by Tertiary limestones and tuffs and an inner line of Pliocene and Pleistocene volcanoes some of which are still active. Late Tertiary and Quaternary sediments of local derivation mantle the flanks of Cuba, Hispaniola and Puerto Rico and fill the trough separating the outer and inner arcs of the Lesser Antilles.

The *oceanic trenches* developed on the Atlantic side of the Caribbean arc (Fig. 9.6) are partly filled by sediment and are characterised by negative bouguer anomalies. The island of Barbados which projects within the trench-zone reveals over 5 km of folded Eocene to Miocene deposits beginning with flysch-like clastics and ending with fine detrital sediments and carbonate rocks. Up to 12 km of sediment may be contained in the southern portions of the trench which lay closest to the sourcelands of the Caribbean mobile belt in Venezuela. The *Puerto Rico trench*, remote from the source, contains little sediment, although it is thought to have been in existence since Eocene times. This old-established crustal feature remains conspicuously unstable. The abyssal plain which makes its floor lies some 2.5 km below the adjacent ocean-floor and more than 8 km below sea level. The thickness of the crust varies rapidly and high negative bouguer anomalies show that isostatic equilibrium has not been attained. Shallow-focus earthquakes are frequent close to its inner (southern) wall where the Benioff zone approaches the surface. The mobility of the continental margin in the Caribbean sector may be due to the effects of relative movement between the crustal plates which incorporate the cratons of North and South America. The Caribbean arc, the effective boundary of the western Atlantic plate off central America, stands out far to the east of the plate boundary to north and south and shields one or more small plates on its concave side.

The thick crustal belt which supports the Caribbean islands appears to incorporate continental material. Cuba may indeed be underlain by a Palaeozoic basement similar to that beneath the neighbouring Florida peninsula. The dominance of andesitic lavas, pyroclastics and intrusives in the younger centres of the Lesser Antilles has been cited as evidence for the occurrence of a continental basement. The older volcanic assemblages of the Greater Antilles, however, are oceanic in character. A 'basement complex' of metamorphosed igneous rocks apparently of early Cretaceous or pre-Cretaceous age is of ophiolitic character, including large masses of serpentinised nickeliferous peridotite together with basic (sometimes spilitic) and keratophyric lavas metamorphosed in greenschist or amphibolite facies. In the volcanic sequences which overlie this basement complex, lavas (usually basic but including keratophyres and andesites) predominate towards the base and pyroclastics towards the top. The earliest (Jurassic to mid-Cretaceous) sediments interbedded with the volcanics are limestones and contain no detrital quartz. Only in the latest Cretaceous and

Eocene do clastic sediments make their appearance: a feature which, with the increasing abundance of pyroclastics, suggests the emergence of subaerial volcanoes. The main clastic sediments date from late Eocene to Miocene times, when the dormant volcanic cores of the large islands shed erosion debris into marginal basins.

In the Greater Antilles and in the outer arc of the Lesser Antilles, the bulk of the igneous rocks belong either to the serpentine-spilite-keratophyre group or to a calc-alkaline suite including high-alumina basic rocks, andesitic lavas and pyroclastics and small tonalitic or dioritic intrusions. The more silicic members of this suite predominate in the relatively small centres of the inner arc in the Lesser Antilles where recent volcanic activity has involved the explosive eruption of *nuées ardentes.*

(c) *South America.* The Atlantic margin of South America is almost devoid of seismic activity. The earliest structures related to this margin are narrow basins corresponding to those of west Africa and, like the latter, often containing evaporites followed by marine strata. Oblique lines of alkaline igneous centres traverse the Brazilian shield near the continental margin (see Chapter 8, p. 236).

Younger sedimentary basins at the continental margin are related to the drainage pattern of the continent. The broad shelf off Argentina is occupied by several east-south-east basins containing up to 8 km of sediment supplied by rivers draining the Andes. These basins are set transverse to the coast and, at least the larger, extend to the edge of the continental shelf. The shelf alongside the positive Brazilian shield is narrow and carries little sediment, whereas the broad Amazon basin is extended seaward by thick submarine deposits.

(d) *The Scotia arc.* At the southern tip of South America, the Atlantic and Pacific Oceans meet. The junction of the oceanic crustal plates is marked by the hairpin Scotia arc, which, like the Antillean arc, stands out far to the east, with its convex side facing the Atlantic. A glance at the map (Fig. 9.2) shows that the southern tip of South America in Tierra del Fuego swings round to the east and, on the other side of Drake Passage, the Antarctic Peninsula of Graham Land makes a similar eastward swing. The eastern tips of these two lands are connected by a shallow submarine loop on which stand a number of widely spaced islands and rocks.

The predominantly andesitic and Pleistocene *South Sandwich Islands,* which form the central part of the arc, appear to mark the junction of Atlantic and Pacific crustal plates. The oceanic trench on their east side and the westward dip of the associated seismic zone suggest that the Atlantic plate is overridden at this junction. Other islands of the Scotia arc, including South Georgia and the South Orkneys and South Shetlands incorporate folded and metamorphosed sediments and volcanics dating back at least to Mesozoic times and associated with Andean-type granites. They are best regarded as displaced and modified fragments of the Andean mobile belt, linked to their moorings via the deflected tips of Tierra del Fuego and the Antarctic peninsula. Disruption, involving transcurrent faulting, was probably post-Aptian according to Frakes. Preliminary magnetic surveys of the Scotia Sea have revealed an old oceanic anomaly pattern suggesting that the Sea opened along a north-easterly ridge which became inactive before the end of the Tertiary.

6 The eastern margin

Along the north-eastern continental margins, several narrow fault-bounded marine troughs received Mesozoic sediments ranging from Triassic to Cretaceous. The narrow Mesozoic outcrops of East Greenland lie within the Caledonides, those of north-west Britain near the Caledonian front. These outcrops are truncated obliquely by the present continental slope, whose trend is clockwise from that of the early troughs. It may be inferred that Europe and Greenland parted along a fracture slightly oblique to the early Mesozoic troughs.

The bordering land-areas of the North Atlantic are made largely of Palaeozoic or Precambrian rocks whose grain is oblique to their margins. The continental slope appears to be defined by faults in some places: notably along the northern and western sides of the Iberian peninsula. A fragment of continental crust made of Precambrian gneisses and granulites and separated from the continental margin by a sediment-filled oceanic basement is seen in the Rockall Bank. Southward from the mouth of the English Channel, the slope is scored by submarine canyons. Rocks of basement aspect appear, from records of continuous reflection profiles (Stride *et al.*, 1969), to crop out far from land and even near the base of the continental slope. These features suggest that the margin of Europe is not primarily a constructional feature. To the south of the Mediterranean area, as we have seen (Chapter 8) the African continent is bordered at many points by Mesozoic–Tertiary basin-deposits arranged parallel to the continental margin.

Sedimentary sequences. Reflection-profiles and other evidence suggest that late Mesozoic and Tertiary sediments form a veneer only a few hundred metres thick on many parts of the shelf off the European coast and that the successions are broken by at least one unconformity (late Cretaceous–early Eocene). Much greater thicknesses are recorded in several basins oblique to the continental margin. Apart from the North Sea basin (see below), the principal basins are that of the *Western Approaches* off the English Channel where chalk, limestones and siltstones (mainly Upper Cretaceous to Miocene) reach thicknesses approaching 4 km; the *Aquitaine basin* which, at its landward end, contains up to 4 km of Jurassic, Cretaceous and Tertiary strata; and the *Lisbon trough* which is perhaps of similar thickness.

The explanation for the restriction of deposition along the European margin appears to reflect an eastward tilt of the continental shoulder. The great rivers north of the Alpine-Carpathian mountains open into shallow epicontinental seas which are separated by a rim of Palaeozoic massifs from the Atlantic. The Baltic Sea, the North Sea and their forerunners acted as sediment-traps through the Mesozoic and Tertiary eras. The total thickness of sediment in the *North Sea basin* immediately north of the Dutch coast runs up to more than 6 km, while Tertiary sediments alone exceed 3 km in the region between southern Norway and northern England. Considerable reserves of oil and natural gas are held in the

North Sea succession where broad upwarps, faults and salt-domes derived from Permo-Triassic evaporites provide oil-traps (Fig. 3.14).

Igneous activity at the continental margin. Marginal igneous activity in the northern Atlantic was concentrated at points where the Iceland-Faroes ridge meets the adjacent continents — in East Greenland south of Scoresby Sound and in north-western Britain. Both of the marginal sub-provinces are classic areas from which fundamental concepts concerning the evolution and emplacement of magmas have been derived. Their general characters are summarised in Table 9.4 and only a few points call for further comment.

Table 9.4. TERTIARY IGNEOUS ACTIVITY IN THE MARGINAL PROVINCES OF THE NORTH-EAST ATLANTIC

	NORTH-WEST BRITAIN	EAST GREENLAND
Age	Mainly Eocene, lavas give ages up to 70 m.y., central intrusions ages of about 55 m.y.	First lavas underlie early Eocene, probably mainly Palaeocene
Tectonic setting	Most plutonic centres and lava-fields lie on a north—south lineament oblique to continental margin	Plutonic centres and lava-fields lie on a lineament parallel to coast and continental margin: lineament monoclinal with downwarp to ocean
Scale	Lineament extends 500 km: maximum thickness of lavas 3 km	Lineament extends for 1500 km: maximum thickness of lavas over 7 km
Lavas	Dominantly basalts, including olivine-basalts, tholeiites, feldspar-rich basalts. Rhyolitic and trachytic lavas mostly near central intrusions	Dominantly basalts, including olivine-basalts, tholeiites Rhyolites, remnants of strongly alkaline lavas recorded from a few localities
Central Intrusions	Layered gabbros Layered ultrabasic-basic associations Granites Gabbro-granite associations Ring-complex of many kinds, local evidence of association with surface calderas (Mull). Central intrusions emplaced in lava-pile or underlying rocks, younger than most of the lavas	Layered gabbros, for example Skaergaard Gabbro-granite associations Alkaline associations including nepheline-syenites Central intrusions emplaced in lavas or underlying rocks, younger than most of the lavas
Dyke swarms	Basalts and dolerites predominate. Swarms related to igneous centres. mostly strike NW—SE, oblique to principal lineament and to continental margin	Dolerites predominate. Swarm localised along monoclinal flexure, parallel to continental margin
Petrological character	Calc-alkaline	Calc-alkaline and alkaline

Vulcanicity in both regions is thought to have begun early in the Palaeogene, that is, not long after the separation of the continental masses began. The igneous assemblages are thus older than those of Iceland but broadly contemporary with those of the Faroes (see Table 9.3). The association of vulcanicity with tectonic disturbance at the continental margin is emphasised by the concentration of activity along certain lineaments. A monoclinal downwarp defines the coast of East Greenland for some 800 km. The Eocene plateau-lavas thicken (Plate VI) seaward and may be downwarped over the monocline by as much as 10 km. A coast-parallel dyke swarm is emplaced in fractures arranged radially with respect to the centre of curvature and is most dense in the monocline. The structural relationships of the igneous tract of north-west Britain are more complex, for the lava-fields and most of the plutonic centres are concentrated along a line oblique to the continental margin and close to an old-established zone of fracturing defined in Mesozoic or earlier times (see p. 271). Local dyke swarms have a north-west trend and their alignment with respect to the main igneous lineament suggests that sinistral transcurrent movement took place along this lineament.

Basic rocks are overwhelmingly predominant, forming by far the greater part of the lava-piles and being represented in many of the central complexes. As was

Plate VI. Tertiary plateau-basalts at the continental margin in East Greenland. The flat-laying lavas on the south side of Gaase Fjord rest on an irregular basement of Caledonian gneisses seen towards the left; the highest tops are at 1900m. (Reproduced by permission of the Director, Greenland Geological Survey.)

realised at an early stage in the investigation of the British province, basic suites with differing chemical characteristics occur in close association and from this discovery arose the concept of differing 'magma-types' available at depth. In the island of Mull, Bailey and others (1924) recognised rocks related to two principal magma-types: the Plateau Type, and the Non-porphyritic Central Type. These types – renamed by Kennedy the *olivine-basalt* (*alkali-basalt* of Tilley) and *tholeiitic* types respectively – have proved to be of world-wide importance, and the problem of their relationship is still open to discussion (see, for example, Yoder and Tilley, 1962). Many minor igneous types in north-west Britain appear to be derivatives of one or other magma-type, modified by fractional crystallisation, by hybridisation or by assimilation. Rocks of alkaline cast, mugearites, trachytes etc., are regarded as derivatives of the olivine-basalt magma-type, while ultrabasic cumulates and acidic rocks, rhyolites, felsites, granophyres, granites, are linked to the tholeiitic magma-type.

The status of plutonic acid rocks in the central complexes has aroused much discussion. Although the proportion of acid rocks in the province as a whole is probably not more than 10 per cent (much the same as it is in Iceland), the acid:basic proportions in the plutonic centres expressed in terms of area of outcrop are almost 2:1; moreover, rocks of intermediate composition which might be expected as members of a differentiation series are represented only in trivial amounts. These anomalies led Holmes (for example, 1936) to conclude long ago that many of the granites and granophyres represented crustal material remelted by the ascending basic magma. Strong positive bouguer anomalies over the outcrops of granites in Skye, Arran and elsewhere, suggest that these bodies are indeed only cappings on columnar basic intrusions and strontium isotope abundances and certain other chemical ratios seem consistent with the derivation of granites in Skye from acid gneisses (Lewisian), or arkosic sandstones (Torridonian), of the Precambrian basement.

The Iberian pensinsula and the Bay of Biscay. Between the peninsulas of Brittany and Spain, the Bay of Biscay lies mainly at depths of over 1000 m and is apparently underlain by a crust of oceanic type some 11 km in thickness. The awkward shape of this embayment spoils the otherwise excellent fit of North America against Europe. Early reconstructions by du Toit improved the fit by rotating the Iberian pensinsula clockwise to close up the Bay of Biscay. Bullard *et al.* (1965) obtained a good fit of the 1000 m contour against the Grand Banks off Newfoundland by a clockwise rotation of 39°. These procedures (which, of course, imply that the Iberian pensinsula rotated anticlockwise at some stage in the opening of the Atlantic) are supported by palaeomagnetic studies, since the Lisbon volcanics, a suite of Eocene basalts, appear to have a palaeomagnetic orientation significantly different from that of other European rocks of the same age; partial closure of the Bay of Biscay restores the palaeomagnetic direction to an orientation like that of rocks from other parts of the craton of Europe. We saw in the last chapter that displacement of small blocks was a feature of the development of the Mediterranean area and the rotation of the Iberian peninsula may fall into place as an adjustment to complex plate movements in this region.

III The Arctic Ocean

The Arctic area is, not unnaturally, little known to geologists and we do not propose to discuss its history. The mid-Atlantic ridge has been traced by Soviet geologists between North Greenland and Spitsbergen, and thence across the polar region to terminate near the New Siberian Islands. Over at least the eastern part of its course it is bordered by symmetrical linear magnetic anomalies. The extent of deep waters underlain by crust of oceanic type is limited and on the tectonic chart of Atlasov *et al.* (see Hope, 1964) the entire tract between Greenland, Arctic Canada and eastern Asia is shown as being underlain by crustal material structurally continuous with that of the neighbouring lands.

IV The Indian Ocean

1 The ocean floor

The Indian and Southern Oceans together occupy an irregular basin opening southward to connect with both the Atlantic and the Pacific. The shape of the basin is determined by *mid-oceanic ridges* which form a system shaped like an inverted Y (Fig. 9.7). The stalk of the Y, running south-eastward from the Gulf

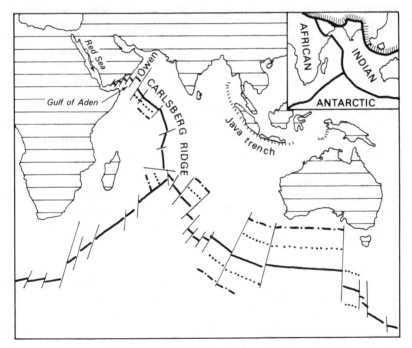

Fig. 9.7. Structural elements of the Indian Ocean with (inset) the boundaries of the principal crustal plates

of Aden to a point north-east of Mauritius, is the *Carlsberg Ridge*, a well-defined feature with the characteristic topography and seismicity of a mid-ocean ridge. Linear magnetic anomalies on either side of this ridge are closely spaced and indicate a spreading rate of not more than 1.5 cm/year. Near Latitude 20 S, the ridge appears to divide: one branch, rather poorly known over much of its course, extends south-westward between South Africa and Antarctica to link up with the mid-Atlantic ridge, while the other curves eastward to pass between Australia and Antarctica, entering the Pacific Ocean in the vicinity of the Antarctic Circle. A pattern of rather broad magnetic stripes mapped over a considerable area south of Australia indicates an average spreading rate of 3.5 cm/year and correlation with the Pacific anomaly pattern suggests that the full width of oceanic crust separating Australia and Antarctica has been formed over the past 35 m.y., that is, since Oligocene times.

Topographical surveys (Heezen and Tharp, 1964) have revealed a conspicuous north-north-east 'grain' produced by the alignment of many narrow aseismic ridges in the Indian Ocean. Some of these features appear to mark transform faults but others are considered to incorporate slivers of continental material.

The mid-oceanic ridges of the Indian Ocean define the boundaries of three outwardly-moving crustal plates. The western plate (Fig. 9.7) includes the whole continent of Africa and the eastern part of the South Atlantic and is almost (though not wholly) encircled by mid-oceanic ridges; this plate is apparently extending towards the east, the south and the west. The southern plate, which includes Antarctica and much of the Southern Ocean, also has a large growing edge. The northern plate is complex, incorporating both peninsular India and Australia, together with the oceanic floor of the Arabian Sea, the northern Indian Ocean and the Tasman Sea. The arrangement of oceanic magnetic anomalies and the alignment of transform faults indicate north-north-easterly migration resulting from the addition of new material at the southern and eastern boundaries, a migration corresponding to that inferred for peninsular India (pp. 7–8). The northern border of the north-eastern plate, defined by the Himalayan mobile belt and its extension into the East Indies has acted as a destructive plate margin.

2 *Continental fragments*

Several island-groups in the western part of the Indian Ocean are strikingly different from the general run of oceanic islands; they have metamorphic or granitic foundations of continental type, which appear to be far older than the surrounding oceanic crust (cf. Rockall Bank, p. 271).

In *Madagascar*, the largest of these continental fragments, a gneissose basement (locally showing metamorphism of granulite facies and incorporating anorthositic intrusions) yields radiometric dates of 500–600 m.y. and incorporates polycylic gneisses probably over 2000 m.y. It is overlain by a late Palaeozoic and Mesozoic cover of Karroo and shallow-water sediments laid down largely in fault-bounded basins. The geological make-up of Madagascar is almost identical with that of eastern Africa and it is clearly to be regarded as a continental fragment. Although the Mozambique Channel which separates it from Africa is more than 2000 m in depth, there is some doubt as to whether it is floored by normal oceanic crust. The *Seychelles*, which represent the high point of an aseismic ridge some 2000 km in length, are made largely of granites giving radiometric dates of just over 500 m.y. These rocks have affinities with those of the Mozambique belt in eastern Africa and appear to represent a continental sliver. Seismic refraction surveys indicate that the Moho lies at depths of some 30 km below them, but rises to only 8.2 km between the islands and the continent of Africa. Other 'microcontinents' may include the Maldive-Laccadive ridge off the south-west coast of India.

3 *Igneous rocks of the ocean floor and oceanic islands*

The limited data suggest that basalts are the dominant rocks represented in the ocean floor. Metabasalts of greenschist facies and serpentines have been obtained from localities near the Carlsberg ridge where faulting reveals rocks from deeper levels. Oceanic islands are rather sparsely distributed and, as in the Atlantic, are predominantly basaltic. Olivine-basalts of mildly alkaline types occupy much of Reunion, an active centre formed more than 2 m.y. ago. Rather similar basalts in the Kerguelen archipelago of the Southern Ocean are associated with trachytes and are intruded by small bodies of syenite, diorite and gabbro. Strongly alkaline rocks are represented in Mauritius.

4 *Sedimentary deposits*

Evidence concerning the distribution of sediment on the floor of the Indian Ocean is derived from seismic studies controlled by study of deep-sea cores which, of course, usually penetrate only the younger parts of the cover (Ewing *et al.*, 1969. Very large areas of the Indian Ocean floor along, and to either side of, the mid-oceanic ridges have turned out to be almost bare of sediment. The rates of accumulation, as calculated from the thicknesses deposited since the beginning of the last (Brunhes) episode of normal magnetization, are not unduly slow and the paucity of sediment is therefore in good agreement with the assumption that the crust adjacent to the ridges is of recent formation. The thickest accumulations far from land occur in the Southern Ocean where biological activity is favoured by the uprise of nutrient-rich currents from deep levels. The sediment here is of pelagic type — core samples consist largely of diatomaceous ooze, much of the sea-floor being below the level of carbonate accumulation. The measured thicknesses are well below a kilometre, and in the more barren tropical regions where red clay predominates they are usually less than 100 metres. Rather greater thicknesses have been recorded on submarine plateaux rising above the level at which carbonate is dissolved.

   Thick sequences of terrigenous sediments are, as would be expected, located in the marginal parts of the ocean-basin. Over 40 per cent of sediment in the ocean is concentrated in the *Indus and Ganges cones*, the submarine extensions of the deltas built by the great rivers draining the Himalayas (Chapter 8, p. 251). These cones appear on topographical maps as huge aprons scored by submarine channels, which extend over the oceanic crust as far as, or (for the Ganges cone) well beyond, the southern tip of India. Ewing *et al.* (1969) calculate that the total volume of the Ganges cone is over 700 000 km³. Assuming that the main Himalayan source-region was elevated not earlier than mid-Miocene, these figures imply average rates of accumulation of about 17 cm per 1000 years. The material of the cones is well-stratified, providing many reflecting surfaces picked up by seismic profiling, and is presumed to consist mainly of turbidites carried from the delta-front into deep water by turbidity currents. It may be noted that the sediment-mass of the Ganges cone has a north—south elongation and that transport by turbidity-currents is mainly axial.

Along the eastern margin of Africa, the bulk of the terrigenous sediment occurs in the *Somali basin* off East Africa, and the *Madagascar basin* extending southward from the Mozambique Channel. These basins, floored mainly by oceanic crust, flank Mesozoic marginal basins (p. 233) and appear to contain up to 2.5 km of Tertiary detrital sediments including turbidites. Small sediment-filled basins lie on the western side of the Australian continent but, interestingly, there is little sediment on the ocean floor off the East Indies where sediment is trapped by the Java trench.

V The Indonesian Arcs

The south-eastern salient of continental Asia is fringed by a tangle of islands which resolve themselves into a number of seismically and volcanically active arcs (Fig. 9.8). The Tertiary fold-belt of Burma and Malaya is continued more or less directly by the Indonesian islands between Sumatra and Timor. In this sector, a volcanic arc subject to explosive eruptions such as that of Krakatoa (1893) is flanked on the south by oceanic trenches and sediment-filled troughs and is associated with a northward-dipping Benioff zone. A lack of isostatic equilibrium is indicated by strong negative gravity anomalies over the trenches and by positive anomalies over parts of the volcanic arc. East of Timor, the arc appears to loop sharply back on itself towards the north-west, coming to flank the Pacific in so doing and linking up with the Philippines where the trenches lie on the east side of the volcanic arc and the Benioff zone dips westward. Another arc, skirting New Guinea and passing through the Solomon Islands and New Hebrides, flanks the continental mass of Australia and New Guinea and in this arc, the dip of the Benioff zone is locally north-eastward.

The complex relationships outlined above reflect the adjustments necessitated by the meeting of three crustal plates. The Indian Ocean plate, advancing northward from its spreading axis in the Southern Ocean, is made of oceanic crust in the west (where it is overridden by the Indonesian arc) but partly of continental crust in the east (where in places it overrides the E. Pacific plate). The continent of mainland Asia, with its extensions in Malaya, West Borneo and the Sunda Shelf, is in contact with both Indian Ocean and Pacific plates. The extremely sharp curvatures of the structural trends suggest crumpling and dis-tortion of the entire arc system as it was pushed northward by the enlargement of the Indian Ocean plate.

The geological history of the islands suggests that the fringing arc environ-ment has persisted at least since the end of the Palaeozoic era. Volcanic groups

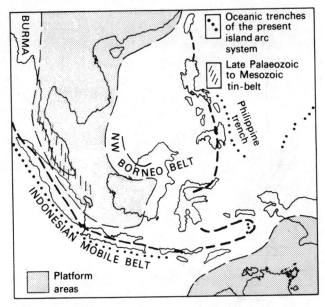

Fig. 9.8. South-east Asia in Tertiary times showing the mobile zones of Indonesia and Malaysia bordering a more stable (often emergent) block. The late Palaeozoic to Mesozoic tin-belt indicated here is located on two lines of granitic batholiths

including both ophiolite-chert and andesite-dacite assemblages, form large parts of the sequence in the volcanic arcs. Some of the associated sediments are of typical shelf facies, but the majority are detrital formations laid down in trenches and other basins flanking the volcanic arcs. A typical association of Cretaceous—Eocene turbidites with cherts and ophiolites, indicating an oceanic trench environment is seen, for example, in north-eastern Borneo. Deep-water Mesozoic sediments including cherts, pelagic limestones and manganese nodules may represent trench-sediments remote from sources of terrigenous detritus. Persistent instability is indicated by many unconformities and by conspicuous folding.

VI The Red Sea and Gulf of Aden

The Red Sea and the Gulf of Aden, though insignificant in terms of area, are of interest for two reasons: first, they are connected both with the typically oceanic Carlsberg Ridge and with the typically continental rift-valley system of East Africa; and, second, they appear to be embryonic oceans in which sea-floor spreading has only recently begun (Fig. 9.9). Both are bordered by high cratonic continental blocks which tilt gently away from their shores. Both are traversed by central zones of irregular topography and pronounced seismicity. Both are

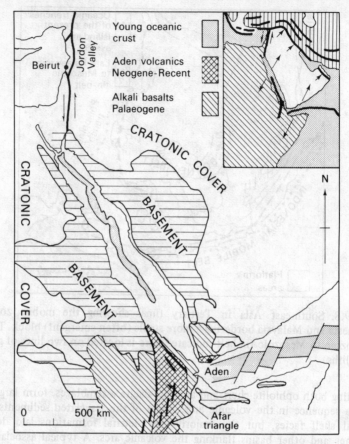

Fig. 9.9. The Red Sea and Gulf of Aden with (*inset*) the relationships of the Arabian plate. The Aden volcanics are intermediate in character between the alkali-basalts which preceded them and the oceanic tholeiites which were erupted beneath the central tracts of the new seas

regions of high heat flow and are characterised by continuing or recent igneous activity.

The Carlsberg Ridge, approaching from the open ocean, is diverted into the Gulf of Aden by many transform faults and continues as the seismic zone of the Gulf. This zone and the corresponding structure in the southern part of the Red Sea, meet the northern extremity of the Ethiopian rift-valley in the 'Afar triangle', a low-lying plain underlain by thick Tertiary and post-Tertiary basalts.

The warping of erosion surfaces suggests that up-doming of the continental crust in early Tertiary times produced a broad swell elongated along the line of the Ethiopian rift and with its centre not far from the triple point at which the rift meets the Red Sea and Gulf of Aden. Flood-basalts of late Cretaceous to Miocene age are distributed around this meeting-point, both on the Arabian and on the

African side of the sea. The early history of the gulfs thus had much in common with that of the rift-valleys, involving arching of the crust and vulcanism near the culmination of the arch.

The scarps which face the marine gulfs have been regarded as fault-controlled structures. Gass and Gibson (1969) among others, however, prefer to regard them as downwarps modified by minor faulting and erosion. Mesozoic sediments are preserved here and there on and below the scarps, while an unconformable mid-Tertiary sequence is confined to the floor of the depression. Miocene evaporites along the borders of the Red Sea provide an interesting parallel with the Mesozoic evaporites laid down in the initial basins of the Atlantic coasts (p. 232). The structural and stratigraphical evidence suggests that by Miocene times the Red Sea and Gulf of Aden had been established as narrow troughs at or a little below sea level; the downwarping of their shoulders may have been the result of drastic thinning of the crust beneath these troughs.

In the southern part of the Red Sea an axial tract some 50 km broad is floored by dense and highly magnetic rocks, shows linear magnetic anomalies and has a high heat flow. A broader and more continuous tract with similar properties underlies the Gulf of Aden. These tracts are traversed by the zones of seismic activity already mentioned and are regarded as strips of oceanic crust generated along mid-oceanic ridges (Laughton, 1966). Their narrowness suggests that sea-floor spreading, as in the adjacent part of the Indian Ocean, was relatively sluggish. Igneous activity in and on the borders of the new seas is recorded by tholeiitic flood-basalts and basic dyke swarms in the Afar depression, by volcanic islands (possibly sited on transform faults) in the southern Red Sea and by varied central complexes which include both acid and alkaline components near the southern tip of Arabia. A remarkable feature apparently related to a high geothermal gradient and to the uprise of juvenile material is the occurrence beneath the deep waters of the Red Sea of a layer some 4 km in thickness with distinctive seismic velocities which is known from boreholes to consist of evaporites. Other sea-floor sediments carry abnormal concentrations of zinc, copper and manganese, and pools of hot brines with similar metal contents have been located in the depths of the Red Sea.

The history of the Red Sea and Gulf of Aden falls into two phases: (i) an initial (early Tertiary) phase of vertical uplift perhaps leading to 'necking' of the crust and development of a linear depression, and (ii) a late (mid-Tertiary to Recent) phase of horizontal separation of continental fragments associated with the generation of new crustal material. The opening of the seas appears to be most readily explained in terms of a north-eastward drift of the Arabian peninsula towards the mobile belt of the Middle East (Fig. 9.9). Reconstruction of the African-Arabian block on this hypothesis produces a good fit apart from some overlap in the Afar triangle (which may by floored by oceanic crust). It will be noted that the lines of movement, while approximately parallel to the conspicuous transform faults in the Gulf of Aden, are oblique to the long axes of both the Red Sea and the Gulf of Aden. Relative movement between Arabia and the Mediterranean zone is indicated by a major sinistral wrench-fault along the Dead Sea—Jordan valley (Fig. 9.9). Freund puts the post-Cretaceous displacement on this structure at 105 km and estimates the rate of displacement over the last 10 m.y. at about 0.4 cm/year.

10

The Pacific Ocean and its Island Arcs

I Introduction

The Pacific Ocean, unlike the Atlantic and Indian Oceans, has a geological history extending back through the entire Phanerozoic eon and possibly well into the Precambrian. It differs, too, from the new oceans in its structural relationships. Mobile belts — orogenic mountain-tracts or systems of island arcs and trenches — border the ocean at almost every point and its internal structural pattern is by no means everywhere concordant with the continental margins.

Although the basin of the Pacific is an old-established structure, it is floored by comparatively young crustal material; a large part of this crust appears to be no older than Tertiary and most of the remainder is Mesozoic. Consumption of the older oceanic crust along the peripheral mobile belts has provided the means of disposal of the surplus and maintained the balance on a global scale (Fig. 9.1); the *Circum-Pacific mobile belt* has thus played an essential role in the evolution of the Pacific basin. We shall here deal with the ocean basin itself and with the northern and western periphery, leaving the eastern portion of the encircling mobile belt to be mentioned in Chapter 11.

II The Ocean Basin

1 The mid-oceanic ridge

The mid-oceanic ridge of the Pacific is unusually well known, largely thanks to the studies of groups based in the United States. The principal ridge enters the ocean from Antarctic waters south of New Zealand and turns northward near Latitude 120 W to form the *East Pacific Rise*. Two short spurs — one off Chile and the second in the vicinity of the Galapagos Islands — approach the margin of

South America and may represent minor spreading axes. The main ridge continues northward and is diverted by means of displacements on transform faults into the Gulf of California whence it skirts the border of North America, finally disappearing south of Alaska (Fig. 10.1). The conspicuous asymmetry of the ocean basin with respect to the East Pacific rise appears to be due in part to the overriding of the oceanic plate generated on its eastern side by the advancing continental rafts of the Americas. An inactive ridge located beneath the *Darwin Rise* of the west Pacific may have acted as a spreading axis in Tertiary times.

The East Pacific rise is revealed by topographical profiles as an exceptionally broad submarine ridge cleft by a central rift and descending outward with gentle slopes to the basin-floor; its breadth is due to the high rate of sea-floor spreading (Table 10.1). The ridge is offset by major transform faults of westerly or west-north-west trend, some of which are marked by conspicuous submarine scarps. Offsets amounting in some instances to several hundred kilometres have the effect of diverting the ridge from its normal course; the tract which parallels the western edge of North America, for example, is made up of short ridge-segments running north-north-east, progressively shifted northwestward by transform faults.

2 General structure

Linear magnetic anomalies parallel to and symmetrical about the mid-oceanic ridge have been mapped over much of the southern and eastern Pacific floor. The symmetrical distribution of crustal strips characterised by alternately normal and reversed magnetization is well illustrated in Fig. 10.2 in which a single magnetic profile across the Pacific Antarctic ridge is shown above the theoretical arrangement of crustal strips required to produce such a profile.

The magnetic profile illustrated in Fig. 10.2, on which certain anomalies have been dated by reference to volcanic sequences, has been used as a standard for comparison with other profiles across the East Pacific rise and also with profiles across the mid-Atlantic and Indian Ocean ridges. Making allowance for differences in the rate of spreading, details of profiles taken at points many thousands of kilometres apart agree with remarkable accuracy, allowing geophysicists to identify, correlate and assign approximate dates to anomalies over much of the Pacific floor.

The profile illustrated in Fig. 10.2 indicates a spreading rate of about 4.5 cm/yr; similar or higher rates are indicated by other profiles, the figures decreasing northward and southwards in a manner which suggests that the Pacific plates rotated outward about a north–south axis (Table 10.1). The east–west trend of the larger transform faults is consistent with opening about such an axis but, as will be seen, many of the active faults close to south-western North America have a north-west trend apparently related to a late change in the direction of crustal movement.

At a spreading rate of over 4 cm/yr, implying the annual addition of at least 8 cm of new crustal material, little more than 100 m.y. would be required for the generation of the whole floor of the eastern Pacific. The oldest rocks might be expected in the north-west of the basin, the region most remote from the

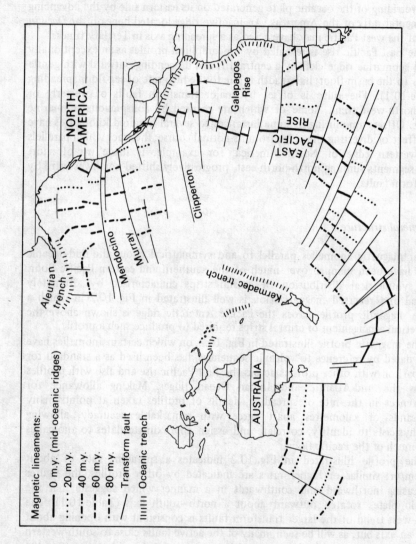

Fig. 10.1. The structural elements of the eastern Pacific Ocean

Table 10.1. RATES OF SPREADING IN THE PACIFIC
(based on Le Pichon, 1968)

Latitude	Longitude	Spreading rate (cm/yr)
48 N	127 W	2.9
17 S	113 W	6.0
40 S	112 W	5.1
45 S	112 W	5.1
48 S	113 W	4.7
51 S	117 W	4.9
58 S	149 W	3.9
58 S	149 W	3.7
60 S	150 W	4.0
63 S	167 W	2.3
65 S	170 W	2.0
65 S	174 W	2.8

spreading axis, and indeed the distribution of sediment suggests (p. 287) that parts of this region are floored by Mesozoic rocks.

The structure of the sea-floor in the north-eastern corner of the Pacific is exceptionally complex (Fig. 10.3). The East Pacific Rise, approaching central America from the south, is dismembered by north-westerly transform faults which divert it into the *Gulf of California*. In the Gulf, a number of north-easterly submarine troughs are linked by north-westerly scarps. Seismicity is concentrated along these features and analysis of recent earthquake movements indicates that the scarp features follow transform faults. The magnetic anomalies of the sea-floor in the southern part of the Gulf resemble those of the eastern Pacific and this part of the Gulf is regarded as a young (post-late Miocene) ocean formed by oblique opening along lines parallel to the transform faults. Continued opening along these lines may be expected to detach Baja California and the coastal tract north of it from the rest of North America.

From the northern tip of the Gulf of California the *San Andreas Fault* cuts northwestward through the continent for more than 1000 km before emerging again at the continental margin. This famous dislocation has been the site of dextral wrench-movements for over 150 m.y. The total displacement is of the order of 500 km and the matching of features of several ages indicates that the rate of movement on the fault has increased greatly since Eocene times. J. T. Wilson has suggested that the San Andreas Fault acts as a transform fault linking the dismembered ridge-segments in the Gulf of California with mid-oceanic ridges to the north. In alignment and sense of movement it agrees with the transform faults in the floor of the Gulf and according to Tobin and Sykes its northern end turns westward into the Mendocino transform fracture-zone (Fig. 10.3). The San Andreas fault in its present very active state thus functions as an accommodation structure facilitating sea-floor spreading; its long history of dextral displacement, dating back to a time when North America lay far to the east of its present site, shows that it is an old lineament possibly initiated in a different crustal environment.

North of the Mendocino fracture-zone the *Juan de Fuca and Gorda ridges*, linked by a transform fault, border a narrow eastern plate of oceanic crust

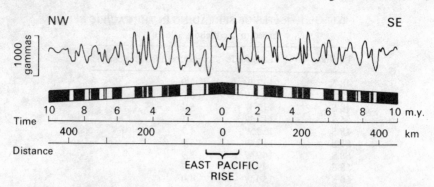

Fig. 10.2. Magnetic profile across the East Pacific Rise (the Eltanin 19 Profile at 51°S) illustrating the symmetry about the spreading centre. The diagram beneath the profile shows the sequence of normally magnetised (black) and reversely magnetised crustal strips (after Heirtzler, 1968, and others)

flanking the continent. Seismic activity along the ridges, and the symmetry of the magnetic anomalies on either side of them, indicate that new material is being added to this plate which consequently moves eastward to plunge beneath the advancing continent. The anomalies on either side of the Juan de Fuca and Gorda ridges show considerable irregularities which have been ascribed to distortion resulting from the late changes of movement-direction already referred to.

Off Vancouver Island, the Juan de Fuca ridge terminates against a transform fault which links it with the Aleutian Island arc. No mid-oceanic ridge has been identified north of this point. The distribution of magnetic anomalies (Fig. 10.1) shows that, contrary to the usual arrangement, the crustal strips nearest to the arc are younger than those lying nearer the centre of the ocean. It appears, therefore, that both the eastern oceanic plate and the mid-oceanic ridge in this region have been overrun and the advancing continent has begun to encroach upon the western oceanic plate. A sharp bend in the linear anomalies south of the Aleutians (the 'great magnetic bight' of some authors) may indicate the former meeting of three oceanic plates two of which have disappeared beneath the overriding material. Towards the Asiatic side of the basin, an angular discordance between linear anomalies off Japan and the Japanese oceanic trenches appears to indicate the overriding of the western oceanic plate.

3 The sedimentary cover

The sediment-cover in the Pacific basin is seldom as much as a kilometre in thickness and often no more than 100 metres. The East Pacific Rise, as might be expected, is almost bare. Some wedges of terrigenous sediment – probably mainly turbidites – fringe the continental margin off North America, but elsewhere the bulk of the cover consists either of pelagic sediments or of detritus contributed from ridges and island groups within the basin.

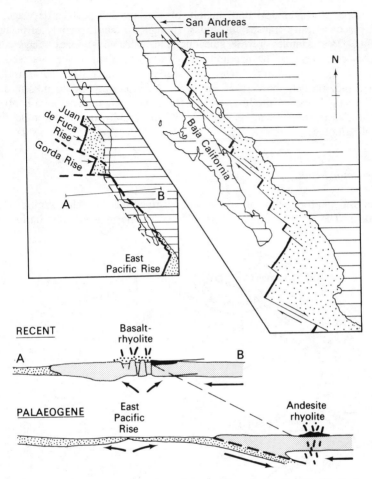

Fig. 10.3. The mid-oceanic ridge system of the eastern Pacific in relation to the continent of North America; the stippled areas east of the ridge mark small oceanic plates partly overridden by the continent. The accompanying diagrammatic sections (after McKee, 1971) illustrate possible stages in the westward advance of the continent

Preliminary sampling and refraction profile surveys recorded by Ewing and his collaborators have led to the identification of two main stratigraphical units with differing accoustical properties whose boundary is thought to be close to the Mesozoic–Tertiary boundary. The lower unit (seldom more than 300 m) is almost confined to the western part of the basin. It is accoustically opaque and consists of well-bedded sediment, probably largely detrital material laid down by turbidity currents. The greatest thicknesses occur near island groups and over the Darwin Rise which, with its numerous atolls and seamounts, probably provided the bulk of the detritus.

The upper stratigraphical unit, mainly of Tertiary and post-Tertiary age, is generally accoustically transparent and is thought to consist mainly of pelagic sediment. Core samples reveal siliceous and calcareous oozes, clays and occasional chert layers. The maximum thicknesses of about 600 m are seen in three east–west zones, one roughly equatorial and the other two near the northern and southern borders of the basin which correspond to present-day zones of high biological productivity. In the intervening regions there is often less than 100 m of sediment on the basaltic floor, and manganese nodules are widely distributed.

4 The oceanic islands

Two classic petrographic provinces are distinguished in the Pacific area (Fig. 10.4). The *oceanic province* coincides with the main oceanic basin and

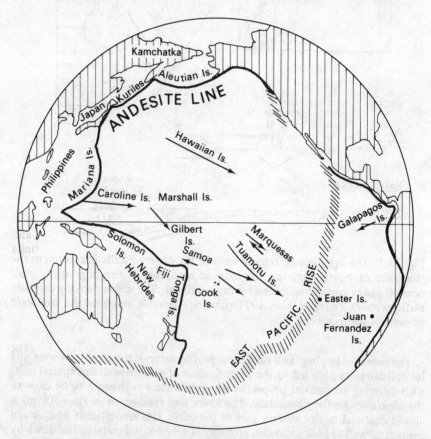

Fig. 10.4. The principal magmatic provinces of the Pacific region; the main igneous lineaments of the oceanic province are marked by the arrows which point towards the most recently active centres

covers the mid-oceanic ridge and island groups such as the Hawaiian Islands. Like the corresponding provinces of the Atlantic and Indian oceans, it is characterised by basalts and their associates. The *Circum-Pacific province* follows the peripheral mobile belt and is characterised by basalts along with abundant andesitic and rhyolitic lavas and pyroclastics. Its boundary against the oceanic province proper is the *andesite line* which follows the oceanic side of the island arcs and skirts the coast of the Americas. The style of igneous activity in the Circum-Pacific province is dictated by the processes of partial melting of material overridden by the crustal plates which border the ocean (see later, section IV).

The volcanic islands of the oceanic province form clusters often arranged along lineaments. There are few such clusters in the northern Pacific or in the vicinity of the East Pacific Rise — Easter Island is one of the few centres on the rise itself, while the Galapagos and Juan Fernandez groups lie close to the eastern spurs of the ridge system. Menard has pointed out that few islands stand on crust which is less than about 10 m.y. in age. Off western North America, *seamounts* which represent evolving volcanoes increase in volume with distance from the Juan de Fuca ridge and Menard infers that some 10 m.y. may be required for the build-up of a submarine volcanic pile before the top breaks surface as an island. The scarcity of islands near the mid-oceanic ridge may be connected with the unusually high rate of sea-floor spreading which allowed little time for the build-up of volcanic centres.

Most of the Pacific volcanic islands are scattered over the western part of the ocean where the crust dates back to Mesozoic or early Tertiary times. These islands, together with large numbers of seamounts, atolls and guyots, are strung out along NW—SE lines oblique to the regional magnetic anomaly pattern. These volcanic lineaments, some of which are 1000 km in length, have a roughly constant alignment over an enormous area (Fig. 10.4). Although not far from parallel to the present direction of drift of the main Pacific plate, they are obviously oblique to the large east—west transform faults; the Hawaiian lineament, for example, is crossed obliquely by the continuation of the Murray fracture zone.

Many volcanic groups include centres in several stages of evolution, from seamounts which have not yet broken surface, through active volcanoes and extinct, dissected volcanoes, to worn-down atolls and guyots. Mid-Cretaceous corals have been dredged from guyots near the Hawaiian ridge, indicating that extinction and submergence of early volcanoes in the group took place some 100 m.y. ago. The dating of rocks from the subaerial parts of the islands suggests, however, that each huge subaerial shield-volcano was built up quite rapidly: the rocks of Hawaii itself, for example, all appear to be less than a million years in age.

In almost every island cluster, the centres which show signs of most recent activity are located at the ends of the lineaments closest to the East Pacific Rise and are followed outward by centres in more advanced stages of evolution. Only in the Samoan group is the arrangement reversed, the active cones lying at the western end of the lineament. The general consistency of the arrangement suggests that new centres were initiated at the end of the lineament closest to the axis of sea-floor spreading and carried passively outward on the rigid crust:

we may note that magmatic source-regions and volcanoes remained coupled during drift.

The typical oceanic assemblages of the Pacific islands need not be considered at length. The Hawaiian islands are made up of coalescing shield-volcanoes whose broad tholeiitic foundations differ from the 'oceanic tholeiites' of the sea-floor in being relatively poor in alumina and soda. The uppermost parts of the volcanoes, formed during and after the establishment of calderas, exhibit a variety of alkaline basalts, mugearites, hawaiites and trachytes generally regarded as differentiates of basaltic parent-magmas which form no more than about five per cent of the volume of the volcanoes.

III The Eastern Pacific Margin

The eastern Pacific is flanked by the Rockies and Andes, which have separated the American cratons from the ancestral Pacific throughout Phanerozoic times. A seismic zone extending for more than 10 000 km along or close to the continental margin performs the same function today.

The continental shelf off South and Central America is narrow and is bordered over most of its length by an oceanic trench or a sediment-filled trough. The *Peru—Chile trench* forms a deep furrow from Latitude 4°N to 40°S and the associated seismic zone and negative gravity anomaly extend beyond it both to north and south. Medium and deep-focus earthquakes in the seismic zone define a Benioff zone dipping steeply beneath the overriding continent.

From the Gulf of California northward, the nature of the margin changes. Seismic activity is limited to shallow-focus earthquakes taking place along the mid-oceanic ridge and its transform faults, including the San Andreas fault. There are today no deep-focus earthquakes and no oceanic trench which might indicate an advance of North America over the Pacific floor although (p. 286) the structure of the ocean floor clearly indicates that a broad oceanic tract was overridden in Tertiary times. The present motion of the Pacific crust (p. 285) appears to be north-westward away from the continent, carrying the continental strip west of the San Andreas fault with it. The change in plate motion indicated by changes in the alignment of transform faults appears to be post-Miocene in age. It was, in part at least, connected with the overrunning of the East Pacific Rise and was accompanied by a corresponding change in the style of volcanic activity in south-western North America (see Chapter 11 and Fig. 10.3).

IV The Western Pacific Margin and the Island Arc Systems

From Alaska to New Zealand, the western Pacific is bordered by a festoon of island arcs. Many of these are rooted in the continental masses to the west, but their central parts stand far out from land, sheltering small seas floored at least in part by crust of oceanic or transitional type. From Japan southward, the festoon is doubled and the outer arcs of the Izu-Bonin, Mariana and Tonga Islands lie more than 1000 km from any major land-mass.

The convex sides of the arcs are flanked by oceanic trenches and generally (though not invariably) face the Pacific. The Circum-Pacific belt of seismicity defines a series of movement-surfaces usually (though again not invariably) dipping steeply towards the bordering regions from the Pacific side of the arcs. The andesite line falls close to the line at which these surfaces intersect the sea-floor: that is, to the effective perimeter of the Pacific oceanic plate.

The western Pacific border-zone is a region of altogether exceptional seismic activity and is subject to earthquakes at very deep levels. Both these features may be attributed to the relatively fast rates at which the collision of oceanic and bordering crustal plates is taking place. Differential movements along the zone of detachment are estimated by Le Pichon to reach rates of up to 9 cm a year (Table 1.1). It will be noted that the leading edges of the overriding plates, unlike those of the American plates, are largely submerged and are made in part of crustal material which is of oceanic or transitional character. The Asiatic and Australian continental masses lie well back from the present mobile zone and are themselves almost aseismic, though they are formed in many places of complexes which were mobile up till Mesozoic or Tertiary times.

The island arcs which fringe the continents (disregarding for the moment the outer Marianas system) have a number of geological features in common. Although Precambrian basements of continental type have not been identified with certainty, their histories stretch back for well over 100 m.y. to Mesozoic or even to early Palaeozoic times. Volcanic activity and mobility have characterised the arcs throughout their life-span and regional metamorphism (sometimes associated with granite-formation) has affected large areas. Geophysical evidence shows that most of the arcs form welts in which the crust is thicker than that on either side, though it seldom reaches more than about 20 km. A crustal layer resembling the 'granitic' layer of continental regions has been distinguished beneath some arcs, but others appear to be underlain by material more closely resembling the oceanic crust.

This mixture of features has led to suggestions that continental crustal material is seen in process of evolution in the island arcs and that the welding of such material to adjacent continents may provide a means of *continental accretion*. A radically different interpretation favoured by certain workers, notably in the U.S.S.R., attributes transitional characters in certain arcs to *oceanisation* of continental material (p. 209). The small seas sheltered behind the Kamchatka-Kurile arcs, for example, are considered to result from fairly recent submergence following on the increase in mass resulting from ocean-isation of the underlying crust.

1 The Japanese arcs

The complex Japanese arcs, exceptionally well documented by the studies of Japanese geologists and geophysicists, provide a good introduction to the features of the western Pacific margin (Figs. 10.5, 10.6).

The northern island of Hokkaido stands at the meeting-point of the Kurile and Sakhalin arcs approaching from the north. From the southern side of this island the main Japanese arc is seismically and volcanically active north of a

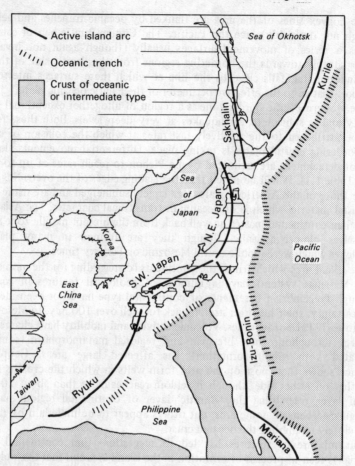

Fig. 10.5. Island arcs and associated structures of the north-west Pacific

transverse feature known as the *Fossa Magna* in central Honshu, but inactive
further south. The active Izu-Bonin arc of small islands takes off southward from
the vicinity of the Fossa Magna and, in the extreme south-west, the Ryuku arc
provides a link with Taiwan.

On the convex (Pacific) side of the arcs, an oceanic trench extends
continuously from the vicinity of the Kuriles to the eastern side of the Bonin
Islands while a less clearly marked trench borders the Ryuku arc. Off Honshu,
linear magnetic anomalies mapped over the Pacific floor appear to be obliquely
truncated near the trench, presumably as a result of the overriding of the Pacific
plate along the Benioff zone associated with this trench. The shallow seas which
separate southern Japan from the mainland of Asia are underlain by crust
apparently of continental type, while beneath the adjacent islands the Moho lies
at 30–35 km. Further north, however, the floor of the Japan Sea lies at oceanic
depths of up to 4 km and the underlying crust is thin and lacking in a 'granitic'

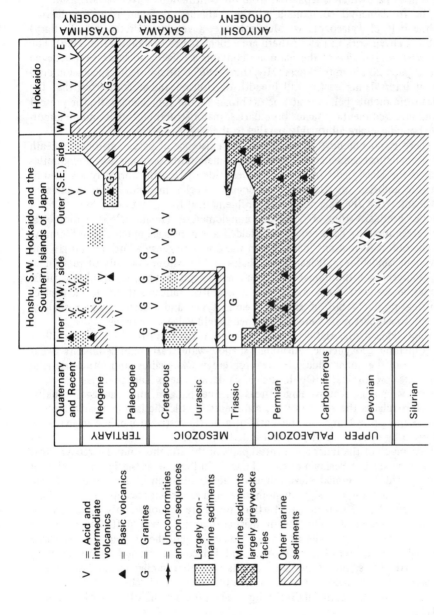

Fig. 10.6. Chart summarising the history of deposition, igneous activity and tectonic activity in Japan (after Matsushita, in: Takai et al. (1963))

layer; beneath the adjacent parts of Honshu, the Moho seldom reaches depths of more than 30 km. There is therefore a structural contrast between the southern tip of the main arc, which is more or less continuous with continental Asia, and the central part which is separated from the continent by a true ocean basin.

The fundamental component of the main arc in Japan is a consolidated mobile belt of Palaeozoic to Mesozoic age; a younger (Cretaceous–Tertiary) mobile belt appears in the northern arc which crosses Hokkaido (Fig. 10.6). The *Chichibu geosyncline* of the main arc contains a Palaeozoic succession believed to approach 20 km in thickness (Miyashiro, 1967). Rocks of this succession crop out at intervals across the full breadth of Honshu and it is inferred that the Palaeozoic mobile belt was at least as broad as, and coincident with, the present island arc. Sedimentary facies boundaries, major structural trends and metamorphic boundaries are all roughly parallel to the arc.

There is no unequivocal record of the existence of basement-rocks beneath the geosynclinal sequence although highly-metamorphosed gneisses and granites such as the Hida gneiss on the north-west side of Honshu, formerly assigned to the Precambrian by Japanese geologists, might incorporate a reactivated basement. With the exception of a problematical K–Ar date of 596 m.y. for a granite boulder from a Permian conglomerate (Imai, 1963) radiometric age-determinations have not so far yielded apparent ages much above 500 m.y.

The oldest formations identified in the Chichibu geosyncline are Silurian and Devonian sandstones, shales, black shales and limestones (locally of reef facies and sometimes bituminous) associated with keratophyres, keratophyric tuffs, dacites, andesites and acid tuffs. This assemblage indicates that volcanic centres emitting products not of simple oceanic types and tracts of water shallow enough to support reefs were in existence in mid-Palaeozoic times.

Upper Palaeozoic, and especially Permian, rocks form the bulk of the fill of the Chichibu geosyncline. Although shallow-water carbonate sediments occur locally on the inner side of the arc, greywacke-pelites and other detrital sediments predominate. Ophiolitic basis lavas, cherts and pyroclastics, with serpentine pods, occupy large areas in the Sangun and Sanbagawa belts. Miyashiro places the source of the sediments in volcanic islands near the outer side of the present arc, on the basis of a north-westward increase in maturity indicated by increases in Al_2O_3 and in K:Na ratios.

Over much of the inner and central parts of the arc, the youngest geosynclinal formations are late Permian or early Triassic. In these areas the Triassic *Akiyoshi orogeny* led to regional elevation and the termination of marine deposition. Post-orogenic (late Triassic) conglomerates and sandstones occupy intermontane depressions; later Mesozoic and Tertiary rocks are predominantly volcanics and non-marine sediments. The site of marine deposition was shifted in Mesozoic times to the outer (southern) side of Honshu where the *Shimanto geosyncline* received a thick series of Mesozoic and Palaeogene detrital sediments with occasional cherts, limestones and basalts. A rather similar assemblage of late Jurassic and Cretaceous greywackes, pillow-lavas and cherts, followed by Palaeogene paralic sediments including coals, provides the oldest geological unit in Hokkaido.

The effects of orogenic activity in the main arc are expressed by complex fold-structures and by the development of tracts of regional metamorphism and granite-formation. Thanks to the work of Miyashiro and his colleagues, Japan has provided the type examples of the *paired metamorphic belts* which characterise a number of the Pacific island arcs (Fig. 10.7). Of each pair of belts, one is marked by a low-pressure facies series with abundant granites and the other by a high-pressure facies series. In Honshu, older paired metamorphic belts on the site of the Chichibu geosyncline were developed during the Akyoshi orogeny and younger belts during the late Mesozoic orogeny. In both pairs, the high-pressure belt lies towards the outer (Pacific) side of the arc. A Tertiary pair of metamorphic belts in Hokkaido appears to show a reverse arrangement with the high-pressure belt on the inner (western) side. The features of the metamorphic belts as outlined by Miyashiro and summarised in Table 10.2.

In the belts characterised by the *low-pressure facies series*, the type of metamorphism (the *Abukumu type* of Miyashiro) suggests an unusually steep geothermal gradient and is similar to the Buchan type developed locally in the British Caledonides (p. 47) and to the characteristic metamorphism of the European Hercynides (Chapter 3). Abundant granites, yielding ages in the range 110–50 m.y., include early concordant and often gneissose types, perhaps formed largely *in situ* and intrusive granites which extend into less-metamorphic terrains on the north-western side of the Abukumu belt and may be associated with subaerial rhyolitic tuffs and ignimbrites.

The Sangun and Sanbagawa belts characterised by the *high-pressure facies series* are those in which complexes of ophiolitic lavas, cherts and serpentines were emplaced during the filling of the geosyncline. The characteristic minerals of low grades, stilpnomelane, prehnite and pumpellyite are rather widely developed and glaucophane, indicative of a higher grade, is somewhat patchy. Granites are almost entirely absent.

The paired metamorphic belts of Honshu and the southern islands (Fig. 10.7) have been interpreted by reference to the island-arc setting – a convenient summary is that of Miyashiro (1967). The geosynclinal sediments and ophiolites of the Sangun and Sanbagawa belts are considered to have accumulated in troughs analogous to the present oceanic trench and high-pressure metamorphism is thought to have taken place as the crustal tract was overridden by the embryonic island arc. The low-pressure, high-temperature metamorphism and granite-formation in the Hida and Ryoka-Abukumu belts resulted from the rise of volatiles and partial melts expelled from the disturbed mantle and overridden crustal plate with consequent influx of heat into the arc advancing above the Benioff zone. This attractive hypothesis follows Fyfe in accepting the down-thrusting of a cool crustal slab at a Benioff zone as the main cause of metamorphism in the high-pressure facies series. Metamorphism of this type is very widely developed in the Circum-Pacific and Alpine belts formed in Mesozoic and Tertiary times, but is extremely uncommon in older terrains, an anomaly which is possibly due to the high rates of crustal consumption achieved in the widely-spaced mobile belts of the late Phanerozoic system.

2 The Kuriles and Kamchatka

From northern Japan, the zone of active seismicity follows an arcuate course north-eastward along the oceanic trench which skirts the Kuriles (Fig. 10.5). Off Kamchatka, it abruptly changes direction to follow the Aleutian trench

Table 10.2. THE PAIRED METAMORPHIC BELTS OF JAPAN
(based mainly on Miyashiro)

Age and Name	Metamorphic facies series	Location	Relationships	Characteristic features
The older pair: metamorphism ended in Triassic				
1 Hida metamorphic belt	low-pressure facies series	north-west Honshu, adjacent part of Japan Sea	the broad belt of the high-pressure facies series lies on the Pacific side of the arc relative to the low-pressure facies series	Gneisses, schists and granites
2 Sangun metamorphic belt	high-pressure facies series	central parts of western Honshu, Kyushu and Shikoku		Recrystallisation patchy, large areas show no metamorphism. Common mineral assemblages in prehnite-pumpellyite, glaucophane-schist and actinolite-greenschist facies. No granites other than those related to belt 3. Basic volcanics and serpentines abundant

The younger pair: metamorphism ended in Cretaceous

3 Ryoka-Abukumu metamorphic belt	low-pressure facies series	south-eastern part of Honshu, Kyushu and Shikoku	the belt of the high-pressure facies series lies on the Pacific side of the arc relative to the low-pressure facies series	Thorough recrystallisation throughout the belt, zones of metamorphism symmetrical about an axis of high-grade rocks. Syntectonic and post-tectonic granites abundant in and to the north of the belt. Characteristic minerals of pelites: (chlorite):biotite:andalusite and cordierite: sillimanite: (orthoclase)
4 Sanbagawa metamorphic belt	high-pressure facies series	south-east of Ryoka-Abukumu belt		Recrystallisation patchy, grade increases towards Ryoka-Abukumu belt. No granites. Characteristic minerals of pelites:chlorite-muscovite-stilpnomelane-albite:(lawsonite-jadeite). Glaucophane locally in basic rocks. Basic volcanics and serpentines abundant.

The Hokkaido pair:

5 Kamuikotan metamorphic belt	high-pressure facies series	west-central Hokkaido southern Sakhalin	the belt of the high-pressure facies series is remote from the Pacific — the relationship is the reverse of that shown by the other pairs	Lawsonite, jadeite, glaucophane, pumpellyite are recorded. Serpentines abundant
6 Hidaka metamorphic belt	low-pressure facies series	central Hokkaido		

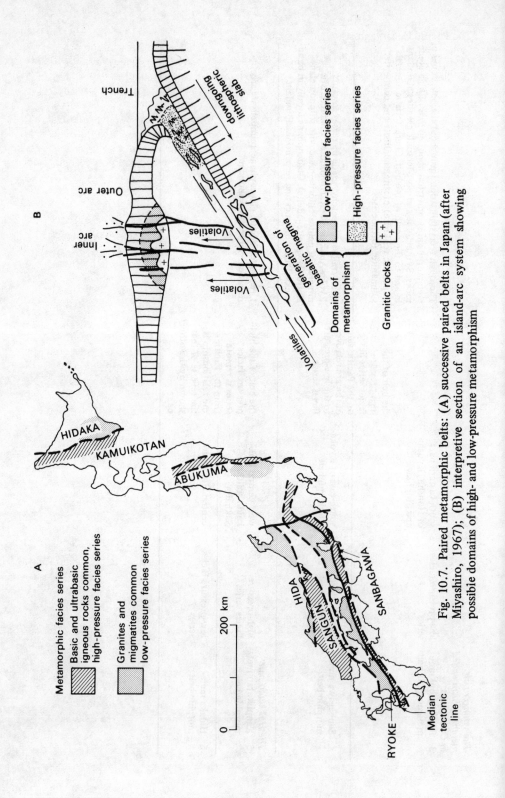

Fig. 10.7. Paired metamorphic belts: (A) successive paired belts in Japan (after Miyashiro, 1967); (B) interpretive section of an island-arc system showing possible domains of high- and low-pressure metamorphism

(Fig. 10.4). The seismic zone which dips steeply north-westward beneath the Kuriles extends to great depths — earthquakes with foci at depths of 600 km have been recorded — and perhaps represents the earth's most active seismic region at the present day.

The island arcs off northern Asia form a pattern of intersecting structures built up over a considerable period. Intense volcanic activity and important vertical movements are thought to have dominated the tectonic regime. The *Kamchatka peninsula* is underlain by crust of continental type of up to 40 km in thickness. Cretaceous and older rocks in the central and southern parts of the peninsula form paired metamorphic belts — an outer (eastern) belt characterised by low-temperature assemblages, and an inner belt including migmatites, granites and knotted andalusite-schists. Kamchatka appears to be linked to the mainland by continental crust extending beneath the northern Sea of Okhotsk: Soviet geologists consider that the sea was formed only in Quaternary times by downwarping of a land-mass. The *Kuriles* stand on a crustal welt up to 30 km thick but are separated from the aseismic arc of Sakhalin by a deep basin floored by thin oceanic crust (Fig. 10.5). The formation of this basin is assigned by Goryatchev (1962) and others to a comparatively recent episode of oceanisation of continental crust.

The present form of the arcs is attributed largely to differential vertical movements of Pliocene and later date. Two major uplifted zones forming eastern Kamchatka and the Greater Kuriles are flanked by downwarped zones and transverse warps are expressed by the pattern of embayments and peninsulas along the east coast of Kamchatka. Kamchatka Cape, one of the upwarps of this system, marks the meeting-point of the Kamchatka and Aleutian arcs. Down-warping in parts of the transverse troughs is continuing today at a rate of 4 mm per year; high bouguer anomalies suggest that dense material underlies them and Goryatchev considers them to represent zones in which oceanisation of continental crust is taking place.

The geological history of the Kuriles is recorded by assemblages of volcanic rocks, volcanigenic sediments and associated intrusives extending back to late Cretaceous times (Table 10.3); sediment derived from continental source-areas is

Table 10.3. THE SUCCESSION OF THE KURILE ISLANDS
(based on Markhinin)

Age	Location	Rocks	Thickness (metres)
QUATERNARY	Greater Kurile ridge (west side of arc)	andesite, basalt (dacite)	2000
PLIOCENE		basalt, andesite, tuff, pumice, clastic sediment: granitoid intrusions	1600
LATE MIOCENE		basic pyroclastics and clastic sediments: granitoid and gabbroic intrusions	2300
MID-MIOCENE		basic breccias, tuffs, lavas, sandstones: granitoid intrusions	900
EARLY MIOCENE		propylitised andesites, rhyolites	2400
PALAEOGENE	Lesser Kurile ridge	andesites, tuffs, basic lavas, gabbros	500
LATE CRETACEOUS		andesites, basalts, tuffs, agglomerates	1800

almost absent. The style of vulcanicity is andesitic; andesites, rhyolites and related pyroclastics, with acid intrusives, recur together with basalts throughout the succession. Markhinin points out that the average composition of the volcanics (characterised by rather low alkalies and high alumina and lime) is close to Poldervaarts' average composition of young fold-belts and concludes that the arc represents a zone of continental-type crust built up by vulcanism since Cretaceous times. A close connection between igneous activity and tectonic regime is indicated by the fact that the aseismic part of the peninsula, north of the junction with the Aleutian arc, has no active volcanoes.

3 The Aleutian arc

The Alaska peninsula and the islands alongside it form the oldest part of the Aleutian arc and are joined to the northern continental regions by a thick crustal layer underlying the shallow part of the Bering Sea and incorporating deformed Mesozoic or pre-Mesozoic rocks. This old part of the arc is double, the volcanic ridge of the peninsula being paralleled by an outer (southern) string of islands made largely of sedimentary rocks. The central and western portions of the Aleutian arc form a single line of small andesitic volcanic islands revealing no rocks older than Tertiary. These portions, though marked by a crustal welt, are separated from the adjacent continents by the deep *Aleutian basin* of the Bering Sea which is floored, beneath a thick cover of young volcanics and sediments, by a 6 km crust of oceanic type. The western end of the arc joins the Kamchatka peninsula at one of the transverse upwarps previously mentioned (p. 299). The *Aleutian trench* skirts the Pacific border of the arc and is associated with a zone of high seismicity dipping northward.

The strong disharmony between the magnetic anomaly pattern of the north Pacific floor and the arc-trench system (p. 286) suggests that a broad tract of oceanic crust has been overridden at the Aleutian Benioff zone. Some authorities regard the Aleutian basin as a former part of the Pacific, isolated by the development of the Benioff zone in Tertiary times. Others regard the floor of the basin as a downwarped and oceanised tract of crust, formerly of continental type.

Near the root of the Alaska peninsula, Triassic and early Jurassic sediments — the oldest rocks of the arc — are penetrated by a narrow granodiorite and granite batholith dated at 160 m.y., which extends parallel to the tectonic trend far to the east. This axial batholith is flanked on either side by great thicknesses of flysch-like late Mesozoic greywackes and slates associated with serpentines and gabbros. Those in the southern tract, which appear (contorted, weakly metamorphosed and invaded by early Tertiary quartz-diorites) in Kodiak Island and other islands of the outer arc, are believed to have accumulated near the foot of the Mesozoic continental slope, possibly in an early oceanic trench. It will be seen that large-scale granite-emplacement began in the eastern part of the arc well back in Mesozoic times and that accretions to the continental crust have, on the interpretation mentioned above, amounted to only the width of the outer island arc in 100 m.y.

The central and western parts of the island arc and trench are considered to be entirely Tertiary and post-Tertiary (early reports of a Carboniferous flora on Adak Island are now discounted). In Eocene times, andesitic, acidic and basaltic volcanics were erupted from many centres and quartz-diorite plutons were emplaced in the volcanic centres. Volcaniclastic sediments accumulated around their feet, especially after phases of isostatic uplift. In late Tertiary and Pleistocene times basaltic shield-volcanoes, andesitic cones and a variety of mildly alkaline assemblages were formed. At no stage in the long history of the arc were strong compressive stresses effective: only the flysch-formations of the east are strongly folded and their structures may be partly due to slumping. The tectonic development of the arc appears to have been dominated by vertical movements and by horizontal displacements on faults. The western end of the arc lies almost parallel to the present direction of movement on the Pacific floor and dextral slip, which allows adjustment to this movement, has been recorded in a number of recent earthquakes.

4 The Philippine, Mariana and Tonga arcs

Of the two island arc systems taking off southward from Japan the more westerly links up via the Philippines with the tangled Indonesian complex (Fig. 10.5). The main island group rises from a core of folded and metamorphosed pre-Triassic rocks. A lower division in this basement is rich in chlorite-schists and amphibolites, derived from basic volcanics, while an upper division of phyllites and slates represent metagreywackes. In the Mesozoic and Tertiary successions which overlie the basement, a tortuous central zone of abundant volcanics is distinguished from eastern and western zones in which reef-limestones are interbedded with clastic sediments. The Mesozoic deposits of the central tract include ophiolitic spilites, cherts and peridotites, with deep-water greywackes and manganiferous sediments. Tertiary rocks record the onset of andesitic volcanicity and the concomitant build-up of a sub-aerial arc with the accumulation of basalts, andesitic and dacitic tuffs, together with greywackes, conglomerates, arkoses, coal measures and limestones.

The more easterly arc, which leaves Japan in the vicinity of the Fossa Magna, is marked by the Izu, Bonin and Mariana Islands and terminates about 1000 km north of New Guinea. Far to the south-west the islands of the Tonga and Kermadec groups define a second arc in a wholly oceanic setting, which is linked by a transform fault to arcs defined by the New Hebrides, the Solomon Islands and New Britain. The broad Philippine Sea, which separates the Philippine and Mariana arcs, is floored by oceanic crust which appears to form an independent plate overriding the Pacific floor and overridden in turn by the Philippine arc at the trench on its eastern side. The South Fiji and Tasman Seas which are sheltered behind the Tonga arc, on the other hand, appear to have acted as portions of the large plate incorporating Australia and India (Fig. 10.4). Oceanic trenches flank these arcs on their Pacific sides and seismic zones dip westward away from the ocean. Exceptionally high rates of relative movement are recorded along the Tonga-Kermadec tract which overrides the west Pacific plate (Table 1.1). To the north, however, the symmetry of the system is abruptly

reversed — the New Hebrides, the Solomon Island and New Britain are flanked
by oceanic trenches to the southwest and the associated seismic zones are
thought to dip towards the Pacific.

Both the inner and outer arc systems appear to have been initiated at least in
early Tertiary times; basalts and andesites metamorphosed to low grades include
Upper Cretaceous components in the Solomon Islands; in Fiji, some 10 km of
Eocene-early Miocene volcanics, and sediments showing low-grade metamorphism, underlie the more recent volcanics; and in Bonin, Eocene foraminiferal
limestones are recorded. Palaeogene volcanic groups include andesitic lavas and
pyroclastics along with basalts; and similar associations recur in the Neogene and
post-Tertiary volcanics. The occurrence of andesites in a wholly oceanic
environment emphasises again the importance of tectonic setting in the genesis
of these rocks. Sedimentary formations other than reef-limestones and volcanigenic detrital sediments are of little importance; modest thicknesses of ooze,
clay and volcanic detritus are ponded in shallow troughs to the west of the arcs
but the distance from continental areas has limited the supply of continental
detritus.

The Tonga-Kermadec area has played an important role in the development
of ideas on plate tectonics. From the mode of propagation of body-waves from
deep-focus earthquakes beneath Fiji, Oliver and Isaaks (1967) have recognised an
anomalous zone about 100 km thick in the asthenosphere beneath the Tonga
arc. The top of this zone coincides with the westward-dipping Benioff zone of
earthquake-foci and can be followed to depths of 700 km (Fig. 10.8). The
properties of the anomalous zone, which extends far into the mantle, suggest
that its strength approaches that of lithosphere material and Oliver and Isaaks
suggest that 'the lithosphere has been thrust, or dragged, or has settled beneath
the Tonga arc on a large scale' (1967, p. 4259). The lithospheric slab is regarded
as the leading edge of the Pacific plate descending into the mantle along the
Benioff zone. The depth of the deepest earthquake indicates approximately the
level at which its identity is lost and may be expected to vary with the rate of
advance.

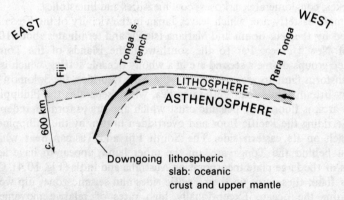

Fig. 10.8 The relationships of the lithospheric slab overridden at the Kermadec
trench (based on Oliver and Isaaks, 1967)

5 New Zealand and New Caledonia

New Zealand, like Japan, occupies the site of an old-established island arc system which appears to have been situated at the border of the Pacific plate for several hundred million years. Although the islands are bordered on the west today by the Tasman Sea with an oceanic crustal structure, it seems likely that in late Palaeozoic and Mesozoic times they lay closer to the Australian continental mass.

The North Island of New Zealand lies at the confluence of two arcs. The western arc, coursing southeastward through New Caledonia, swings through New Zealand in a great sigmoidal curve and finally turns sharply east to the Chatham Islands (Fig. 10.9). This sigmoidal belt marks the site of the late Palaeozoic and early Mesozoic *New Zealand geosyncline*. The Tonga-Kermadec arc meets North Island in the extreme north-east and is continued by a line of active and extinct volcanic centres near the eastern side of New Zealand and by a trench and a seismic zone extending part-way down the east coast. The *Macquarie ridge* taking off southward from the south-west tip of South Island may represent a continuation of this arc but is distinguished by the eastward dip of the associated seismic zone. Seismic activity in New Zealand today is concentrated along the opposed Benioff zones of the Macquarie and Kermadec ridges and along the dextral *Alpine fault* which links them, imparting a clockwise motion to the region as a whole. The New Zealand geosyncline is now partially dismembered by the Alpine fault.

The geological history of New Zealand falls into three stages as follows:

Stage 1: Formation of a metamorphic complex with associated granites which constituted a western 'foreland', or geanticlinal massif, during the evolution of the New Zealand geosyncline: late Precambrian to mid-Palaeozoic?

Stage 2: Initiation and infilling of the New Zealand geosyncline followed by the *Rangitata orogeny* and development of the Alpine fault: late Palaeozoic and Mesozoic.

Stage 3: Late-orogenic and post-orogenic vulcanism and sedimentation and development of the Kermadec-Macquarie arc system: late Cretaceous to Recent.

The *Palaeozoic metamorphic complex*, formed during the first stage, occupies much of the westernmost part of New Zealand. It consists of altered andesitic and basaltic volcanics, greywackes and argillaceous rocks which have yielded a few fossils of Cambrian, Ordovician and Devonian ages and many include late Precambrian rocks. The assemblage suggests an unstable island arc environment; detrital zircons giving radiometric ages of 1350 m.y. indicate that sourcelands of continental type were available for erosion.

Of the many granites in the Palaeozoic complex, a large number appear to date from the Rangitata orogeny and have given isotopic ages of 120–100 m.y. The extensive Mesozoic plutonism which they indicate makes it inappropriate to regard the complex as a foreland to the New Zealand geosyncline: it exhibits, indeed, some features of a belt of low-pressure metamorphism paired with that geosyncline and suffered fairly widespread recrystallisation in Mesozoic times.

A conspicuous lineament, the *median tectonic line,* separates the Palaeozoic

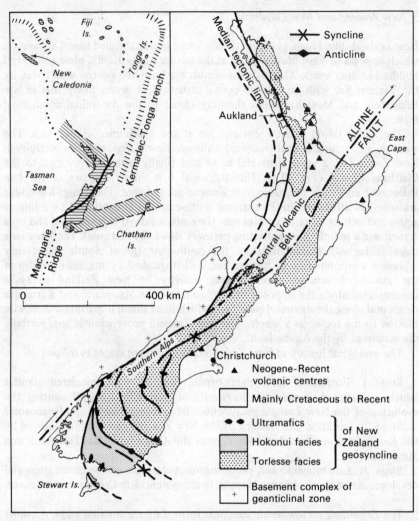

Fig. 10.9. Geological structure of New Zealand (based mainly on compilations of Fleming, 1970). The inset shows the relationships of the New Zealand geosyncline (oblique lines) with the neighbouring arcs and trenches

massif from the rocks of the *New Zealand geosyncline* which occupy most of the rest of the country and which reappear in similar facies in New Caledonia. Two facies belts are distinguished: the narrow and discontinuous *Hokonui facies* composed mainly of shallow-water deposits in the west, and the broader tract of the *Torlesse, alpine or axial facies* characterised by deeper-water sediments in the east. The line which separates the two facies belts is marked by strong magnetic anomalies, by frequent outcrops of Permian pillow-lavas, by serpentines and other ultrabasic rocks and by local developments of high-pressure metamorphic assemblages. This important lineament is regarded by some authors as the approximate eastern margin of continental crust and the trace of the Benioff zone during the evolution of the New Zealand geosyncline. All the structures mentioned above curve into and are displaced along the Alpine fault.

The rocks of *Hokonui facies* are contained in synclinal tracts which swing away from the Alpine fault in western North Island and southern South Island. Permian, Triassic and Jurassic formations thicken eastward to thicknesses estimated at up to 18 km. Permian pillow-lavas, with their associated ultrabasics, make a considerable proportion of this total on the eastern synclinal limbs and are followed by shallow-water tuffaceous sandstones, shales and shelly limestones containing non-marine intercalations towards the top.

The rocks of the *Torlesse facies* are largely of eugeosynclinal type and include poorly fossiliferous turbidites, black pelites, cherts, pillow-lavas and manganiferous sediments suggestive of deep-water deposition. The bulk of the sequence, possibly amounting to over 20 km, appears to be of Permian, Triassic or Jurassic age, though older Palaeozoic formations may also be represented. Syn-orogenic Lower Cretaceous greywackes and pillow-lavas are the youngest strata affected by the Rangitata orogeny. The Upper Cretaceous, unconformable on folded rocks, includes non-marine conglomerates, breccias, sandstones and coal measures.

Orogenic deformation connected with the orogeny took place over a short time-span during the Cretaceous period and ended with the initiation of the Alpine fault. Early nappe-like folds refolded by more upright structures have been recognised in the Torlesse facies belt. The fold-traces curve sigmoidally into the fault and the folds themselves tighten as they approach it. Metamorphism is generally of low grades and granitic intrusions are almost lacking. An exceptional range of low-temperature metamorphic assemblages is developed, especially in the tuffaceous sediments of the Hokonui facies belt where zeolites are present in abundance. The principal assemblages in the Torlesse facies belt, long studied by Turner and his colleagues, fall in the prehnite-pumpellyite and greenschist facies, reaching amphibolite facies in the lowest parts of the succession. The style of metamorphism in the New Zealand geosyncline approaches that of the high-pressure facies series. The occurrence of lawsonite and more locally glaucophane near the junction of Hokonui and Torlesse facies, provides, with the granite-rich terrain to the west of the geosyncline, an analogy with the paired metamorphic belts of Japan.

The Alpine fault. The north-easterly fault which disrupts the New Zealand geosyncline has been intermittently active since late Mesozoic times and is subject to frequent movements today. Matching of the displaced boundary between the Torlesse and Hokonui facies indicates a dextral displacement of some 450 km since Cretaceous times. The fault links active Benioff zones to north and south of New Zealand and represents an accommodation structure allowing adjustment to the complex movement of the plates which meet at these zones. The cumulative effects of displacement have been a clockwise rotation of New Zealand and a tightening of the curvature of the inactive New Zealand geosyncline.

Tertiary and Recent events The early Cretaceous Rangitata orogeny raised the backbone of New Zealand above sea-level and blocked out the present land-masses. Volcanic activity continued through much of the Tertiary era along the eastern border of South Island, close to the Pacific margin where basaltic and andesitic cones were developed. In North Island, rhyolites, ignimbrites, andesites and basalts were erupted in Neogene and Recent times along the still-active *central volcanic belt* (Fig. 10.9). Vertical movements, associated with limited folding, culminated in the mid-Pleistocene *Kaikoura 'orogeny'* and led to the accumulation of coarse sediments in intermontane and marginal tracts as well as to the development and elevation of successive generations of erosional features. Uplift of the block east of the Alpine fault has been estimated at an astonishing 18 km (Suggate, 1963).

An even more disturbed history is recorded in *New Caledonia* where an enormous series of Palaeogene basalts and flysch-like sediments are conspicuously folded and metamorphosed to varying degrees. Overlying early Tertiary rocks are huge slabs of *peridotite and serpentine* on whose lateritised surfaces residual deposits of nickel and subsidiary chromite are developed. These ultramafic slabs have thicknesses of up to about 2 km and have been variously interpreted as surface extrusions, laccolithic intrusions and portions of an overthrust oceanic plate. Regional metamorphism appears to have taken place in the late Palaeogene at about the time of emplacement of the ultrabasic rocks. It is of a high-pressure type characterised by aragonite-pumpellyite at low grades and by lawsonite-glaucophane and epidote-garnet at higher grades. From the known stability ranges of these minerals and from their distribution in the north-west of the island, Brothers (1970) has concluded that the highest ratio of total pressure to temperature was attained in the aragonite-pumpellyite zone which is developed at stratigraphically and structurally high levels. This anomaly is attributed by Brothers to excessive pore water pressures resulting from the seal provided by the overlying serpentines.

11

The Cordilleran and Andean Mobile Belts and the North American Crato

I Introduction

When viewed in relation to the tangle of island arcs and small seas which characterise the western Pacific margin, the continental border-zone flanking the eastern Pacific arrests the eye by its simplicity (Fig. 11.1). From Alaska to Mexico and from the Caribbean to Cape Horn two broad mountain-belts 8000–10 000 km in length rise abruptly above the narrow continental shelf to separate the ocean from the large cratons of North and South America. These mountain-belts mark a continental margin which has been mobile throughout the Phanerozoic eon, but a margin characterised by features which can hardly be matched in other mobile belts past or present.

In late Precambrian times, North and South America belonged to different supercontinents and although they were probably in contact, their relative positions were not those which they occupy today. With the opening of the Atlantic, both continents were propelled westward at the leading edges of the west Atlantic plates. The arcs and small seas of *Mexico and the Caribbean* which link them today were developed or modified during the period of drift, as was the *Scotia arc* which links South America and Antarctica. Despite the enormous lateral migrations of the American continents during the Phanerozoic eon no collision with other continental plates took place and the style of mobility was dictated by the effects of uncoupling of crustal plates at the continent–ocean interface. Here we shall briefly examine the history and structure of the mobile belt and deal with the post-Palaeozoic evolution of the North American craton. The Palaeozoic history of this craton has been touched on in Chapter 4 and the history of the South American and Antarctic cratons in Chapters 5, 6 and 8.

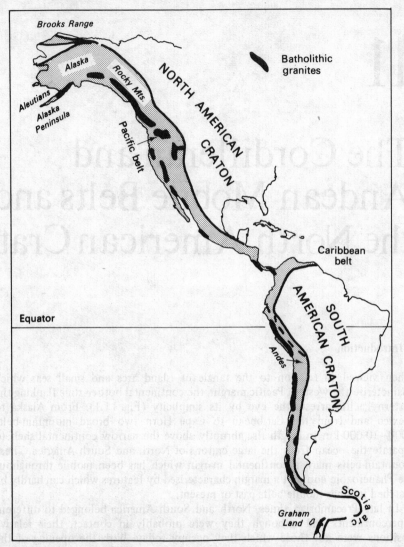

Fig. 11.1. The mobile belt of the western Americas showing the principal granitic batholiths

II The Mobile Belt in North America

1 Tectonic zones

The greater part of the mountain tract of western North America was subjected to strong orogenic disturbances in late Mesozoic to early Tertiary times and

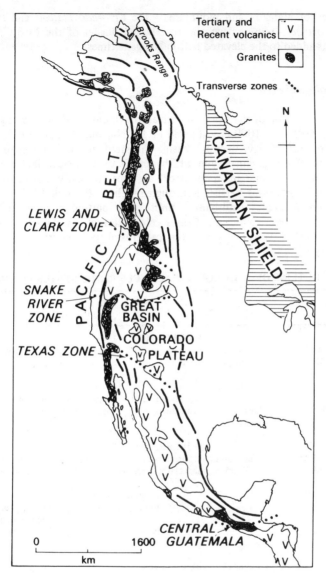

Fig. 11.2. The western mobile belts of North America (based on P. B. King)

formed a highland region throughout the Tertiary and Quaternary eras. This *Cordilleran belt* is commonly distinguished from a narrower *Pacific belt* to the west in which the principal orogenic disturbances were mid-Tertiary or later (Fig. 11.2). In the Cordilleran belt itself, a western tectonic zone incorporating a cover of *eugeosynclinal type* and an eastern zone incorporating a cover of *miogeosynclinal type* are distinguished. In the south-western United States a broad area of plateaux and mountains – the *Colorado Plateau* and *Basin and Range province* – representing the reactivated border of the North American craton was added to the elevated belt in Neogene times.

2 The geological cycle

The western portion of the North American continent has evolved as a geological entity for well over 1000 m.y. (Table 11.1). The oldest supracrustal divisions related to this entity occupy basins initiated in mid-Proterozoic times. Formations of geosynclinal type which appear to have been laid down at an active continental margin date back at least to early Palaeozoic times. Geosynclinal conditions persisted through much of the belt till late Mesozoic times when orogenic disturbance, uplift and plutonism were felt almost everywhere. Late-orogenic conditions characterised by largely non-marine sedimentation, by volcanicity and vertical crustal movements continue to the present day.

Table 11.1. OUTLINE OF THE GEOLOGICAL CYCLE IN THE WESTERN MOBILE BELT

Supracrustal sequence			Orogenic and plutonic events
		PACIFIC BELT	
QUATERNARY AND NEOGENE		late-orogenic formations mainly subaerial volcanics	(mainly vertical movements) granite plutons
MAINLY PALAEOGENE	younger older (Franciscan)	} eugeosynclinal } formations	folding and metamorphism, high-pressure series
		CORDILLERAN BELT	
QUATERNARY AND TERTIARY		late-orogenic formations, abundant subaerial volcanics	(mainly vertical movements), mid-Tertiary granite plutons
END-MESOZOIC		Laramide, strong in miogeosyncline Nevadan, strong in eugeosyncline	} principal orogenic } phases, main granite } batholiths
MESOZOIC PALAEOZOIC	younger older	} eugeosynclinal } and miogeosynclinal } cover formations	several orogenic events including mid-Palaeozoic (Antler) and end-Palaeozoic events (granite plutons in Colorado)
INFRACAMBRIAN AND LATE PROTEROZOIC MID-PROTEROZOIC	younger older	} 'pre-geosynclinal' } formations	small granite plutons in Canada (E. Kootenay)

3 The basement

The basement-complexes, which were in existence when the western belt was first defined, are over 1300 m.y. in age. Archaean and early Proterozoic rocks are widely exposed in the reactivated cratonic border-regions of Colorado, Wyoming and Montana and appear in a number of inliers and uplifted blocks within the western (eugeosynclinal) zone of the Cordilleran belt. These basement rocks crop out sporadically almost to the continental margin in Death Valley, California and in Central America. There are, however, few outcrops of crystalline basement within the Pacific mobile belt which may have developed, at least in part, on a foundation of oceanic crust; we may recall that portions of the Aleutian arc appear to have been derived from a late Mesozoic trough-filling laid down near the foot of the then continental slope (p. 300).

Radiometric dating has led to the recognition of two principal age-groupings in the basement. An *Archaean tectonic province*, perhaps an extension of the Superior province, occupies parts of Montana and Wyoming. Granites and metamorphosed supracrustal assemblages, some traversed by Proterozoic dyke swarms, are represented in this province together with the Stillwater igneous complex. *Proterozoic provinces* yielding ages in the range 1800–1300 m.y. flank the Archaean tract on either side; the north-east—south-west boundaries of these provinces are paralleled by Precambrian shear-zones as well as by Phanerozoic lineaments such as that marked by the Colorado mineral belt (p. 323).

The tectonic grain of the basement is truncated obliquely by the Cordilleran mobile belt. Since this belt appears to mark a continental margin defined in Proterozoic times, its discordance raises the possibility that the continental margin originated as a fracture feature during a mid-Proterozoic phase of continental disruption. Such an interpretation implies that the Pacific Ocean might have been formed by sea-floor spreading towards the end of the Svecofennide chelogenic cycle. We have noted in Part I (p. 83) that the North American continent was subjected to east—west extension at about this time, during the eruption of the Keweenawan lavas.

III The Northern and Central Cordillera

1 Proterozoic developments

The oldest formations which may be assigned to the cover-succession of the western mobile belt have given minimum ages of the order of 1300 m.y. They include the *Belt Supergroup* and its equivalent the *Purcell Supergroup* in the northern Rockies, and the comparable *Apache* and *Grand Canyon Series* of U.S.A. The principal outcrops of the Belt-Purcell Supergroup line up in a tract bordered on the south-west by the transverse Lewis and Clark line (Fig. 11.2). Fine-grained detrital sediments predominate through much of the succession, which reaches a maximum thickness of at least 13 km near the axis of the basin. Lateral and vertical variations are shown by the distribution of carbonaceous siltstones (representing a rather deep-water facies) and of shallower-water dolomitic carbonate-rocks containing stromatolites. There are indications that sediment was supplied to the basin both from the eastern craton and from sourcelands to the west, that is, within, or to the west of, the developing mobile belt. Apart from its exceptional thickness, there is little to suggest that the Beltian accumulated in a zone of marked crustal mobility and it is possible that the supergroup was laid down on a stable continental margin during the opening phase of development of the Pacific. Local deformation and metamorphism affect the Purcell Supergroup which is invaded by some small granites dated at 800 m.y.

The later Precambrian divisions of the mobile belt are commonly unconformable on the mid-Proterozoic groups and may be separated from the earliest Palaeozoic units by only minor breaks. Boulder-beds of glacial origin at a number of localities probably correspond with the late Proterozoic or Infracambrian glacigene horizons of other continents (p. 156). Most of the late Precambrian divisions consist of detrital sediments, some, like the *Windermere Series* of Canada, being coarse clastics apparently derived from elevated sources both within the mobile belt and on the foreland. Acid-intermediate volcanics are locally present. Indications of plutonic activity round about the beginning of the Palaeozoic era are provided by granites dated at 580–550 m.y. which crop out in Colorado well to the east of the present orogenic front.

2 The Palaeozoic and early Mesozoic cover

By the early Palaeozoic, the pattern of crustal mobility that was to dominate the evolution of the Cordilleran belt for several hundred million years had been established. Andesitic-rhyolitic volcanic assemblages appeared sporadically in the eugeosynclinal zone from Cambrian times onward, suggesting that uncoupling of continental and oceanic crustal plates, with the consequent development of a Benioff zone, had taken place at or before the end of the Proterozoic era. A contrast between the eugeosynclinal belt characterised by volcanism and the more easterly (external) miogeosynclinal zone in which the cover-sequence was almost wholly sedimentary was maintained from Cambrian times onward; the junction between these zones changed its position only slightly till the ending of geosynclinal conditions in the late Mesozoic.

The northern Cordillera (Fig. 11.3). The eugeosynclinal zone in Canada and Alaska exhibits a varied assemblage of volcanic and sedimentary rocks following, in some places, with only minor breaks on the late Proterozoic groups and reaching thicknesses which are believed locally to exceed 20 km. Volcanic rocks and cherts occur at many levels and are especially important in the Devonian–Mississippian and Permian–Triassic; lines of serpentines are located along several old-established dislocations. The Upper Palaeozoic sediments are predominantly fine detrital deposits and limestones. Persistant instability during Mesozoic times led to the definition in Canada of two positive tracts (now occupied by the crystalline rocks referred to on p. 318) which shed debris into the intervening

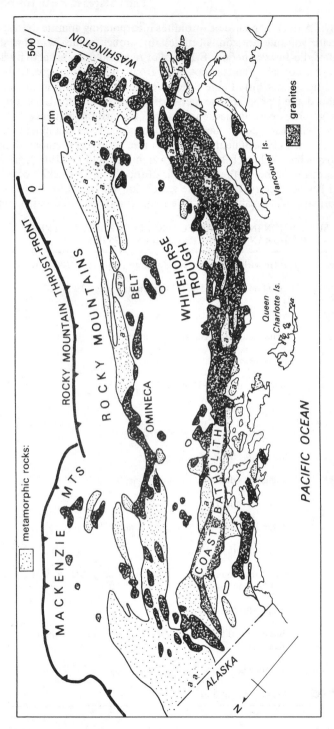

Fig. 11.3. The Cordilleran belt in Canada showing granites (dark shading) and metamorphic complexes (light shading): a = amphibolite facies; b = blueschist facies; no letter = greenschist facies (based on Metamorphic Map of the Canadian Cordillera)

Whitehorse trough. A thick sequence of turbidites incorporating granitic detritus and associated with volcanic horizons occupies this trough, which was almost filled by the end of the Jurassic period. Evidence of repeated orogenic activity is given by early Palaeozoic granites not far from the Pacific coast: by pre-Devonian phyllites and schists in the Brooks Range and elsewhere in Alaska: by locally developed pre-Upper Triassic metamorphic rocks in Canada: and by the abundance of granitic debris in clastic formations through the sequence.

In the miogeosynclinal zone, the cover-succession from Cambrian to Cretaceous is almost wholly sedimentary and is more nearly continuous than that of the eugeosyncline: Table 11.2, based on a summary by Fox (1969), gives details of the succession in the southern foothills of the Canadian Rockies where extensive drilling has been carried out in the search for oil. It will be seen that the Palaeozoic formations are mainly carbonate-rocks and fine detrital sediments

Table 11.2. SUCCESSION IN THE MIOGEOSYNCLINAL ZONE, SOUTHERN
CANADIAN ROCKIES (based on Fox, 1969)

U. CRETACEOUS	shale with sandstone, conglomerate, coal, max. 3000–4000 m
L. CRETACEOUS	conglomerate, sandstone, pyroclastics, max. 1000 m
	(disconformity)
JURASSIC	shale, sandstone (coal), max. 500 m
	(disconformity)
TRIASSIC	siltstone, shale, 0–1000 m
	(disconformity)
PERMIAN	carbonates quartzite, shale, 0–300 m
	(disconformity)
PENNSYLVANIAN	arenaceous dolomite and limestone, 0–500 m
	(disconformity)
MISSISSIPPIAN	dolomite, limestone, shale, chert, max. 1300 m
	(disconformity)
DEVONIAN	limestone, dolomite (partly of reef facies), siltstone, shale, with mid-Devonian disconformity, max. 1000 m
	(major disconformity)
CAMBRIAN	dolomite and limestone above shale and basal sandstone, max. 350 m
←———————— *(unconformity)* ————————→	
MID-PROTEROZOIC	Beltian Supergroup

of modest thicknesses; some of these rocks, especially the massive, rubbly and biohermal Devonian carbonate-rocks, are important oil-reservoirs. The Mesozoic formations, principally detrital or clastic sediments culminating in several kilometres of (partly non-marine) sandstones, shales and conglomerates of Cretaceous age, are *clastic wedges* laid down in connection with the Mesozoic orogenic disturbances in the eugeosynclinal zone. The general pattern recorded in Table 11.2 – an early phase of sedimentation in stable shallow-water conditions followed by the build-up of clastic wedges from westerly sources – is significantly varied in the Brooks Range of northern Alaska where Upper Devonian strata rest on a metamorphic complex including early Palaeozoic rocks, and late Palaeozoic granites are exposed.

The nature of the boundary between miogeosynclinal and eugeosynclinal zones during the period of deposition is seldom clear. In Canada, the older Palaeozoic rocks appear to interfinger across the boundary; the younger units are not often in contact and may have been deposited in separate basins. On the east, the miogeosynclinal succession passes with no marked change of facies into the thinner platform-deposits of the foreland.

The central Cordillera. In the United States, the pre-Tertiary rocks and structures of the mobile belt are heavily masked by late-orogenic volcanics and are disrupted by late-orogenic faulting (Fig. 11.4). The eugeosynclinal zone was subject to volcanicity in every period from Cambrian to Jurassic, with maximum activity in Permian and Triassic times (Table 11.3). Spilites and keratophyres recur at several levels and serpentines are distributed along dislocations, notably in the Klamath Mountains and Sierra Nevada of California. Andesitic and rhyolitic assemblages are even more widely distributed, both geographically and in time, suggesting the consumption of crustal material at a subduction zone over a very long period.

In the western part of the eugeosynclinal zone, the cover-sequence extends, though with many local breaks, from Lower Palaeozoic up to Lower Cretaceous and reaches maximum thicknesses of well over 9 km. Coarse clastic formations incorporating granitic as well as volcanic debris are seen at several horizons from Upper Cambrian upwards and some appear to have been derived from sourcelands within or to the west of the mobile belt. Syn-orogenic sediments of flysch facies are represented by Upper Jurassic and Lower Cretaceous conglomerates and turbidites laid down in restricted basins between rising land-masses in south-west Oregon.

In the eastern part of the eugeosynclinal zone in Nevada, deposition was halted in mid-Palaeozoic times by major disturbances referred to as the Antler orogeny (Table 11.1). In the course of these disturbances, early Palaeozoic cherts and volcanics were transported eastward for almost 100 km on the low-angle *Roberts thrust* over the corresponding succession of miogeosynclinal facies, while folds with westerly vergence, associated with medium-grade metamorphism, were formed further west. In Nevada a north-north-east '*Antler orogenic belt*' some 800 km in length was consolidated during these movements and thereafter functioned as a positive unit between eugeosynclinal and miogeosynclinal troughs.

Fig. 11.4. Locality map of the central Cordillera

Table 11.3. GEOSYNCLINAL VOLCANICITY IN THE CORDILLERAN BELT, U.S.A.
(based on Gilluly, 1965)

JURASSIC	andesites, rhyolites, dacites, basalts, pyroclastics	Washington–California
TRIASSIC	andesites, rhyolites, dacites, pyroclastics	Washington–California
PERMIAN	spilite-keratophyre volcanics, andesitic, dacitic, rhyolitic, volcanics	Washington–Oregon Nevada–California
PENNSYLVANIAN	andesitic, spilitic and trachytic volcanics	Washington–N. California
MISSISSIPPIAN	andesitic volcanics andesitic, basaltic, keratophyric and trachytic volcanics	Nevada N. California
DEVONIAN	basaltic and andesitic volcanics quartz-keratophyres	Oregon–California N. California
SILURIAN	siliceous tuffs	N. California, C. Nevada
ORDOVICIAN	andesites, basaltic pillow-lavas, pyroclastics, siliceous tuffs	Central Navada
CAMBRIAN	andesites, basaltic pillow-lavas, pyroclastics	Central Nevada

The miogeosynclinal zone is at its broadest west of the Colorado plateau and merges into the foreland to north and south. Shallow-water limestones and dolomites form the bulk of the Palaeozoic succession. Although their lithology suggests a fairly stable environment, marked variations in thickness indicate that the basement on which they accumulated was subject to differential vertical movements: in Idaho and Montana, there were long periods of non-deposition, near Las Vegas, a more complete sequence reached 5500 m, while in north-west Utah the Upper Palaeozoic alone reached 9000 m. A clastic wedge of Mississippian age in Nevada was derived from erosion of the Antler orogenic belt.

In Jurassic times, deposition in the old miogeosyncline was halted by the effects of orogeny in the internal zone and a new foredeep-basin came into existence further east, on what had been the foreland region in Wyoming and the neighbouring states. This *Rocky Mountains geosyncline* received Cretaceous and Palaeocene clastic sediments (the earlier marine, the later non-marine) whose total volume is estimated at about 4 000 000 km^3. Gilluly (1963) calculates that in order to supply this bulk of syn-orogenic detritus, the newly-formed Cordilleran belt would have had to be eroded to an average depth of 8 km.

3 The main orogenic phases

As we have already seen, the eugeosynclinal portion of the Cordilleran belt had been subject to repeated disturbances, some of them orogenic in character, from early Palaeozoic to mid-Mesozoic times. In the temperate words of P. B. King (1969, p. 68) 'it is difficult to make any meaningful generalizations as to orogenic climaxes' and not all structures or granites can be assigned to specific stages of activity. Nevertheless, it seems generally agreed that the late Mesozoic and early Tertiary periods saw the terminal stages of orogeny in the Cordilleran belt. The rocks of the *eugeosynclinal* zone had been folded and elevated well before the end of the Mesozoic era during the *Nevadan orogeny*, though they continued to receive intrusive granites till Palaeogene times. The structures of the *miogeosynclinal* zone date mainly from the *Laramide orogeny* which took place at or after the end of the Mesozoic. We see here a record of the migration of orogenic activity from internal to external orogenic zones comparable with that which became familiar in our survey of the Alpine-Himalayan belt. The variations in style of activity also followed a familiar pattern; the eugeosynclinal zone and its eastern border-tract were distinguished by extensive regional metamorphism and by emplacement of enormous granite batholiths, the miogeosynclinal zone by lack of metamorphism, by smaller (though not negligible) granites and by *décollement* tectonics. The reactivated cratonic zone showed a distinctive style which will be discussed separately.

The northern Cordillera. The *internal (western) zone* of the Canadian Cordillera is occupied by two narrow tracts of crystalline rocks separated by the broader zone of little-altered Mesozoic and late Palaeozoic cover-rocks in the Whitehorse trough (pp. 313–14). Rather broad folds broken by dislocations are displayed by the competent volcanic groups of the trough and the total amount of shortening across the structure appears to have been small. The Coast geanticline which separates the Pacific and Cordilleran belts is occupied by granite (the Coast batholith) associated with small areas of gneisses and metasediments. The more easterly crystalline tract, the *Omineca geanticline* separating cover-rocks of eugeosynclinal and miogeosynclinal facies, includes both granites and metamorphic rocks, the latter being the more extensive. Early geologists assigned many of these crystalline rocks to the Precambrian and regarded the Omineca tract as an ancient barrier between two Phanerozoic troughs. More recent work has shown that Palaeozoic and even Mesozoic metamorphic rocks appear within the crystalline complexes. In the *Shuswap* complex, near the Canadian border, a sequence including Palaeozoic and Proterozoic members, mostly showing metamorphism of greenschist facies, overlies an infrastructure of gneiss domes which could represent the basement. In the Eastern Alaska Range, pre-Palaeozoic complexes locally retain evidence of granulite facies metamorphism, possibly dating from a basement cycle. The evidence of boulders in conglomerates, with limited radiometric evidence, suggests that metamorphism and granite-emplacement were repeated over a considerable period in the Coast and Omineca tracts, which were zones of persistent high heat flow.

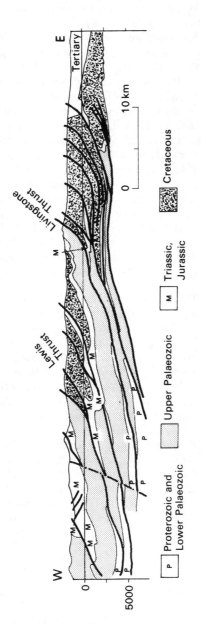

Fig. 11.5. Cross-section of the miogeosynclinal zone of the northern Cordillera in Alberta (based on Fox, 1969)

The *external (eastern) zone* of the Canadian Cordillera forms the *Rocky Mountains* and their eastern foothills in which the principal structures are of late Cretaceous or Tertiary age. The basement is not exposed in this zone, where the miogeosynclinal cover-rocks are stacked in thrust-slices and tight folds with north-eastward vergence (Fig. 11.5). Seismic studies suggest that the top of the basement descends smoothly westward toward the mobile belt and it appears that the cover-sequence was pushed eastward as packets of folds and schuppen above a *décollement* surface; shortening of over 150 km is implied by the observed reduplications. A ·thrust-front marked by major dislocations (the *Lewis thrust* has a minium displacement of 30 km) separates the external zone from the less strongly disturbed autochthonous cover of the craton.

Sedimentation of *late-orogenic* facies began early in the internal (western) regions where the first 'successor basins' developed in late Jurassic times. Clastic sediments (largely non-marine) mixed with volcanic components accumulated in these intermontane basins in Cretaceous and early Tertiary times, reaching considerable thicknesses in the Alaskan sector, and were themselves quite strongly folded. In the external (eastern) region, Tertiary piedmont wedges spread eastward over the craton. Volcanicity in the northern Cordillera was trivial compared with that in the central sector (p. 326). Block movements along rejuvenated or newly formed dislocations such as the *Rocky Mountains—Tintina 'trench'* in Canada and *Fairweather fault* in Alaska appear to have been dominated by transcurrent (predominately dextral) displacements.

The central Cordillera. The eugeosynclinal zone of the central Cordillera was invaded towards the end of the Mesozoic times by the *Sierra Nevada and Idaho batholiths* and their satellites. The cover-rocks closely associated with them are commonly metamorphosed in amphibolite facies. Around the north-west of the Idaho batholith, for example, the Belt Series is represented by coarse schists, gneisses, quartzites and marbles, while Palaeozoic strata show similar characters against the Sierra Nevada batholith. Further from the batholiths, and especially in the younger divisions, cover-rocks, though strongly folded, are non-metamorphic or of low grades. A series of eastward-dipping thrusts marks their boundary with the Pacific belt in California.

The structures developed in the *miogeosynclinal zone*, like those of the northern Rockies, are related to the detachment of the cover from the relatively undisturbed basement. In Idaho and Wyoming, thrust slices and overturned folds with easterly vergence are the principal structures. Further south, large low-angle thrusts carry Palaeozoic or Proterozoic cover-formations over the Cretaceous fill of the Rocky Mountains geosyncline. Minor thrusts and steep reverse faults are seen also in the rejuvenated craton, especially around the borders of the Colorado plateau.

4 Granites of the Cordilleran belt

A glance at the map (Fig. 11.1) shows that outcrops of granite extend with only short gaps along the Cordilleran belt from the peninsula of Alaska to that of

California and from northern Peru through Bolivia into Chile. The great bulk of this granitic material was emplaced over a period of a few tens of millions of years in late Mesozoic times during an episode which seems almost without parallel in geological history.

In North America, the granites of the western mobile belt were concentrated in elongated *batholiths* up to 1000 km in length, which are aligned in the tectonic grain. Smaller *plutons*, which are generally ovoid or circular in plan, are more widely scattered; some are satellites of the great batholiths and may represent cupolas rising from their buried extensions. The batholiths themselves are composite structures made up of numerous units which are chemically distinct or outlined by concentric foliation patterns. The suite includes gabbroic and dioritic components and granites in the strict sense, but by far the most abundant types are granodioritic. Geophysical studies suggest that the batholiths pass downward in some places into rocks with higher densities and seismic velocities, which are presumably more basic in average composition.

Most of the smaller plutons are obviously intrusive and are either surrounded by hornfelsic metamorphic aureoles or by little-altered rocks. The great batholiths are flanked by metamorphic complexes of regional extent, including schists and gneisses. It seems probable that these complexes are not contact-aureoles in the strict sense but are products of high regional geothermal gradients in the tracts through which the batholiths rose. The predominance of hornblende–hornfels assemblages in the aureoles around many of the smaller and later plutons, the preservation of small age-differences in the minerals of adjacent plutons, the style of mineralisation and the stratigraphical evidence that granitic sourcelands were repeatedly available for erosion in Mesozoic times all suggest that emplacement took place at shallow depths – Evernden and Kistler (1970) suggest that the cover on many granitic bodies was no more than about 5 km in thickness.

Although in a general sense the batholiths and plutons of the Cordilleran belt are members of one great suite, significant variations in time and space have been revealed. A compositional variation, irrespective of age, is indicated in the United States by a distinction between a region (mostly to the west of the great batholiths) characterised by quartz-diorites and a region, including the batholiths and land to the east of them, characterised by granodiorites. Some authors see in this variation an expression of variations in the depth at which magma was generated in relation to an eastward-dipping Benioff zone. Studies of strontium isotope abundances suggest that some at least of the granites are derived from mantle material.

Setting aside the volumetrically trivial Palaeozoic granites, the earliest stage of plutonism is recorded by small Triassic granites. Jurassic granites form significant parts of some major batholiths and some smaller bodies mostly between the batholiths and the Pacific. By far the greater part of the Sierra Nevada and Idaho batholiths as well as much of the Coast and Alaska batholiths and innumerable smaller plutons are of Cretaceous age. Eocene (Laramide) granites form the eastern part of the Coast batholith and numerous small plutons in the miogeosynclinal zone and reactivated craton of the central Cordillera. Mid-Tertiary plutons are more or less confined to the miogeosynclinal zone and reactivated craton where sub-volcanic porphyries, rhyolites and other minor intrusions also occur. The age-variations suggest an irregular eastward displacement of the sites of plutonism with time, possibly related once again to an eastward increase in depth of the magma-source.

The plutonic cycle recorded by the granite series in the western mobile belt spanned a total period (Triassic–mid-Tertiary) of about 180 m.y. reaching its peak in the late Mesozoic. Evernden and Kistler have suggested that emplacement in California took place in several pulses, lasting 10–20 m.y.,

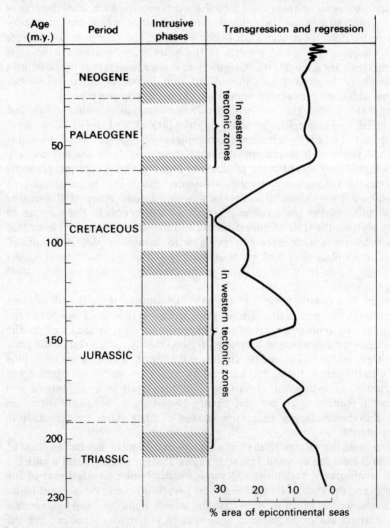

Fig. 11.6. Phases of granite-emplacement in the central Cordillera set against the record of transgression and regression on the North American craton (modified from Evernden and Kistler, 1971)

separated by intervals of about 30 m.y.; although this plutonic cycle coincided with a prolonged marine transgression on the North American craton, each intrusive pulse was reflected in a temporary reversal of the transgression as shown schematically in Fig. 11.6. This relationship illustrates in a striking way the coincidence of granite-emplacement and regional uplift which is apparent in the terminal stages of many orogenic cycles.

5 Mineralisation

The rich late-orogenic mineral-deposits of the Cordillera may be mentioned here, since some are related to late Cretaceous and Tertiary granites. Many, however, were related to the Tertiary volcanic activity which has still to be dealt with (p. 326). *Gold* is of major importance in California (the *Mother lode* belt on the south-west side of the Sierra Nevada pluton) and occurs in many smaller deposits from Alaska to Mexico. Its wide distribution reminds us that the Circum-Pacific belt as a whole constitutes the world's only major post-Archaean gold-province. Of other deposits, *lead-zinc-silver* assemblages of various kinds occur in association with high-level plutons and sub-volcanic intrusions, mostly in the external zone and reactivated craton. These deposits may be derived in part from older mineral provinces and are certainly influenced by basement structures such as the Precambrian shear-zones which parallel the *Colorado mineral belt* (p. 311). *Tin*, though a very minor component, is widely distributed and is of considerable importance in South America. *Copper* occurs in large deposits from Buete (Montana) to the Mexican border and in the Chilean Andes. It is associated with volcanic centres (*porphyry copper*) and will be mentioned again later (p. 336).

IV The Pacific Belt

1 General outline

In the Pacific belt (Fig. 11.2) marine sedimentation continued, often in small unstable basins flanked by islands, till well on in Tertiary times, while the internal zone of the Cordilleran belt to the east was rising as a mountain-region. Folding and faulting took place sporadically, the general conditions being those of extreme instability; the main phases of elevation were delayed till late Neogene and Quaternary times.

   In the much-discussed Californian sector, the oldest exposed cover-rocks are later Jurassic to early Palaeogene greywackes, shales and cherts, together with basic lavas and serpentine pods. Submarine basic volcanics of Eocene age reach over 15 km in thickness in Oregon and Washington. In the Coast Ranges of California the eugeosynclinal assemblage constitutes the *Franciscan Formation* which is thought, like the corresponding assemblage of Alaska (p. 300), to have accumulated on an oceanic basement near the foot of the continental slope. The younger cover-units represent the fills of short-lived unstable basins, predominantly marine at low levels and towards the coast, predominantly non-marine at high (Pliocene and Quaternary) levels and towards the interior.The *Ventura basin* of southern California contains 4–5 km of late Neogene turbidites.

The structural style of the Pacific belt is disruptive. The north-eastern margin of the belt is marked both in California and in Alaska by eastward-dipping thrusts on which units of the Cordilleran belt override the Pacific belt. In Alaska, folds of south-westward vergence, stacked in many thrust-slices, suggest that the continental plate overrode the Pacific Margin. All these structures are traversed by large transcurrent faults foremost among which is the *San Andreas fault* with dextral displacements of several hundred kilometres (Fig. 10.3) and by normal faults resulting from vertical block-movements.

2 The Franciscan blueschist assemblage

The Franciscan Formation displays a chaotic structural style characterised by numerous *mélange zones* in which tectonically-defined blocks up to a few hundred metres in diameter are scattered through a less competent meta-sedimentary matrix. Most of these blocks consist of coarse metamorphic rocks of blueschist, amphibolite or eclogite facies derived from basic or ultrabasic parents. Some are of higher metamorphic grade than their matrix and have a 'rind' in which actinolite–chlorite–talc assemblages replace their characteristic minerals. They are usually regarded as tectonically-emplaced 'pips' derived from deeper levels.

The metasedimentary formation in which the blocks lie show a metamorphic zoning defined by three assemblages developed in greywackes and shales:

    (a) (low grade) *Laumontite zone:* laumontite + albite ± calcite
    (b) *Pumpellyite zone:* pumpellyite and/or lawsonsite + albite + calcite
    (c) *Lawsonite zone:* lawsonite (± pumpellyite) + albite ± aragonite: lawsonite + jadeitic pyroxene ± aragonite

The mineral assemblages, both of the formation as a whole and of the contained blocks are of high-pressure facies series. The regional grade increases eastward. The lawsonite zone lies close to the bounding thrusts on which units of the Cordilleran belt override the Franciscan, and Blake and others (1967) have concluded that the grade increases upward to the thrusts. These relationships, with the wide distribution of higher-grade blocks, have been interpreted in various ways. Early suggestions involving metasomatism associated with the emplacement of ultrabasic intrusions in the formation have been largely abandoned in favour of hypotheses involving high pressures of one sort or another. Deep burial resulting from downdrag at the marginal Benioff zone, with subsequent retrieval of slices and pips by isostatic recovery is favoured by Ernst (Fig. 11.7A). The generation of tectonic 'overpressures' connected with the relative movements of colliding plates in conditions of high pore water pressures is favoured by Blake and his colleagues (Fig. 11.7B).

The distinctive feature of the California blueschist assemblages is the chaotic structural style as a result of which components of differing characters are

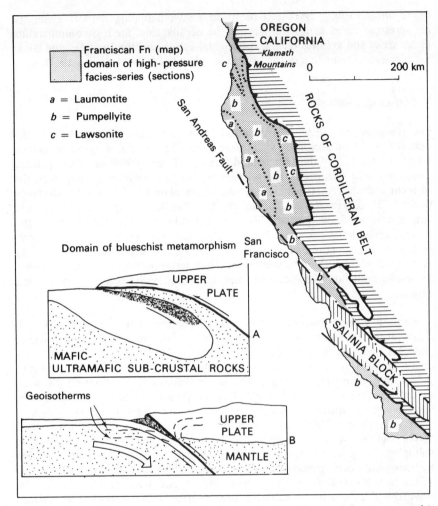

Fig. 11.7. The Franciscan Formation in California showing the metamorphic zones defined by laumonite, pumpellyite and lawsonite. The diagrammatic sections (A) and (B) illustrate two interpretations of the setting of the blueschist metamorphic assemblage: (A. after Ernst, 1971. B. after Blake *et al.*, 1969)

jumbled together. Whereas metamorphic minerals from the Franciscan matrix give apparent ages in the range 135–80 m.y. (consistent with the stratigraphical age of the formation) some of the included blocks give ages of 150–140 m.y. and may have been derived from an older blueschist complex as 'pips' squirted up during the phases of downdrag and retrieval.

The time-span of metamorphism of high-pressure facies series in the Pacific belt is broadly the same as that of metamorphism and granite-emplacement in the internal zone of the Cordilleran belt to the east. These two units constitute *paired metamorphic belts* related to an eastward-dipping Benioff zone: the high-pressure facies series forming on the oceanic side, the higher-temperature facies series and granites on the continental side where the Benioff zone lay at deep levels and remelting of the overridden plate had begun (see Fig. 10.7).

V Late-Orogenic Stages

The Neogene and post-Tertiary periods were distinguished, especially in the central Cordilleran sector, by widespread basic magmatism and by block-faulting affecting much of the reactivated craton. These activities were perhaps connected with changes in the relationships of the continental mass with the adjacent oceanic crustal plates. The east Pacific plate which formerly intervened between the continental margin and the East Pacific Rise appears to have been largely overrun in late Tertiary times and the old-established oceanic trench and Benioff zone at the continental margin went out of action when the movements of the remaining oceanic plate took on a new coast-parallel direction (Fig. 10.3). As a result of these changes, western North America passed into a realm affected by transcurrent fault-displacements and block-faulting related to an extensional regime.

Tertiary—Recent volcanism. The northern sector of the Cordilleran belt is characterised by a comparatively minor suite of late-orogenic volcanics, mainly andesites and basalts. The central and southern Cordillera are distinguished by enormous piles of subaerial lavas and pyroclastics on which stand some imposing volcanoes of quite recent origin. The lower (Palaeocene to Eocene) divisions, extending as far east as the Colorado Plateau, are predominantly andesitic. They are moderately disturbed, are penetrated by late granites and by high-level porphyries, rhyolites and other acid differentiates and are frequently mineralised in the vicinity of these intrusives. In contrast to these typical late-orogenic volcanics, the principal *Neogene and Quaternary volcanics* are plateau-basalts with rhyolites, acid ignimbrites and tuffs which have affinities with anorogenic volcanics elsewhere. Only in western Oregon and Washington, close to the still-active Pacific belt, do andesites predominate in the Miocene and younger volcanics; these rocks interfinger along the crest of the Cascades with very different basaltic assemblages. The *Columbia River Basalts*, mainly Miocene tholeiites, fill a shallow crustal downwarp and are followed by Quaternary olivine-basalts. The *Snake River Basalts* (Pliocene—Pleistocene) occupy a transverse downwarp of east—west trend which is thought by some to be sited on a tract of crust lacking a granitic layer. Rhyolites, dacites, ignimbrites and other acid volcanics formed by differentiation and/or remelting of crustal material appear at centres within the lava-plateaux and form extensive spreads towards the east.

Structural disturbances in Neogene and post-Tertiary times were of two principal kinds also related to a change of crustal regime. *Transcurrent displacements* on the dextral San Andreas fault and its associates speeded up in the mid-Tertiary (p. 285). Differential *vertical movements* led to the elevation of mountain blocks separated by troughs which filled up with post-orogenic sediments; much of the vertical movement is of quite recent date, uplift of the Sierra Nevada for example being largely Pleistocene.

VI The North American Craton

1 The main cratonic region

After the stabilisation of the Appalachian and Ouachita belts, almost the whole North American continent, with the exception of the western mobile belt and a zone fringing the Canadian Arctic, entered on a phase of cratonic development. Much of the continent, in fact, remained almost continuously above sea level, subject to eroson or receiving deposits of continental facies. Epicontinental seas flooded parts of the interior during a late Mesozoic transgression (Fig. 11.6) and encroached repeatedly on the continental margins where predominantly marine successions accumulated.

Variations in geological history from Mesozoic times onward allow one to distinguish a number of differing crustal regimes as follows:

(1) *The Canadian shield* remained, on balance, a positive region receiving only thin deposits, or subject to erosion. In the large downwarp of *Hudson Bay*, the basement surface was progressively depressed to depths of up to two kilometres. Igneous activity in the shield was on a minor scale throughout Phanerozoic times, but a few alkaline ring-complexes and kimberlites were emplaced in the basement.

(2) *The Appalachian belt* (with the exception of its easternmost part) began the Mesozoic era as a mountain-tract and was subjected to successive cycles of degradation, rejuvenation and erosion. In late Mesozoic times the *Schooley peneplain* formed from degradation of the original mountains was arched to heights reaching well over a kilometre, producing the highland block from which the present mountains were carved. Two Tertiary erosion-surfaces were developed at lower levels.

(3) *The Atlantic margin* was defined by the development of basins of subsidence early in the Mesozoic and was characterised thereafter by thick sedimentation of coastal-plain, shelf and slope types as described on p. 266. Basic volcanism in late Triassic times (p. 114) was associated with the emplacement of a basic dyke swarm (dated at 200–190 m.y.) whose members extend for some 2000 km in the marginal tract. May (1971) suggests that this swarm, with others of similar age in West Africa and northern South America, form a radial set converging on a region in the vicinity of the Blake plateau and Senegal basin. Assuming the dykes to be tensional features, stress-trajectories are constructed by May which suggest that a crustal dome was formed immediately prior to the break-up of the continents (Fig. 11.8). Small igneous centres of other types are located along lineaments oblique to the American coast – the *White Mountain magma-series* of syenites and alkali granites is mid-Mesozoic, the *Monteregian suite* of syenites and nepheline-syenites is late Mesozoic. In the far north, a patch of early Tertiary plateau-basalts corresponding to those of West Greenland adheres to the coast of Baffin Island.

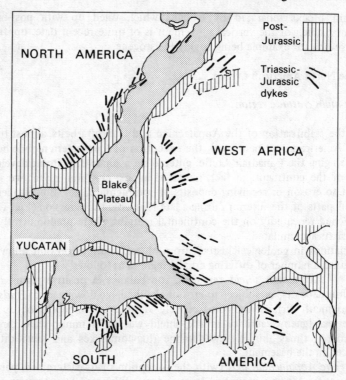

Fig. 11.8. Dyke swarms emplaced in relation to the rupturing in early Mesozoic times, of the American—African supercontinent (based on May, 1971)

(4) The *Gulf Coast* at the south-east border of the craton was a region of subsidence and sedimentation throughout Mesozoic and post-Mesozoic times. Shelf, delta and coastal-plain deposits of this region smothered much of the Ouachita belt which was consolidated at the end of the Palaeozoic era.

(5) The *interior lowlands* of the United States in which a cratonic Palaeozoic cover had accumulated on a Precambrian basement (Chapter 4), remained as a generally low-lying stable unit through the remainder of the Phanerozoic eon. The eastern part of the lowlands has retained little Mesozoic or Tertiary sediment but the western part, along the whole border of the Cordillera, exhibits an almost continuous cover of non-marine and shallow-water marine sediments: volcanics are almost absent except in the reactivated region which is dealt with separately (p. 329).

The sediments which accumulated on the western part of the craton in Triassic and Jurassic times include relatively thin sequences of non-marine red-beds with evaporites. In parts of Colorado, late Jurassic sandstones of the Morrison Formation are impregnated with uranium minerals. The effects of Nevadan orogeny are reflected in the appearance of thick wedges of Cretaceous sandstones, red-beds and shales thinning eastward from the mountain-front (p. 317). A marine transgression through late Mesozoic times ensured that the finer distal sediments were mainly of marine facies; minor regressions corresponding to pulses of granite-intrusion in the Cordilleran belt punctuated this transgression (Fig. 11.6). From late Cretaceous times onward the sea retreated from the continental interior and the Tertiary and Quaternary deposits in the western part of the craton took the form of non-marine sandstones, red-beds and shales in which some celebrated mammalian faunas are contained.

Igneous activity outside the reactivated region discussed below was expressed mainly by the emplacement of plugs and dykes of syenite, nepheline-syenite or mica-peridotite and (in Utah and Arizona) of kimberlitic pipes carrying eclogite nodules. Alkaline bodies ranging in age from mid-Palaeozoic to late Cretaceous are concentrated on an east—west lineament from Kansas to West Virginia.

2 The reactivated cratonic zone

The portion of the craton flanking the western mobile belt in Utah, Colorado, Arizona and New Mexico became, through Mesozoic and Tertiary times, progressively more involved in the instability characteristic of the western belt and was finally incorporated in the rising orogenic belt. From late Palaeozoic times onward, the distribution of sediment was influenced by the presence of positive blocks in which the Precambrian basement remained at or near the surface. The late-orogenic Palaeogene sediments derived from the rising mountain-tract, especially, showed conspicuous variations of thickness and facies due to this cause. Igneous activity began in Tertiary times with the emission of varied volcanics and the emplacement of granitic plutons with concomitant mineralisation in many places.

It was thus from a structure which was already inhomogeneous that the *Basin and Range province* of Nevada was developed in Neogene times, when the tectonic regime changed once more. A system of graben and horsts with an irregular north—south grain was developed in the province, where the graben are spaced at intervals of 25—30 km and are themselves of similar breadths. By Pleistocene times, systems of fault-troughs separated by mountain blocks had been formed through the whole width of the Cordilleran range.

The *Colorado Plateau* forms a node in the tracery of faults and appears to have been outlined before the ending of the Laramide orogeny. Thrusts formed during the orogeny curve around the plateau as did some Palaeogene basins and uplifts. The plateau appears to have stood at a low level in early Tertiary times and was elevated only in the late Neogene, forcing the Colorado River and tributaries rising beyond its north-eastern border to entrench themselves in canyons up to 1500 m in depth.

The style of the late Tertiary and Quaternary structures is dominated by *normal faulting* and appears to have been related to an extensional regime. From geometrical considerations, Stewart (1971) estimates extension in the Basin and Range province at 50—100 km. Many geological and geophysical anomalies

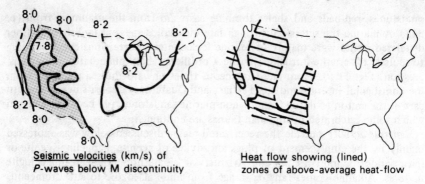

Fig. 11.9. Geophysical anomalies in the western United States

characterise the region. The dominance of a basic-acid volcanic association in the more recent volcanics has been mentioned. The heat flow is high and hot spring activity still continues (Fig. 11.9). The crust is of variable but usually small thickness; beneath the Great Valley of California and parts of the Basin and Range province the recorded thicknesses are less than 25 km which would be small even for an old stable region: even the figures of around 40 km for the Sierra Nevada and Colorado Plateau are relatively modest.

Clearly, there has been no marked regional thickening of crust adjacent to the mobile belt, indeed, extension during the phase of normal faulting may have resulted in a thinning of the crust. Despite the absence of crustal roots, much of the central Cordillera and reactivated craton stands 1–2 km above sea level, forming a vast highland region which appears to be at least partly in isostatic balance. The lower crust and uppermost mantle beneath this region have anomalous properties (Fig. 11.9) – velocities of P waves are well below the common figure of 8.0 km/sec in the mantle and mantle rocks may also be of less than average density, providing an explanation of the buoyancy of the overlying crust. The remarkable combination of features displayed by the central Cordillera has been attributed by Menard and others to the operation of a buried 'mid-oceanic ridge' – the East Pacific Rise which, as we saw, was overrun by North America in Neogene times. It appears, however, that at least one branch of the Rise now runs up the Gulf of California (Fig. 10.3) and some authors envisage an extensional regime resulting from right-lateral movement of the Pacific floor past the American continent. Either regime could reasonably account for the extensional structures, the basic vulcanicity and the high heat flow.

VII The Arctic Mobile Belt

The Arctic border of North America and Greenland is occupied by a tract of thick Phanerozoic cover-rocks which show fold-systems of various ages. Over much of this tract, the basement includes late Precambrian rocks subjected to orogenic folding during the formation of the *Carolinidian belt*. The older Phanerozoic cover-units, up to 6 km in thickness (Cambrian or Ordovician to Devonian) show a detrital facies of greywackes, shales and volcanics in northern Ellesmere Island and parts of Peary Land passing southward into a miogeosynclinal carbonate facies. All were folded in mid-Palaeozoic times to produce the *Innuitian fold-belt* (Fig. 2.1). Regional metamorphism affected the northern (internal) zone in which a few granite plutons were also emplaced.

The fold-complexes of Palaeozoic age at the Arctic border were effectively stabilised before the resumption of deposition in Pennsylvanian times; in character and timing they have something in common with the early metamorphic complexes of the Brooks Range in Alaska, which were incorporated in the western Cordillera (p. 314). Along the borders of Arctic Canada, the principal later Phanerozoic structure was the large *Sverdrup basin* whose deposits range with scarcely a break from Pennsylvanian to Eocene and reach total thicknesses of nine kilometres. The late Palaeozoic members, largely carbonate rocks, are followed by evaporites, sandstones and shales apparently built out by northward-flowing rivers into the Arctic Ocean. Tertiary folding and thrusting affected the deposits of Ellesmere and Axel Heiberg Islands at the east end of the basin and salt-dome structures are quite widely developed. The Sverdrup basin, despite its remoteness, is a potential source of oil.

VIII Mexico and Central America

The greater part of Mexico, Guatemala and Honduras is occupied by the *southern Cordillera* which are similar in essentials to the central Cordillera already described (Table 11.4). In these regions, the north-eastern front of the

Table 11.4. TECTONIC ZONES OF THE SOUTHERN CORDILLERA

south-west (Pacific margin)

1 *Pacific belt:* eugeosynclinal late-Mesozoic assemblages resembling the Franciscan Formation

2 *Internal zones of Cordilleran belt:* crystalline basement and Cretaceous granite batholith of Baja California: ? basement geanticline of Sierra Madre del Sur: remnants of Mesozoic eugeosynclinal assemblage

3 *External zones of Cordilleran belt:* miogeosynclinal Mesozoic cover (carbonate rocks, clastic wedges, evaporites) folded and overlain by marine and non-marine clastic sediments of Tertiary-Quaternary successor basins: late-orogenic Tertiary andesites, ignimbrites and pyroclastics with lead-zinc-silver mineralisation in Sierra Madre Occidental

4 *Marginal zone of Cordilleran belt:* orogenic front marked by rapid north-eastward decrease in folding and by uplifts of basement

5 *Foreland zone:* little-disturbed Mesozoic cover of shelf facies including evaporites overlain by clastic wedges and detrital Tertiary-Quaternary coastal-plain sediments

north-east (Caribbean margin)

Cordillera truncates the Ouachita belt, and the Mesozoic cover appears to rest on a basement of folded and metamorphosed Palaeozoic and late Precambrian rocks, a few samples of which have yielded radiometric ages of as much as 1000 m.y. The mobile belt is evidently of considerable antiquity and the crustal regime is continental. The narrow tract of southern Central America, on the other hand, is built up of Mesozoic and post-Mesozoic rocks which are thought to rest on oceanic crust. This tract is part of the *Antillean belt* whose other branch forms the Caribbean arc (Fig. 9.6).

Two east—west lineaments, oblique to the tectonic grain, cross the southern Cordillera close to their union with the Central American belt (Fig. 11.2). The transverse *Mexican volcanic belt* following an east—west course immediately south of Mexico City is marked by huge cones of olivine-basalt and related rocks formed since late Neogene times and including most of the active centres of Mexico. This young volcanic lineament emerges on the Pacific coast close to the eastern termination of the trench and seismic zone which flank Central America. A few hundred kilometres to the south, an irregular geanticlinal tract occupied by the Precambrian and Palaeozoic basement extends from the Sierra Madre del Sur in Mexico to eastern Honduras. The tectonic grain in this tract is obliquely truncated by the Pacific coast and towards the east end of the lineament basement metasediments, granites, ultramafic rocks and cover-sediments are jumbled together by steep faults which appear to pass eastward into the fault-system defining the Cayman trench (Fig. 9.6). The obvious disruption, the relationship with oceanic structures, and the apparent absence of pre-Mesozoic rocks south of the geanticline suggest that the early Phanerozoic continent terminated close to this lineament.

The central American tract which links the southern Cordillera with the Andes is flanked on both sides by ocean basins and is built on a basement of Mesozoic eugeosynclinal volcanics and detrital sediments. Mobility and sporadic vulcanicity has continued up till the present day, with consequent generation and accumulation of erosional debris. The tract appears to have been represented until Neogene times by island groups various distributed and it was not until the Pliocene that an unbroken isthmus emerged.

The southern border of the Caribbean Sea is fringed by a mobile belt in which continental influences are once more dominant and where the principal structures are of early Tertiary age. The *Caribbean belt* of northern Venezuela and Trinidad is backed by the Guyana shield from which it is separated by the *Eastern Venezuela basin* (p. 246). Along the northern border of the land, a sliver of metamorphosed Mesozoic eugeosynclinal supracrustal rocks lies on the seaward side of an east—west dislocation whose effect is to detach the more mature Caribbean belt from the still-active volcanic arc of the Lesser Antilles. The branches of the Antillean belt, with the Caribbean basin which they frame, appear to form a small crustal plate moving independently against the larger continental and oceanic regions which surround it.

IX The Andes

1 Introduction

The western mobile belt of South America is, like that of North America, founded on an old-established continental margin which appears to have been active through much of Phanerozoic time. Successions of Palaeozoic (and perhaps late Precambrian) strata record subsidence and thick marine sedimentation contrasting with the history of more restricted sedimentation on the South American craton to the east. The Mesozoic succession in the mobile belt is predominantly marine and, in the northern and southern sectors, includes assemblages of eugeosynclinal types: it contrasts with the predominantly continental Gondwana successions on the craton. Episodes of orogenic disturbance were widely recorded in mid-Palaeozoic and late Palaeozoic times. Similar episodes in late Mesozoic times were associated with the emplacement of enormous granitic batholiths and were followed by phases of widespread intermediate and acid vulcanicity and by repeated uplift. An oceanic trench off the Pacific coast and an eastward-dipping seismic zone subject to deep-focus earthquakes show that movement between the South American continent and the South Pacific floor still continues.

Where the Andean belt is flanked by the stable continental mass of South America, it consists in most places of at least two parallel ranges separated by less elevated plateaux or sediment-filled troughs. From northern Peru northward and in southern Chile and Tierra del Fuego, the more westerly range incorporates Mesozoic assemblages of eugeosynclinal type which contrast with assemblages of shallow-water sediments further east, but no general correlation has been made between the ranges of the present mountains and any system of eugosynclinal and miogeosynclinal zones. An eastern *sub-Andean zone* bordering the high ranges and characterised by relatively simple structures forms a distinct unit through most of the belt.

A crystalline basement of continental type is exposed at many localities even as far west as the present coastline. Much of this basement represents early Palaeozoic fold-complexes, but high-grade metamorphic rocks assigned to the Precambrian are recognised in the coastal region of Chile and elsewhere. In Colombia, basement rocks of the central range have yielded radiometric dates of about 1300 m.y. There seems little evidence for any westward build-out of the

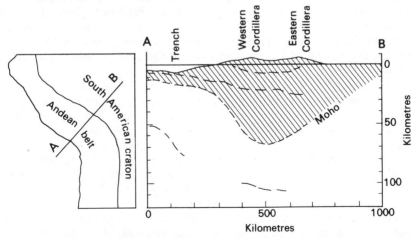

Fig. 11.10. Crustal section of the Andes, based on seismological data (after James, 1971)

continent since Palaeozoic times. Seismic and gravity data suggest that in the Peruvian sector, at least, the crust thickens rapidly from about 30 km near the coast to 60–70 km beneath the western Cordillera: the Andes, in contrast to the North American Cordillera, appear to have a substantial crustal root (Fig. 11.10).

To north and south the Andean chains splay out in a number of curved structures linking up with other units of the Circum-Pacific belt. The westernmost ranges in the north turn into the Isthmus of Panama, while the eastern range terminates against the Caribbean belt of Venezuela. In Argentina the oblique Precambrian Pampean, Patagonian and Deseado massifs mark ancient ridges which swing out south-eastward from the main belt and are separated by tracts containing Palaeozoic basin-sequences. The most southerly of these basin-tracts follows the curve of Tierra del Fuego and links up with the Scotia arc. These complex distal belts record changes in relationships at the edges of the continental craton.

2 Palaeozoic and Mesozoic deposition

Throughout the Palaeozoic eras, troughs within the Andean belt received thick sedimentary sequences in which detrital material predominated, while the cratonic regions of deposition fringing the Precambrian shields and massifs received thinner and less complete sequences of shelf facies (see Chapters 5 and 6). Glacigene sediments, among which marine tillites are conspicuous, are recorded in Silurian-Devonian and Pennsylvanian-Permian groups in Bolivia.

The Lower Palaeozoic and Devonian formations are principally of marine facies. Widespread unconformities below, within or above the Devonian appear to correspond to phases of mid-Palaeozoic orogenic activity resulting in the elevation of certain parts of the mobile belt: the late Palaeozoic formations, especially those along the eastern side of the belt, are largely of non-marine facies. Apart from some greenstones in the Devonian of the Venezuelan Andes, there is little evidence of volcanicity until late Palaeozoic times. The early Palaeozoic assemblages are commonly phyllitic or slaty and suffered low-grade metamorphism during one or more Palaeozoic orogenic episode. Mid-Palaeozoic granites are recognised in Colombia and late Palaeozoic granites may also be represented.

The onset of Mesozoic deposition marked a new phase in the evolution of the Andes. Igneous rocks of ophiolitic type with cherts and marine clastic deposits are conspicuous in the southern Andes (southern Chile and Tierra del Fuego) and the northern Andes (northern Peru, Ecuador and Colombia) but are absent from the long central sector of the belt. In this central sector, and towards the eastern side of the mobile belt generally, the Triassic and Jurassic consist largely of fine marine detrital sediments and carbonate rocks, interfingering, especially in the north, with red-beds of non-marine facies. Widespread disturbances and upheaval are marked by unconformities above the Jurassic.

The Cretaceous successions in the Andes record an initial extension of the sea over almost the whole area of the mobile belt and onto the northern border of the craton in Venezuela. Later Cretaceous deposits, however, are largely non-marine, reflecting the changes brought about by widespread orogenic disturbances, and are concentrated in intermontane troughs and in the sub-Andean zone. By the close of the Mesozoic era most of the Andean belt had risen above sea level.

3 The orogenic and late-orogenic stages

Although some Palaeozoic orogenic episodes resulted in extensive deformation, low-grade metamorphism and the emplacement of a number of granites, the principal Andean orogenic phases were of late Mesozoic date. The majority of Andean granites (Fig. 11.1)are late Jurassic and Cretaceous and the emplacement of these bodies was followed by general uplift which expelled the sea from most of the area. *Post-Mesozoic sediments* of non-marine facies (some of which carry oil) are concentrated in the sub-Andean zone where huge clastic fans and wedges spread onto the craton, and in fault-bounded intermontane troughs located between the main ranges. From the Altiplano of Bolivia to the Upper Magdalena Valley of Colombia, these graben-like interior troughs received up to 6 km of sedimentary or volcanic debris.

The western ranges of the Andes constitute tectonically elevated tracts in which Precambrian and Palaeozoic crystalline rocks emerge at many points beneath Mesozoic cover-rocks. The huge *granitic batholiths* of these ranges — of which the 1500 km Peruvian batholith is only one — appear to be mainly Cretaceous, though some are as old as Triassic. Folding in the country-rocks is by no means intense and even in the vicinity of the batholiths the layering of the cover-rocks is not always strongly distorted. In the central and eastern ranges, Mesozoic and Tertiary strata invaded by granites somewhat smaller and considerably younger than those of the coastal ranges predominate; folding which took place mainly in Tertiary times, was only locally intense. In Colombia, and probably also elsewhere, the cover appears to have moulded itself irregularly over fractured and elevated segments of the basement. Steep reverse faults separate the uplifted tract from the sub-Andean zone in which Mesozoic and early Tertiary strata show simpler folds and faults.

The structural patterns across the full width of the mobile belt provide little evidence of significant tangential shortening during the late Mesozoic and Tertiary episodes. The great batholiths consist of many intrusive units emplaced within one envelope and appear to have mounted through the country-rocks in a series of pulses. The evidence for differential vertical movements resulting from block-faulting is impressive. In northern Colombia, for example, the basement surface beneath a young sedimentary trough lies some 12 km below the corresponding surface in the adjacent cordillera; subsidence and elevation evidently went on side by side during the late-orogenic period. Regional elevation giving the present highland belt appears to date only from late Neogene times; an early Tertiary erosion-level was upwarped to heights of at least 4 km in the Chilean Andes.

Late-orogenic volcanicity in the Andes continued sporadically from late Cretaceous times to the present day. Andesitic eruptions predominated in the Palaeogene and continue from a number of modern cones in the eastern ranges.

Small intrusions of granite, porphyry and similar rocks occur in the andesitic provinces. In Chile, these bodies are associated with a sulphide mineralisation responsible for large disseminated *porphyry copper* deposits. Isotopic dating of the igneous host and of hydrothermal minerals shows that these deposits range from early Palaeogene (56 m.y.) near the Peruvian border to late Neogene (6 m.y.) 1000 km further south. Mid-Tertiary tin-mineralisation is seen in Bolivia where the eastern Cordillera intersect older tin-provinces.

Volcanicity of a different style is recorded by enormous sheets of ignimbrites and acid pyroclastics which mantle the underlying structure, especially in eastern Chile. These rocks represent very large volumes of acid-intermediate magmas and appear to have been erupted during explosive phases coinciding with the late Tertiary episodes of regional elevation.

X The Antarctic

From the tip of Tierro del Fuego, the complexes forming the Andean belt disintegrate into fragments of late Palaeozoic and younger complexes which form the submarine ridge and islands on the northern and southern limbs of the Scotia arc (p. 270). The peninsula of *Graham Land* which connects the Scotia arc with West Antarctica incorporates late Palaeozoic detrital metasediments and volcanics and late Palaeozoic granites, followed unconformably by folded Mesozoic strata in which elongated granitic batholiths are emplaced. No pre-Carboniferous rocks are exposed at the distal end of the peninsula, but towards its root a crystalline basement resembling the basement of the East Antarctic craton emerges. These relationships suggest that both the Scotia arc and its root in Grahamland were built up during Palaeozoic times to form a unit linking old-established sectors of the Andean belt which were backed by continental cratons.

In the main portion of West Antarctica, the Andean tract parallels the late Precambrian-early Palaeozoic Ross mobile belt which occupies the Trans-antarctic mountains (Part I, p. 168). The oldest deposits of the Andean tract are Carboniferous greywackes and phyllites (the Trinity Peninsula Series), all of which are strongly disturbed and metamorphosed to low grades. Mesozoic sequences, which are unconformable on the Palaeozoic, include a non-marine series of detrital sediments and a thick group of Upper Jurassic andesitic and rhyolitic volcanics. Marine Upper Cretaceous strata record a transgression corresponding to that in the South American sector and mark the final deposits of the geosynclinal stage. These Mesozoic formations, although they are invaded by large Andean granite batholiths of late Mesozoic—Tertiary age, show only moderate tectonic disturbances and no regional metamorphism. Tertiary sediments occur only locally, but Neogene and Quaternary volcanics, mainly of basic composition, form important outcrops in the region of the Edsel Ford Ranges and King Edward VII Peninsula as well as in the distal part of Grahamland and the South Shetland Islands.

XI Retrospect: The Post-Palaeozoic Mobile Belts

The connection between the evolution of the Phanerozoic mobile belts and the plate movements induced by sea-floor spreading has already been referred to many times and is expressed diagrammatically in Fig. 11.11 which shows that

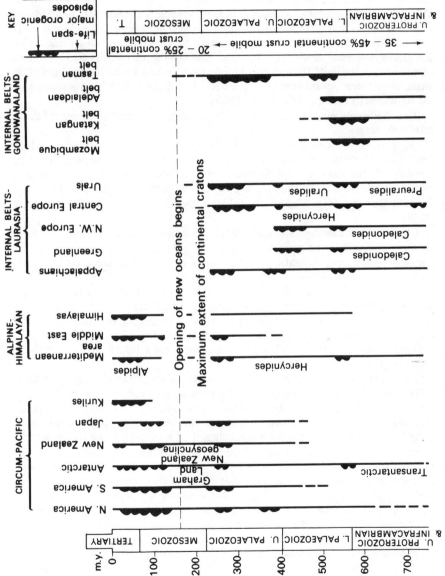

Fig. 11.11 The evolution of some Phanerozoic mobile belts in the supercontinents of Gondwanaland and Laurasia

whereas the *peripheral belts* encircling Laurasia and Gondwanaland remained mobile throughout the Phanerozoic eon, the *internal belts* were stabilised at or before the end of Palaeozoic times. It was within the enormous cratons welded together by the stabilised internal belts that the fracture-systems associated with the opening of the Atlantic, Indian and Southern Oceans were developed (Fig. 7.1). These relationships in space and time suggest, as Sutton has pointed out (1963, 1968), that the Mesozoic and Tertiary episode of continental drift resulting from the opening of the new oceans should be regarded as a stage in an integrated sequence of events — the Grenville chelogenic cycle (Part I) — extending back to the initiation of the mobile belts in late Precambrian times. The siting of the new oceans was predetermined to a large extent by the pattern of the mobile belts (compare Kennedy, 1965).

The chelogenic cycle as a whole can be thought of, rather crudely, as falling into three stages:

(1) From late Proterozoic to mid-Palaeozoic times, a large number of mobile belts were in existence both within and at the margins of the supercontinents. Cratons of continental crust between the mobile belts were small. Simple measurements on a globe showing the continents restored to pre-drift positions indicate that the total lengths of mobile belts active at this stage was at least 90 000 km.

(2) In later Palaeozoic and early Mesozoic times, the internal mobile belts of Laurasia and Gondwanaland were progressively stabilised. The large continental cratons created by stabilisation of these belts were more or less intact, though major fracture-zones began to appear within them.

(3) In later Mesozoic and Tertiary times, the super-continents were disrupted by the opening up of new oceans along some of these fracture-zones. The continental fragments were propelled outward, encroaching on the older Pacific ocean and almost eliminating the Tethys. Belts of orogenic mobility were now reduced to those of the peripheral system which had a total length of no more than 60000 km.

Three general points arise from this time-sequence. First, if orogenic mobile belts represent plate boundaries at which crust has been consumed, it would appear that the consumption of crust took place over shorter lengths of plate-boundary during the later stages of the chelogenic cycle. Assuming a roughly constant rate of sea-floor spreading, this restriction of the consumption-zones would suggest that the rate of consumption at any point was speeded up. Support for this inference comes from the marked increase in the development of the high-pressure metamorphic facies series. As we have noted, blueschists, eclogites and related rocks of this series which are widely distributed in Mesozoic-Tertiary mobile belts are extremely scarce in older belts and although the origin of these rocks is disputed, it is at least possible that tectonic overpressures were involved in their development (pp. 324—6).

The Mesozoic and Tertiary history of the Circum-Pacific mobile belt differed in other ways from that of earlier mobile belts and possibly from the Palaeozoic history of this belt. Compressional structures involving the basement appear to be of very minor importance: the great thrust-structures of the Rockies are developed entirely in the cover. Vertical movements were of greater significance than horizontal shortening, a fact which suggests that the overriding plates were effectively uncoupled from those of the Pacific floor. The enormous volumes of granitic material and of andesitic to rhyolitic volcanics seem to be greater than those generated over a comparable time-span in any older mobile belt. All these features could indicate that the Mesozoic era was the first stage in earth history at which the mobile belts were simplified to a single system.

A second point arising from the contrast between early and late stages of the chelogenic cycles concerns the bodily displacement of mobile belts. It will be evident that the long-lived peripheral mobile belts have not remained in fixed positions on the globe; the Cordilleran and Andean belts of the western Americas, for example, have been propelled westward at the leading edge of the continental cratons by the opening of the Atlantic Ocean. If, for simplicity, we regard the African and European continents as remaining at fixed longitudes, we must assume that the western mobile belts have been displaced by a distance equal to the width of the Atlantic – several thousand kilometres – in less than 200 m.y. The mobile belts originally bordering the Tethys have been brought into juxtaposition by the virtual elimination of the ocean which separated them, and the Indonesia-Philippine arcs have been crowded together by the northward advance of Australia. In addition to this bodily transport of the belts, adjustments between adjacent sectors were evidently necessitated by changes in length of the peripheral system. Rodolpho (1971) summarising some data on this topic, points out that the Circum-Pacific system was extended up to a stage at which it approximated to a great circle (about 40 000 km) when continental drift had reduced the ancestral Pacific to 50 per cent of the earth's surface and subsequently contracted to a small circle of approximately 37 800 km length. Adjustments resulting from these changes appear to have been taken up at the few points – the Aleutian, Caribbean, Scotia and south-west Pacific arcs – at which the mobile belt was not reinforced by a rigid continental craton: extension in the early stages being allowed for by development of new strings of islands, and subsequent contraction by increased curvature of the arcs. The migrating mobile belts retained their characteristic styles of activity while in transit and it therefore appears that the characteristic high heat flow and mantle peculiarities migrated along with, and at the same rate as, the belts themselves. Such an arrangement accords with the idea that mobile belts are located above descending convection currents in the mantle only if one envisages progressive changes in size of the convection-cell.

Where a buoyant continental crustal plate overrides oceanic crust, as at the western margin of the Americas, the uncoupling of crustal units at the plate margin may in itself initiate disturbances in the thermal and tectonic regimes at depth, which migrate with the plate margin. It is less easy to envisage this explanation for the siting of the mobile belt where, as along much of the western Pacific margin, the active plate boundaries separate crustal units of not very different thickness and constitution.

Finally, it will be apparent that the responses of the zones in which consumption of crust has taken place have shown periodic variations over time-spans of the order of 100 m.y., whereas the production of new crust at oceanic ridges is thought to have continued in a more regular way. The long-term orogenic and chelogenic rhythms appear to be adjusted on a global scale to the steadier process of sea-floor spreading and it may be supposed that the advance of downward-moving slabs of crust at Benioff zones sets in motion a train of tectonic and thermal modifications which act relatively slowly, controlling the progression from geosynclinal to orogenic and post-orogenic stages. The termination of mobility has, in some instances, been accompanied by permanent stabilisation of the zone without the development of a new mobile belt in the vicinity. Such a change, seen for example in Devonian times in the Caledonides of Britain, Scandinavia and Greenland, suggests the welding of formerly separate crustal plates into larger units. In other settings, notably in the peripheral belts, activity has been renewed after stabilisation either within the consolidated belt or to one side of it. Such displacements of the mobile zone, clearly indicating the persistance of an active plate margin, suggest that the Benioff zone became unserviceable at the end of each geological cycle — a deterioration possibly associated with the build-up of abnormal distributions of mass resulting from the introduction of lithospheric slabs into relatively dense layers of the mantle.

Bibliography

As with the reference lists in Part 1, the references that follow are intended to supplement the information given in successive chapters as well as to cover some points of detail referred to in the text and figures. They are not comprehensive and certainly do not cover all the sources consulted during the writing of this book; but the information given is intended to provide ways into the literature by the citation of a number of reviews and other general works which have comprehensive bibliographies. Where articles or books are referred to more than once the full reference is given in the first chapter list; later lists refer back to this chapter list.

For regional geological maps, see the list at the beginning of the Bibliographical References in Part I.

1 New Themes in Earth History

Blackett, P. M. S. (1961). Comparison of ancient climates with the ancient latitudes deduced from rock magnetic measurements. *Proc. Roy. Soc. Lond.,* A, **263**, 1

Blackett, P. M. S. *et al.* (1965). A symposium on continental drift. *Phil. Trans. Roy. Soc. Lond.,* A, **258**

Bullard, E. C. (1964). Continental drift. *Q. Jl geol. Soc. Lond.,* **120**, 1

Bullard, E. C., J. E. Everett and A. G. Smith (1965). The fit of the continents around the Atlantic. *In:* Blackett *et al.,* 1965

Creer, K. M. (1965). Palaeomagnetic data from the Gondwanic continents. *In:* Blackett *et al.,* 1965

—— (1970). A review of palaeomagnetism. *Earth. Sci. Rev.,* **6**, 369

Dietz, R. S. (1961). Continent and ocean-basin evolution by spreading of the sea floor. *Nature, Lond.,* **19**, 854

Du Toit, A. L. (1937). *Our Wandering Continents,* Oliver & Boyd

Flinn, D. (1971). On the fit of Greenland and north-west Europe before continental drift. *Proc. geol. Ass.,* **82**, 469

Harland, W. B. (1965). *In:* Blackett *et al.,* 1965

Heirtzler, J. R. (1968). Evidence for ocean floor spreading across the ocean basins; *In: The History of the Earth's Crust* (ed. Phinney) Princeton Univ. Press

Hill, D. (1957). Sequence and distribution of Upper Palaeozoic coral faunas. *Aust. J. Sci.,* **19**, 42

Holmes, A. (1929). A review of the continental drift hypothesis. *Min. Mag.,* **40**, 205, 286, 340

—— (1944). *Principles of Physical Geology.* Nelson

—— (1965). *Principles of Physical Geology.* (2nd Edition) Nelson

Hughes, N. F. (ed.) (1973). Organisms and continents through time. *Spec. Pap. Pal. Ass.,* 12

Kennedy, W. Q. (1946). The Great Glen Fault. *Q. Jl geol. Soc. Lond.,* **102**, 41

Le Pichon, X. (1968). Sea-floor spreading and continental drift. *J. geophys. Res.,* **73**, 3661

Martin, H. (1961). The hypothesis of continental drift in the light of recent advances . . . in Brazil etc. *Trans. geol. Soc. S. Africa, Annexure,* 44

Mason, R. G. and A. D. Raff (1961). Magnetic survey off the west coast of North America, 32°N–42°N latitude. *Bull. geol. Soc. Am.,* 72, 1259

Robinson, P. L. (1972). Palaeoclimatology and continental drift. *Phil. Trans. Roy. Soc. Lond.,* A 272

Runcorn, S. K. (1965). Palaeomagnetic comparison between Europe and North America. *In*: Blackett *et al.,* 1965

Smith, A. G., J. C. Briden and G. E. Drewry (1973). *In*: N. F. Hughes (ed.) 1973 (see above)

Strakhov, N. H. (1967). *Principles of Lithogenesis,* Vol. 1, Consultants Bureau and Oliver and Boyd

Tarling, D. H. and S. K. Runcorn (eds) (1973). *Implications of Continental Drift to the Earth Sciences,* Vols. 1 and 2, Academic Press

Tarling, D. H. and M. P. Tarling (1971). *Continental Drift,* Bell

Taylor, F. B. (1910). Bearing of the Tertiary mountain belt on the origin of the earth's plan. *Bull. geol. Soc. Am.,* **21**, 179

Vine, F. J. and D. H. Matthews (1963). Magnetic anomalies over oceanic ridges, *Nature, Lond.,* **199**, 947

Wegener, A. (1924). *The Origin of Continents and Oceans,* Methuen

Wellman, H. W. (1955). New Zealand Quaternary Tectonics, *Geol. Rundsch.,* **43**, 248

Westoll, T. S. (1965). Geological evidence bearing upon continental drift. *In*: Blackett *et al.,* 1965

2 The Caledonides and their Forelands

Allen, J. R. L. (1964). Studies in fluviatile sedimentation etc. *Sedimentology,* **3**, 163

Bailey, E. B. (1922). The structure of the south-west Highlands of Scotland. *Q. Jl geol. Soc. Lond.,* 78, 82

Barrow, G. (1893). On an intrusion of muscovite-biotite gneiss in the south-eastern Highlands of Scotland and its accompanying metamorphism. *Q. Jl geol. Soc. Lond.,* **49**, 330

Cowie, J. W. (1960). Notes on Lower Cambrian stratigraphy in the Boreal regions. *Int. geol. Congr. 21st Session,* 8, 57

Dewey, J. F. (1969). The evolution of the Appalachian/Caledonian orogen. *Nature, Lond.,* **222**, 124

Dewey, J. F. and R. J. Pankhurst (1970). The evolution of the Scottish Caledonides in relation to their isotopic age pattern. *Trans. Roy. Soc. Edinb.,* **68**, 361

Giletti, B. J., S. Moorbath and R. St. J. Lambert (1961). A geochronological study of the metamorphic complexes of the Scottish Highlands. *Q. Jl geol. Soc. Lond.,* **117**, 233

Haller, J. (1971). *The Caledonides of East Greenland,* Interscience

Harland, W. B. and M. J. S. Rudwick (1964). The great Infra-Cambrian Ice Age. *Sci. Am.,* **211**, 28

Johnson, M. R. W., and F. H. Stewart (eds.) (1963). *The British Caledonides,* Oliver and Boyd

Jones, O. T. (1938). On the evolution of a geosyncline. *Q. Jl geol. Soc. Lond.,* **94**, 1X

Kennedy, W. Q. (1948). On the significance of thermal structure in the Scottish highlands. *Geol. Mag.,* **85**, 229

McIntyre, D. B. (1954). The Moine thrust. *Proc. geol. Assoc.,* **65**, 203

Peach, B. N., J. Horne and others (1907). The geological structure of the north-west Highlands of Scotland. *Mem. geol. Surv. Scot.*

Pitcher, W. S. and A. R. Berger (1972). *The Geology of Donegal,* Interscience

Ramsay, J. G. (1958). Moine-Lewisian relations at Glenelg, Inverness-shire. *Q. Jl geol. Soc. Lond.,* **113**, 487

Read, H. H. (1952). Metamorphism and migmatization in the Ythan Valley, Aberdeenshire. *Trans. Edinb. geol. Soc.,* **15**, 265

Shackleton, R. M. (1958). Downward-facing structures of the Highland Border. *Q. Jl geol. Soc. Lond.,* **113**, 361

Spencer, A. M. (1971). Late Pre-Cambrian glaciation in Scotland. *Mem. geol. Soc. Lond.,* **6**

Størmer, L. (1967). Some aspects of the Caledonian geogyncline and foreland west of the Baltic shield. *Q. Jl geol. Soc. Lond.,* **123**, 183

Strand, T. and O. Kulling (1971). *The Scandinavian Caledonides,* Wiley-Interscience

Watson, J. (1964). Conditions in the metamorphic Caledonides during the period of late orogenic cooling. *Geol. Mag.,* **101**, 457

Williams, G. E. (1969). Characteristics and origin of a Precambrian pediment. *J. Geol.,* **77**, 183

Wood, A. (ed.) (1969). *The Precambrian and Lower Palaeozoic Rocks of Wales,* Univ. of Wales Press

Ziegler, A. M. (1970). Geosynclinal development of the British Isles during the Silurian period. *J. Geol.,* **78**, 445

3 The Hercynides and Uralides with their Forelands

Audley-Charles, M. C. (1970). Triassic palaeogeography of the British Isles. *Q. Jl geol. Soc. Lond.,* **126**, 49

Bogdanoff, A., M. V. Mouratov and N. S. Schatsky (eds) (1964). *Tectonique de L'Europe,* (Explanation of Tectonic Map of Europe) Moscow

Brinkmann, R. (ed.) (1955). Tektonik und Lagerstätten im Rheinishen Schiefer-
gebirge. *geol. Rundsch.*, 44
—— (1960). *Geologic Evolution of Europe*. Enke, Stuttgart
Cloos, H. and A. Rittmann (1939). Zur Einteilung und Benennung der Plutone.
Geol. Rundsch., **30**, 600
Coe, K. (ed.) (1963). *Some Aspects of the Variscan fold belt*, Manchester Univ.
Press
George, T. N. (1958). Lower Carboniferous palaeogeography of the British Isles.
Proc. Yorks. geol. Soc., **31**, 227
Gèze, B. (1949). Etude géologique de la Montagne Noire et des Cévennes
méridionales. *Mem. Soc. Géol. France,* 29
Hall, A. (1971). The relationship between geothermal gradient and the
composition of granitic magmas in orogenic belts. *Contr. Mineral and Petrol,*
32, 186
Hamilton, W. (1970). The Uralides and the motion of the Russian and Siberian
plates. *Bull. geol. Soc. Am.,* **81**, 2553
House, M. R. (1971). Devonian faunal distributions: In Faunal provinces in space
and time. *Geol. J.,* Special Issue, 4, 77
Jung, J. and M. Rogues (1952). Introduction a l'étude zonéographique des
formations crystallophylliennes. *Bull. Serv. Carte geol. France,* 235
Kennedy, W. Q. (1946). The Great Glen Fault. *Q. Jl geol. Soc. Lond.,* **102**, 41
Kent, P. E. (1949). A structural contour map of the buried pre-Permian rocks of
England and Wales. *Proc. geol. Ass.,* **60**, 87
Kossmatt, F. (1927). Gliederung des variskischen Gebirgsbaues. *Abh. sächs. geol.
Landesamsts,* **1**, Leipzig
Nalivkin, D. V., (trans. N. Rast) (1973). *Geology of the U.S.S.R.* Oliver &
Boyd
Oftedahl, C. (1959). Volcanic sequence and magma formation in the Oslo
region. *Geol. Rdsch.,* **48**, 18
Oswald, D. H. (ed.) (1967). International symposium on the Devonian system, 1.
Alberta Soc. Petrol. Geol.
Robinson, P. (1972). See p. 342
Rutten, M. (1969). *The Geology of Western Europe,* Elsevier
Suess, F. E. (1926). *Intrusions- und Wandertektonik im Variszischen Grundge-
birge,* Berlin
Trueman, A. E. (1946). Stratigraphical problems in the Coal Measures of Europe
and North America. *Q. Jl geol. Soc. Lond.,* **102**, xlix
—— (1947). Stratigraphical problems in the coalfields of Great Britain. *Q. Jl
geol. Soc. Lond.,* **103**, lxv
Wallace, P. (1968). The sub-Mesozoic palaeogeology and palaeogeography of
north-eastern France and the Straits of Dover. *Palaeogeog. Palaeoclim. and
Palaeoecol.,* **4**, 241
Zwart, H. J. (1960). Relations between folding and metamorphism in the
Central Pyrenees etc. *Geol. en Mijnb.,* **39**, 163
—— (1967). The duality of orogenic belts. *Geol. en Mijnb.,* **46**, 283
—— (1968). The Palaeozoic crystalline rocks of the Pyrenees in their
structural setting. *Krystallinikum,* **6**, 125

4 The Appalachians and Interior Lowlands of the North American Craton

Balk, R. and T. Barth (1936). Structural and petrologic studies in Dutchess County, New York. *Bull. geol. Soc. Am.*, **47**, 685

Benson, W. N. (1926). The tectonic conditions accompanying the intrusion of basic and ultrabasic plutonic rocks. *Natn Acad. N. Zealand Sci. Mem.*, **19**, 6

Bird, J. M. and J. F. Dewey (1970). Lithosphere plate-continental margin tectonics and the evolution of the Appalachian orogen. *Bull. geol. Soc. Am.*, **58**, 1031

Cloos, E. (1947). Oolite deformation in the South Mountain fold, Maryland. *Bull. geol. Soc. Am.*, **58**, 843

Davis, G. L., G. R. Tilton and G. W. Wetherill (1962). Mineral ages from the Appalachian province in North Carolina and Tennessee. *J. geophys. Res.*, **67**, 1986

Dewey, J. F. and R. J. Pankhurst (1970). See p. 343

Fairbairn, H. W. and P. M. Hurley (1970). Northern Appalachian geochronology as a method for interpreting ages in older orogens. *Eclog. geol. Helv.*, **63**, 83

Faul, H. *et al.*, (1963). Ages of intrusion and metamorphism in the northern Appalachians. *Am. J. Sci.*, **261**, 1

Fisher, G. W. *et al.*, (eds.) (1970). *Studies in Appalachian Geology: Central and Southern*, Interscience

Gwinn, V. E. (1970). *In*: Fisher *et al.*, 1970

Ham, W. E. and J. L. Wilson (1967). Palaeozoic epeirogeny and orogeny in the central United States. *Am. J. Sci.*, **265**, 332

Holland, C. H. (ed.) (1971). *Cambrian of the New World*, Interscience

Kay, M. (1948). North American geosynclines. *Mem. geol. Soc. Am.*, 48

Kennedy, M. J. (1972). The Appalachian province. *Sp Pap. geol. Ass. Canada*, 11

King, P. B. (1954). *Tectonics of Middle North America*, Princeton Univ. Press

Lowry, W. D. (ed.) (1964). Tectonics of the Southern Appalachians. *Virginia Polytech. Inst. geol. Dept. Sci. Mem.*, 1

Rodgers, J. (1964). *In*: Lowry (ed.) (1964)

——— (1967). Chronology of tectonic movements in the Appalachian region of eastern North America. *Am. J. Sci.*, **265**, 408

——— (1970). *The Tectonics of the Appalachians*, Interscience

Thompson, J. B. *et al.*, (1968). Nappes and gneiss domes in west-central New England, *In*: Zen, E-an, *et al.* (eds), 1968

Upadhyay, H. D., J. F. Dewey and E. R. W. Neale (1971). The Betts Cove ophiolite complex, Newfoundland: Appalachian oceanic crust and mantle. *Proc. geol. Ass. Canada.*, **24**, 27

Woodward, H. P. (1957). Chronology of Appalachian folding. *Bull. Am. Ass. Petrol. Geol.*, **41**, 2312

——— (1961). Preliminary subsurface study of south-eastern Appalachian interior plateau. *Bull. Am. Ass. Petrol. Geol.* **45**, 1634

Zartman, R. E. *et al.*, (1967). K—Ar and Rb—Sr ages of some alkalic intrusive rocks from central and eastern United States. *Am. J. Sci.*, **265**, 848

Zen, E-an. (1967). Time and space relationships of the Taconic allochthon and autochthon. *Sp. Pap. geol. Soc. Am.*, 97

Zen, E-an., *et al.*, (eds.) (1968). *Studies of Appalachian Geology: Northern and Maritime*, Interscience

5 Gondwanaland in Late Proterozoic and Early Palaeozoic Times

Brown, D. A., K. S. W. Campbell and K. A. W. Crook (1968). *The Geological Evolution of Australia and New Zealand*, Pergamon

Cahen, L. (1963). Grands traits de l'agencement des éléments du soubassement de l'Afrique centrale. *Annls Soc. géol. Belg.* 85, 183

Cahen, L. and N. J. Snelling (1968). *The Geochronology of Equatorial Africa*, North-Holland

Cahen, L. and J. Lepersonne (1967). The Precambrian of Congo, Rwanda and Burundi, *In*: *The Precambrian*, 3, Interscience

Campana, B. and R. B. Wilson (1955). Tillites and related glacial topography of South Australia, *Eclog. geol. Helv.*, 48, 1

Clifford, T. N. (1967). The Damaran episode in the Upper Proterozoic–Lower Palaeozoic structural history of southern Africa. *Sp. Pap. geol. Soc. Am.*, 92

Compston, W. and P. Arriens (1968). The Precambrian geochronology of Australia. *Can. J. Earth Sci.*, 5, 561

Cooray, P. G. (1962). Charnockites in the Precambrian of Ceylon. *Q. Jl geol. Soc. Lond.*, 118, 239

Crawford and Oliver, (1961). p. 196

Daniels, J. L., W. Skiba and J. Sutton (1965). The deformation of some banded gabbros in the northern Somalia fold-belt. *Q. Jl geol. Soc. Lond.*, 121, 111

Dow, D. B. (1965). Evidence of a late Precambrian glaciation in the Kimberley region of Western Australia. *Geol. Mag.*, 102, 407

Garlick, W. G. (1961). Structural evolution of the copper belt. *In*: Mendelsohn, F. (ed.) (1961).

——— (1972). Sedimentary environment of Zambian copper deposition. *Geol. en Mijnb*, 51, 277

Glaessner, M. F. (1962). Pre-Cambrian fossils, *Biol. Rev.*, 37, 467

Glaessner, M. F. and L. W. Parkin (eds.) (1958). The geology of South Australia. *J. geol. Soc. Aust.*, 5, 1

Glaessner, M. F. and B. Daily (1959). The geology and late Precambrian fauna of the Ediacara fossil reserve. *Rec. S. Aust. Mus.*, 13, 369

Harland, W. B., and M. J. S. Rudwick (1964). See p. 343

Hepworth, J. V. (1972). The Mozambique orogenic belt and its foreland in northern Tanzania. *J. geol. Soc. Lond.*, 128, 461

Holmes, A. (1951). The sequence of orogenic belts in South and Central Africa. *Rep. int. geol. Congr.*, 14 (1948), 14, 254

Johnson, R. L. (1968). Structural history of the western front of the Mozambique belt in north-east Southern Rhodesia. *Bull. geol. Soc. Am.*, 79, 513

Martin, H. (1965). *The Precambrian Geology of South West Africa and Namaqualand.* University of Capetown

Mendelsohn, F. (ed.) (1961). *The Geology of the Northern Rhodesian copper belt,* Macdonald

Rhodes, F. H. T. (1966). The course of evolution. *Proc. geol. Ass.*, 77, 1

Sanders, L. D. (1965). Geology of the contact between the Nyanza shield and the Mozambique belt in western Kenya. *Bull. geol. Surv. Kenya*, 7

Schermerhorn, L. J. G. and W. I. Stanton (1963). Tilloids in the West Congo geosyncline. *Q. Jl geol. Soc. Lond.*, 119, 201

Vail, J. R. (1965). An outline of the geochronology of the late Precambrian formations of eastern central Africa. *Proc. R. Soc. Lond.*, 284, A, 354

Wells, A. T. and others (1970). Geology of the Amadeus basin, Central Australia. *Bull. Aust. Bur. Min. Res.*, 100

6 Gondwanaland in Late Palaeozoic Times

Adie, R. (ed.) (1972). *Antarctic Geology and Geophysics*, I.U.G.S.

Blackett, P. M. S. (1961). See p. 341

Briden, J. C., and E. Irving (1964). Palaeolatitude spectra of sedimentary palaeoclimatic indicators *In*: Nairn (ed.) *Problems of Palaeoclimatology*, Interscience

Brown, D. A. *et al.*, (1968). See p. 346

Creer, K. M. (1965). Palaeomagnetic data from the Gondwanic continents. *Phil. Trans. Roy. Soc.*, A, 258, 27

Du Toit, A. L. (1954). *Geology of South Africa* (3rd Edition), Oliver & Boyd.

Crook, K. A. W. (1969). Contrasts between Atlantic and Pacific geosynclines. *Earth planet. Sci. Lett.*, 5, 429

Evernden, J. F. and J. R. Richards (1962). Potassium—argon ages in eastern Australia. *J. geol. Soc. Aust.*, 9, 1

Frakes, L. A. and J. C. Crowell (1969—71). Late Palaeozoic Glaciation. Part 1 (1969) *Bull. geol. Soc. Am.*, 80, 1007: Part 2 (1970) *Ibid.* 81, 2261: Part 4 (1971) Ibid. 82, 2515

Guppy, D. J. *et al.*, (1958). The geology of the Fitzroy basin, Western Australia. *Bull. Aust. Bur. Min. Res.*, 36

Maack, R. (1957). Uber Vereisungsperioden und Vereisungsspuren in Brasilien, *Geol. Rdsch.*, 45, 547

Martin, H. (1961). See p. 346

Richards, J. R. and others (1966). Isotopic ages of acid igneous rocks in the Cairns hinterland, North Queensland. *Bull. Bur. Min. Res. Aust.*, 88

Robinson, P. L. (1972). See p. 342

Sougy, J. (1962). West African fold belt. *Bull. geol. Soc. Am.*, 73, 871

7 The Alpine-Himalayan Belt and the Eurasian Craton

Ager, D. V. and B. M. Evamy (1963). The geology of the southern French Jura. *Proc. geol. Ass.*, 74, 325

Aubouin, J. (1965). *Geosynclines*, Elsevier

Beloussov, V. V. (1969). Inter-relations between the earth's crust and upper mantle. *Am. geophys. Union*, Monograph 13, 698

Carey, S. W. (1955). The orocline concept in geotectonics. *Proc. R. Soc. Tasmania*, 89, 255

Cloos, H. (1939). Hebung-Spaltung-Vulkanismus. *Geol. Rundsch.*, **30**, 405

Debelmas, J. and M. Lemoine (1970). The western Alps: palaeogeography and structure. *Earth Sci. Rev.*, **6**, 221

Ellenberger, F. (1953). Sur l'extension des faciés briançonnais en Suisse. *Eclog. geol. Helv.*, **45**, 285

Gansser, A. (1964). *Geology of the Himalayas,* Interscience

—— (1966). The Indian Ocean and the Himalayas *Eclog. geol. Helv.*, **59**, 831

Gass, I. G. and D. Masson Smith (1963). The geology and gravity anomalies of the Troodos massif, Cyprus. *Phil. Trans. R. Soc.* A, **255**, 417

Gignoux, M. (1950). *Géologie Stratigraphique,* Masso

Jäger, E., E. Niggli and E. Wenk (1967). Rb–Sr Altersbestimmungen an Glimmern der Zentralalpen. *Beitr. geol. Karte Schweiz.*, NF, 134

Kominskaya, I. P., *et al.,* (1969). Explosion seismology in the U.S.S.R. *Am. geophys. Union,* Monograph **13**, 195

Ksiazkiewicz, M. (1963). Evolution structurale des Carpathes polonaises. *Livre à la Mémoire P. Fallot,* **2**, 529

Kuenen, P. H. and C. I. Migliorini (1950). Turbidity currents as a cause of graded bedding. *J. Geol.,* **58**, 91

Mahel, M. (1968). *Regional Geology of Czechoslovakia,* Part 2, E. Schweizerbart'sche Verlagsbuchhandlung

Mercier, J. (1966). Sur l'existence et l'âge des deux phases régionale de métamorphisme alpin . . . en Macédonie central (Grèce). *Bull. Soc. géol. de France,* **7**, 1014

Nalivkin, D. V. (1973). See p. 344

Oxburgh, E. R. (1968). An outline of the geology of the central Eastern Alps. *Proc. geol. Ass.,* **79**, 1

Ramsay, J. G. (1963). Stratigraphy, structure and metamorphism in the Western Alps. *Proc. geol. Ass.,* **74**, 357

—— (1969). The measurement of strain and displacement in orogenic belts. *Sp. Pap. geol. Soc. Lond.,* **3**, 43

Rutten, M. G. (1969). See p. 344

Smith, A. G. (1971). Alpine deformation and the oceanic areas of the Tethys, Mediterranean and Atlantic. *Bull. geol. Soc. Am.,* **82**, 2039

Spencer, A. M. (1974). Mesozoic and Cenozoic orogenic belts. *Sp. Pap. Geol. Soc. Lond.,* 4

Sutton, J. (1963). Long-term cycles in the evolution of the continents. *Nature, Lond.,* **198**, 731

—— (1968). Development of the continental framework of the Atlantic. *Proc. geol. Ass.,* **79**, 275

Trümpy, R. (1960). Palaeotectonic evolution of the central and western Alps. *Bull. geol. Soc. Am.,* **71**, 843

Vogt, P. R. (1971). Hypotheses on the origin of the Mediterranean basin. *J. geophys. Res.,* **76**, 3207

Wenk, E. (1962). Plagioclas als Indexmineral in der zentralalpen. *Schweiz. miner. petrogr. Mitt.,* **42**, 139

Wunderlich. H. G. (1967). Orogenfront-Verlagerung in Alpen, Apennin und Dinariden und die Einwurzelung strittiger Deckenkomplexe. *Geologie Mijnb.,* **46**, 40

8 Gondwanaland: Disruption of a Supercontinent

Adie, R. (ed.) (1972). See p. 347

Belmonte, Y., P. Hirtz and R. Wenger (1965). The salt basins of the Gabon and the Congo (Brazzaville). See Kennedy 1965

Brown, D. A. *et al.,* (1968). See p. 346

Clifford, T. N. and I. Gass (eds.) (1970). *African Magmatism and Tectonics,* Oliver & Boyd

Cratchley, C. R. and G. P. Jones (1965). An interpretation of the geology and gravity anomalies of the Benue Valley, Nigeria. *Overseas geol. Surveys Geophys., Paper* 1

Dixey, F. (1956). Erosion surfaces of Africa. *Trans. geol. Soc. S. Africa,* **59,** 1–16

Du Toit, A. L. (1954). See p. 347

Falcon, N. L. *et al.,* (eds.) (1970). A discussion on the structure and evolution of the Red Sea etc. *Phil. Trans. R. Soc. Lond.,* A., **267**

Holmes, A. and H. F. Harwood (1937). The petrology of the volcanic area of Bufumbira. *Mem. geol. Surv. Uganda,* 2, pt. 2

Holmes, A. (1965). See p. 342

Hospers, J. (1965). Gravity field and structure of the Niger delta etc. *Bull. geol. Soc. Am.,* **76,** 407

Jacobson, R. R. E., W. N. MacLeod and R. Black (1958). Ring complexes in the younger granite province of northern Nigeria, *Mem. Geol. Soc. Lond.,* 1

Kennedy, W. Q. *et al.,* (1965). *Salt basins around Africa,* Inst. Petrol.

King, B. C. (1970). Vulcanicity and rift tectonics in East Africa. *In*: *African Magmatism and Tectonics,* Oliver & Boyd

King, L. C. (1962). *Morphology of the Earth,* Oliver & Boyd

Martin, H. (1961). See p. 346

McDougall, I. (1963). Potassium–argon age measurements on dolerites from Antarctica and South Africa. *J. geophys. Res.,* **68,** 1535

Plumstead, E. P. (1964). Gondwana floras, geochronology and glaciation in South Africa. *Int. geol. Congr. 22nd session,* **9,** 303

Teichert, C. (1958). Australia and Gondwanaland. *Geol. Rdsch.,* **57,** 562

Vail, J. R. (1968). The southern extension of the East African rift system etc. *Geol. Rdsch.,* **57,** 601

Wadia, D. N. (1957). *Geology of India* (3rd Edition), Macmillan

9 The New Ocean Basins

Adie, R. (ed.) (1972). Antarctic geology and geophysics. *I. U. G. S. Series B,* No. 1

Audley-Charles, M. C., D. J. Carter and J. S. Milsom (1972). Tectonic development of Eastern Indonesia etc. *Nature, Lond. Phys. Sci.,* **239,** 35

Bailey, E. B. *et al.,* (1924). Tertiary and post-Tertiary Geology of Mull etc. *Mem. Geol. Surv. Scotl.*

Baker, I., N. H. Gale and J. Simons (1967). Geochronology of the St Helena volcanoes. *Nature, Lond.,* **215,** 1451

Bullard, E. C., J. E. Everett and A. G. Smith (1965). See p. 341

Curray, J. R. and D. G. Moore (1971). Growth of the Bengal deep-sea fan and denudation of the Himalayas. *Bull. geol. Soc. Am.,* **82**, 563

Delaney, F. M. (ed.) (1971). The geology of the east Atlantic continental margin. *I.G.S. Rep.,* 70/16

Ewing, M. *et al.* (1969). Sediment distribution in the Indian Ocian. *Deep Sea Res.* **16**, 231

Funnell, B. and A. G. Smith (1968). Opening of the Atlantic Ocean. *Nature, Lond.,* **219**, 1328

Gass, I. S. and I. L. Gibson (1969). Structural evolution of the rift zones in the Middle East. *Nature, Lond.,* **221**, 926

Hawkes, D. D. (1962). The structure of the Scotia arc. *Geol. Mag.,* **99**, 85

Heezen, B. C., M. Tharp and M. Ewing (1959). The floors of the oceans: I. the North Atlantic. *Sp. Pap. geol. Soc. Am.,* **65**

Heezen, B. C. and M. Tharp (1965). Tectonic fabric of the Atlantic and Indian Oceans etc. *Phil. Trans. R. Soc. Lond.,* A., **258**, 90

Heirtzler, J. R., X. le Pichon and J. G. Baron (1966). Magnetic anomalies over the Reykjanes Ridge. *Deep-sea Res.,* **13**, 427

Hill, M. N. (ed.) (1966). A discussion concerning the floor of the north-west Indian Ocean. *Phil. Trans. R. Soc. Lond.,* A, **259**

Holmes, A. (1936). The idea of contrasted differentiation. *Geol. Mag.,* **73**, 228

Hope, E. R. (trans.) (1964). A new tectonic chart of the Arctic by I. P. Atlasov and others. Can. Defence Research Board

———(trans.) (1967). A chart of the recent tectonics of the Arctic by A. P. Puminov, ibid.

Katili, J. (1971). A review of geotectonic theories and tectonic maps of Indonesia. *Earth Sci. Rev.,* **7**, 143

Kennedy, W. Q. (1965). The influence of basement structure on the evolution of the coastal basins around Africa. *In: Salt basins around Africa,* Inst. Petrol.

Le Pichon, X. (1968). See p. 342

Laughton, A. S. (1966). The Gulf of Aden. *In:* Hill (ed.) 1966

McElhinney, M. W. (1970). Formation of the Indian Ocean. *Nature, Lond.,* **228**, 977

Mitchell-Thomé R. C. (1970). *Geology of the South Atlantic Islands,* Born-traeger, Berlin

Moorbath, S. and H. Welke (1969). Lead isotope studies on igneous rocks from the Isle of Skye, north-west Scotland. *Earth planet. Sci. Lett.,* **5**, 217

Stride, A. *et al.* (1969). Marine geology of the Atlantic continental margin of Europe. *Phil. Trans. R. Soc. Lond.,* A, **264**, 31

Sutton, J. (1968). See p. 348

Sykes, L. R. and M. Ewing (1965). The seismicity of the Caribbean region. *J. geophys. Res.,* **70**, 5065

Trendall, A. F. (1953). The geology of South Georgia: I. *Sci. Rep. Falklands Is. Depend. Surv.,* 19

Walker, G. P. L. (1960). Zeolite zones and dike distribution in relation to the structure of the basalts of eastern Iceland. *J. geol.,* **68**, 515

———(1965). Evidence of crustal drift from Icelandic geology. *Phil. Trans. Soc. Lond.,* A, **258**, 199

Wilson, J. T. (1963). Evidence from islands on the spreading of ocean floors. *Nature, Lond.,* **197**, 536

—— (1965). A new class of faults and their bearing on continental drift. *Nature, Lond.*, **207**, 343

Yoder, H. S. and C. E. Tilley (1962). Origin of basalt magmas etc., *J. Petrol.*, **3** 342

10 The Pacific Ocean and its Island Arcs

Allen, C. R. (1965). Transcurrent faults in continental areas. *Phil. Trans. R. Soc. Lond.*, A, **258**, 82

Anderson, D. L. (1971). The San Andreas fault. *Continents Adrift (reading from the Scientific American)*, Freeman, 1972

Brothers, R. N. (1970). Lawsonite—albite schists from northernmost New Caledonia. *Cont. Miner. Petrol.*, **25**, 185

Ewing, M. *et al.* North Pacific sediment layers measured by seismic profiling. In Knopoff *et al.* (eds) (1968) (see below)

Fleming, C. A. (1970). The Mesozoic of New Zealand etc. *Q. Jl geol. Soc. Lond.*, **125**, 125

Haile, N. S. (1969). Geosynclinal theory and the organisational pattern of the North-west Borneo geosyncline. *Q. Jl geol. Soc. Lond.*, **124**, 171

Heirtzler, J. R. (1968). See p. 341

Imai, H. (1963). Pre-Tertiary igneous activity. In: Takai *et al.,Geology of Japan*, Univ. of California Press

Isaaks, B., J. Oliver and L. R. Sykes (1968). Seismology and the new global tectonics. *J. geophys. Res.*, **73**, 5855

Karig, D. E. (1971). Structural history of the Mariana island arc system. *Bull. geol. Soc. Am.*, **82**, 323

Knopoff, L., C. L. Drake and J. H. Pembroke (eds.) (1968). The crust and upper mantle of the Pacific area. *Am. geophys. Union*, Geophys. Monograph 12

Landis, C. A. and D. S. Coombs (1967). Metamorphic belts and orogenesis in southern New Zealand. *Tectonophysics*, **4**, 501

Le Pichon, X. (1968) See p. 342

Macdonald, G. A. and H. Kuno (eds. (1962). The crust of the Pacific basin. *Am. geophys. Union*, Geophys. Monograph 6

Menard, H. W. (1969). The deep-ocean floor. *Continents Adrift (*readings from the *Scientific American)*, Freeman (1972)

Miyashiro, A. (1967). Orogeny, regional metamorphism and magmatism in the Japanese Islands. *Dansk. Geol. For.*, **17**, 390

Oliver, J. E., and B. Isaaks (1967). Deep earthquake zones, anomalous structures in the upper mantle, and the lithosphere. *J. geophys. Res.*, **68**, 4259

Suggate, R. P. (1963). The Alpine fault. *Trans. R. Soc., N.Z. (Geology)*. **2**, 105

Sykes, L. R. (1968). Seismological evidence for transform faults, etc. In: Phinney (ed.). *The History of the Earth's Crust*, Princeton Univ. Press

Tobin, D. G. and L. R. Sykes (1968). Seismicity tectonics of the NE Pacific. *J. geophys. Res.*, **73**, 3821

Vine, F. (1968). Magnetic anomalies associated with mid-ocean ridges. In: Phinney (ed.). *The History of the Earth's Crust*, Princeton Univ. Press

Wellman, H. W. (1969). Wrench (transcurrent) fault systems. *Am. geophys. Union,* Geophys. Monograph, **13**, 544

11 The Cordilleran and Andean Mobile Belts and the North American Craton

Blake, M. C., W. P. Irwin and R. G. Coleman (1969). Blueschist facies metamorphism related to regional thrust-faulting. *Tectonophysics,* 8, 237

Cobbing, E. J., and W. S. Pitcher (1972). The coastal batholith of central Peru. *Q. Jl geol. Soc. Lond.,* 128, 421

Ernst, W. G. (1971). Do mineral parageneses reflect high pressure conditions of Franciscan metamorphism? *Am. J. Sci.,* **270**, 81

Evernden, J. F. and R. W. Kistler (1970). Chronology of emplacement of Mesozoic batholithic complexes in California and Western Nevada. *Prof. Pap. U.S. geol. Surv.,* 623

Fox, F. G. (1969). Some principles governing the interpretation of structure in the Rocky Mountain Belt, *Sp. Pap. geol. Soc. Lond.* 3, 23

——— (1972). The Cordilleran province. *Sp. Pap. geol. Ass. Canada,* 11

Gansser, A. (1973). Facts and theories on the Andes. *Q. Jl geol. Soc. Lond.,* 129, 93

Gilluly, J. (1963). The tectonic evolution of the western United States. *Q. Jl geol. Soc. Lond.,* 119, 133

James, D. E. (1971). Andean crustal and upper mantle structure. *J. geophys. Res.,* 76, 3246

Kennedy, W. Q. (1965). See p. 350

King, P. B. (1969). The tectonics of North America etc. *Prof. Pap. U.S. geol. Surv.,* 628

May, P. R. (1971). Pattern of Triassic-Jurassic diabase dykes around the North Atlantic etc. *Bull. geol. Soc. Am.,* 82, 1285

Mitchell, A. H. G. and M. S. Garson (1972). Relationship of porphyry copper and circum-Pacific tin deposits to palaeo-Benioff zones. *Trans. Instn Min Metall,* 81, B10

Noble, J. A. (1970). Metal provinces of the western United States. *Bull. geol. Soc. Am.,* 81, 1607

Stewart, J. H. (1971). Initial deposits in the Cordilleran geosynclines, etc. *Bull. geol. Soc. Am.,* 83, 1345

Sutton, J. (1963). See p. 348

——— (1968). See p. 348

Turneaure, F. S. (1971). The Bolivian tin—silver province. *Econ. geol.,* 66, 215

Wheeler, J. O. *et al.* (1972). The Cordilleran structural province. *Geol. Ass. Can. Sp. Pap.,* 11, 1

Index

Numbers in roman refer to pages in Part I; numbers in italic refer to Part II.